Lecture Notes in Mathematics

Edited by A. Dold and B. Eckmann

1121

Singularities and Constructive Methods for Their Treatment

Proceedings of the Conference
held in Oberwolfach, West Germany,
November 20–26, 1983

Edited by P. Grisvard, W. Wendland and J. R. Whiteman

Springer-Verlag
Berlin Heidelberg New York Tokyo

Editors

Pierre Grisvard
Université de Nice, IMSP, Parc Valrose
06034 Nice Cedex, France

Wolfgang L. Wendland
Technische Hochschule Darmstadt, Fachbereich Mathematik
Schloßgartenstr. 7, 6100 Darmstadt, Federal Republic of Germany

John R. Whiteman
Brunel University, Department of Mathematics and Statistics
Uxbridge, Middlesex UB8 3PH, United Kingdom

AMS Subject Classification (1980): 35B65, 45A05, 45L10, 49G10, 65N30, 65R20, 73C45, 73M05, 76Q05

ISBN 3-540-15219-9 Springer-Verlag Berlin Heidelberg New York Tokyo
ISBN 0-387-15219-9 Springer-Verlag New York Heidelberg Berlin Tokyo

Printing and binding: Beltz Offsetdruck, Hemsbach/Bergstr.
2146/3140-543210

Math Sci
Sep

PREFACE

These proceedings contain the majority of the lectures given at the conference with the same title at the "Mathematisches Forschungsinstitut Oberwolfach", Germany, November 21-25, 1983.

In proposing such a conference the organizers thought that it would be timely to bring together persons from various fields of mathematics and engineering science whose common interests lay in analysing and solving numerically problems involving singularities. The list of invitees was compiled with this aim in mind, in an effort to ensure that expertise in theoretical analysis, numerical analysis, and practical application was represented. This in fact mirrored the similar but complementary interests of the organizers.

A very strong programme of lectures was offered, of which 19 are collected together here. Subjects covered include:
treatment of crack problems, analysis and regularity properties of solutions of problems with singularities, application of finite element and boundary element methods to problems involving singularities with associated error analysis, problems of fluid flow containing singularities, retrieval of physically meaningful parameters such as stress intensity factors.

It has become obvious that in mathematics a field of "singularities" now exists in its own right, that increasingly numerical techniques will be used for problems with singularities which can be two- or three-dimensional and also nonlinear. Many of the problems currently belong to the class of boundary value problems for elliptic partial differential equations with solutions having singularities due to geometric boundary irregularities or discontinuities of the differential or boundary operators. These present increasingly demanding challenges for theoreticians to produce relevant singular forms and regularity results. Corresponding solution procedures are based on appropriate series expansions, asymptotic expansions or corresponding numerical approximate solution procedures.

The methods presented in these proceedings are extremely diverse, and are for the most part applied to individual problems. Thus it seems that an effort should be made in the future to assess such methods comparatively in the various fields where they are applied; such as fluid mechanics, solid mechanics, and crack propagation.

The organizers take this opportunity to thank all participants of the conference for their contributions and to express their gratitude to the Oberwolfach Research Institute and its sponsors for their support.

P. Grisvard, W. Wendland, J.R. Whiteman

C O N T E N T S

<u>List of participants</u>

Dr. A. Beagles
Institute of Computational Mathematics
Brunel University
Uxbridge, Middlesex UB8 3PH
Great Britain

Dr. H. Blum
Institut für Angewandte Mathematik
Universität Bonn
Wegelerstr. 6
5300 Bonn

Dr. M. Blumenfeld
Institut für Mathematik III
Freie Universität Berlin
Arnimallee 2-6
1000 Berlin 33

Prof. Dr. M. Brüning
Mathematisches Institut
Universität Augsburg
Memminger Str. 6
8900 Augsburg

Prof. Dr. H.F. Bueckner
Turbine Engineering
General Electric Company
Schenectady, N.Y. 12309, USA

Prof. Dr. L. Collatz
Institut für Angewandte Mathematik
Universität Hamburg
Bundesstr. 55
2000 Hamburg 13

Dr. M. Costabel
Fachbereich Mathematik
Technische Hochschule Darmstadt
Schloßgartenstr. 7
6100 Darmstadt

Dr. R. Davies
Dept. of Applied Mathematics
University College of Wales
Aberwystwyth, Dyfed SY23 3BZ
Great Britain

Prof. Dr. P. Destuynder
Direction des études et recherches
E.D.F.
Division Mécanique Théorétique
92141 Clamart, France

Prof. Dr. M. Djaoua
E.N.I.T.
B.P. 37/Belvedere
Tunis, Tunisia

Dr. M. Dobrowolski
Institut für Angewandte Mathematik
Universität Bonn
5300 Bonn

Dr. M. Durand
Unité d'Enseignement et de Recherche de Mathématiques
3, Place Victor Hugo
13331 Marseille, France

Dr. G. Dziuk
Lehrstuhl und Institut für Mathematik
TH Aachen
Templergraben 55
5100 Aachen

Prof. Dr. G. Fichera
Academia Nazionale dei Lincei
Via Della Lungara 10
00165 Roma, Italy

Prof. Dr. P. Filippi
Laboratoire de Mécanique et d'Acoustique - CNRS
BP 71
13277 Marseille, France

Prof. Dr. P. Grisvard
I.M.S.P.
Université de Nice
Parc Valrose
06034 Nice, France

Prof. Dr. D. Groß
Fachbereich Mechanik
Technische Hochschule Darmstadt
Hochschulstr. 1
6100 Darmstadt

Dr. E. Hartmann
Institut für Baumechanik uns Statistik
Universität Dortmund AVZ
August Schmidt-Str. GB II
4600 Dortmund 50

Prof. Dr. G.C. Hsiao
Dept. for Mathematical Sciences
University of Delaware
Newark, Delaware 19711, USA

Prof. Dr. R.B. Kellogg
University of Maryland
IPST
College Park, Maryland 20742, USA

Prof. Dr. St. Krenk
Risø National Laboratory
4000 Roskilde, Denmark

Prof. Dr. J. Lewis
University of Illinois at Chicago Circle
P.O. Box 4348
Chicago, IL 60680, USA

Prof. Dr. J. Mason
Mathematics Branch
Swindon
Shrivenham, Wilts. SN68LA
Great Britain

Prof. Dr. M. Moussaoui
U.S.T.H.B.,
B.P. 9
Dar El Beida, Algérie

Prof. Dr. J. Nitsche
Institut für Angewandte Mathematik
Universität Freiburg
Hebelstr. 14
7800 Freiburg

Prof. Dr. J. Osborn
Dept. of Mathematics
University of Maryland
College Park, MD 20742, USA

Dr. N. Papamichael
Dept. of Mathematics
Brunel University
Uxbridge, Middlesex UB8 3PH
Great Britain

Prof. Dr. J.B. Rosser
Mathematical Research Center
University of Wisconsin
Madison, WI 53706, USA

Prof. Dr. A. Schatz
Dept. of Mathematics
Cornell University
Ithaca, New York 14853, USA

Dipl.-Math. T. Schleicher
Fachbereich Mathematik
Technische Hochschule Darmstadt
Schloßgartenstr. 7
6100 Darmstadt

Prof. Dr. E. Schnack
Institut für Techn. Mechanik
und Lehrstuhl für Mechanik
Universität Karlsruhe
7500 Karlsruhe

Prof. Dr. E. Stein
Lehrstuhl für Baumechanik
Universität Hannover
Callinstr. 32
3000 Hannover

Prof. Dr. J. Steinberg
Israel Inst. of Technology
Dept. of Mathematics
Technion City
Haifa, 32000, Israel

Dr. P. Tolksdorf
Institut für Angewandte Mathematik
Universität Bonn
Wegelerstr. 6
5300 Bonn

Prof. Dr. L.B. Wahlbin
Dept. of Mathematics
Cornell University
Ithaca, New York 14853, USA

Prof. Dr. W.L. Wendland
Fachbereich Mathematik
Technische Hochschule Darmstadt
Schloßgartenstr. 7
6100 Darmstadt

Prof. Dr. J. Whiteman
Institute of Computational Math.
Brunel University
Uxbridge, Middlesex UB8 3PH
Great Britain

Prof. Dr. G.R. Wickham
Dept. of Mathematics
University of Manchester
Manchester M13 3PL, Great Britain

Prof. Dr. C. Zenger
Institut für Mathematik
Technische Universität München
Arcisstr. 21
8000 München 2

Dr. L. Zink
Institut für Informatik
Technische Universität München
Postfach 2024
8000 München 2

FINITE ELEMENT METHODS FOR THE SOLUTION OF PROBLEMS WITH ROUGH INPUT DATA

I. Babuška[+] and J.E. Osborn[‡]

1. Introduction

The finite element method is based on the variational or weak formulation of the boundary value problem under consideration. The approximate solution is obtained by restriction of the variational formulation to finite dimensional trial and test spaces.

Accuracy of the approximate solution is achieved by the choice of a trial space with good approximation properties and the choice of a test space which guarantees that the finite element solution is an approximation of the same quality as is the best approximation of the exact solution by the trial functions.

The success of the finite element method as a practical computational tool is related to the special construction of the trial and test functions in terms of "element" trial and test functions (defined on the finite elements) satisfying appropriate constraints at the nodes. This element based structure is the basis for the architecture of existing finite element codes. Usually the "element" trial and test functions are polynomials.

It is well known that when the solution is rough, which will in general be the case when the data is rough, then the use of polynomials as "element" trial functions does not lead to good accuracy. In this paper we will construct "element" trial functions which reflect the properties of the problem under consideration. This will lead to increased accuracy - the maximum possible accuracy - while not changing the structure of the code. In fact, the approach can be implemented by changing only the "element" trial functions employed in the method. The questions of optimal trial functions relative to a class of problems is also addressed.

The abstract framework of the approach is given in Section 2. In Section 3, five examples of various types are presented and in terms of these examples the ideas presented in Section 2 are developed. In Section 4, we make some general comments on the design of finite element methods for problems with rough coefficients.

Detailed proofs of the results in the paper will be presented elsewhere. The ideas in the paper can be generalized in various directions.

Throughout the paper we will use the usual L_p spaces and Sobolev spaces H_1, $\mathring{H}_1$.

2. Variational Approximation Methods

In this section we discuss variational approximation methods and state a basic error estimate.

Let $\mathcal{H}_1, \|\cdot\|_{\mathcal{H}_1}$ and $\mathcal{H}_2, \|\cdot\|_{\mathcal{H}_2}$ be two Hilbert spaces and let $\mathcal{H}_{1,h}, \|\cdot\|_{\mathcal{H}_{1,h}}$, $0 < h \leq 1$, $\mathcal{H}_{2,h}$, $\|\cdot\|_{\mathcal{H}_{2,h}}$, $0 < h \leq 1$, be two one parameter families of Hilbert spaces satisfying

$$\mathcal{H}_1 \subset \mathcal{H}_{1,h}, \|u\|_{\mathcal{H}_{1,h}} = \|u\|_{\mathcal{H}_1} \forall h \text{ and } \forall u \in \mathcal{H}_1 \tag{2.1}$$

and

$$\mathcal{H}_2 \subset \mathcal{H}_{2,h}, \|u\|_{\mathcal{H}_{2,h}} = \|u\|_{\mathcal{H}_2} \forall h \text{ and } \forall u \in \mathcal{H}_2. \tag{2.2}$$

Let $B_h(u,v)$ be a bilinear form on $\mathcal{H}_{1,h} \times \mathcal{H}_{2,h}$ satisfying

$$|B_h(u,v)| \leq M\|u\|_{\mathcal{H}_{1,h}} \|v\|_{\mathcal{H}_{2,h}} \forall u \in \mathcal{H}_{1,h}, \forall v \in \mathcal{H}_{2,h}, \text{ and } \forall h \tag{2.3}$$

(M is independent of h).

Let f be a bounded linear functional on $\mathcal{H}_2$. Corresponding to f we assume that there is a unique $u_0 \in \mathcal{H}_1$ satisfying

$$B_h(u_0,v) = f(v) \quad \forall v \in \mathcal{H}_2 \text{ and } \forall h. \tag{2.4}$$

u_0 can be thought of as the exact solution to a boundary value problem under consideration and is unknown. B_h and f are given input date and are known.

We are interested in approximating u_0 and toward this end we assume we are given or have chosen finite dimensional spaces $S_{1,h} \subset \mathcal{H}_{1,h}$ and $S_{2,h} \subset \mathcal{H}_{2,h}$ with $\dim S_{1,h} = \dim S_{2,h}$ and

$$\inf_{\substack{u \in S_{1,h} \\ \|u\|_{\mathcal{H}_{1,h}}=1}} \sup_{\substack{u \in S_{2,h} \\ \|v\|_{\mathcal{H}_{2,h}}=1}} |B_h(u,v)| \geq \gamma > 0 \quad \forall h \tag{2.5}$$

(γ is independent of h). Then we define u_h by

$$u_h \in S_{1,h}$$
$$B_h(u_h,v) = B_h(u_0,v) \quad \forall v \in S_{2,h} \tag{2.6}$$

and consider u_h to be an approximation to u_0. (2.6) is uniquely solvable. We often denote u_0 be u. Note that u_h is defined for every $u \in H_{1,h}$. Given bases for $S_{1,h}$ and $S_{2,h}$, (2.6) is reduced to a system of linear equations. But (2.6) does not define an algorithm for finding u_h since it depends on the unknown u_0. We note from (2.4), however, that $B_h(u_0,v) = f(v)\,\forall v \in S_{2,h}$ provided

$$S_{2,h} \subset H_2 \ \forall h. \tag{2.7}$$

We now assume that (2.7) holds. (2.6) can then be written

$$u_h \in S_{1,h}$$
$$B_h(u_h,v) = f(v) \ \forall v \in S_{2,h}. \tag{2.8}$$

(2.8) is called a variational approximation method.

Having defined u_h we are interested in an estimate for $\|u_0-u_h\|_{H_{1,h}}$. This is provided by the standard

Theorem. The error u_0-u_h satisfies

$$\|u_0 - u_h\|_{H_{1,h}} \le (1 + \gamma^{-1}M) \inf_{\chi \in S_{1,h}} \|u_0 - \chi\|_{H_{1,h}} \tag{2.9}$$

where M and γ are the constants in (2.3) and (2.5), respectively.

For a complete discussion of variational approximation methods see [1,2].

The spaces $S_{1,h}$ are called trial spaces and the functions in $S_{1,h}$ are called trial or approximating functions. The spaces $S_{2,h}$ are called test spaces and the functions in $S_{2,h}$ are called test functions.

Remarks: 1) In our applications there will be a bilinear form $B(u,v)$ defined on $H_1 \times H_2$ and satisfying $B_h(u,v) = B(u,v)\ \forall u \in H_1,\ v \in H_2$. From (2.4) we see that

$$B(u_0,v) = f(v) \quad \forall v \in H_2. \tag{2.10}$$

(2.10) is the variational formulation of our boundary value problem and H_1, H_2, and B are the spaces and the bilinear form in this formulation.

2) (2.9) suggests we choose $S_{1,h}$ so that $\inf_{\chi \in S_{1,h}} \|u_0 - \chi\|_{H_{1,h}}$ is small, i.e., so that the trial functions have good approximation properties, and, with $S_{1,h}$ so chosen, $S_{2,h}$ can be selected so that (2.5) holds with as large a constant γ as possible.

3) In many applications we can choose $S_{1,h} \subset H_1$, but in others the requirement that $\inf_{\chi \in S_{1,h}} \|u_0 - \chi\|_{H_{1,h}}$ be small leads one to choose $S_{1,h} \not\subset H_1$. The trial space is then nonconforming in the sense that $S_{1,h}$ does not lie in the basic variational space H_1. This fact leads to a use of the family of spaces $H_{1,h}$ and forms B_h.

In certain situations one has $H_1 = H_2$ and one wishes to choose $S_{2,h} = S_{1,h}$. Then, if $S_{1,h}$ is nonconforming, $S_{2,h}$ will be also, and we are led to the use of the family $H_{2,h}$. Note that in this circumstance,

$$B_h(u_0,v) = f(v) \ \forall v \in S_{2,h} \tag{2.11}$$

is not valid. Nonetheless, an approximation u_h can still be defined as in (2.8). This, in fact, is what is done in the class of methods commonly referred to as nonconforming in the finite element literature (see, e.g., Section 4.2 in [4]). The error analysis for such problems does not follow directly from (2.9) and the additional complications in the analysis are due to the fact that (2.11) does not hold.

In the methods discussed in this paper we will always have $S_{2,h} \subset H_2$, i.e., our test spaces will be conforming. We emphasize that the choice of nonconforming trial spaces causes no difficulty in the analysis, provided the test spaces are conforming.

3. Examples

In the section we will discuss the approximate solution of five specific boundary value problems with rough input data. In each example we will use a variational approximation method employing trial functions which reflect the properties of the underlying problem in that they provide an accurate approximation to the unknown solution.

a. A Two Point Boundary Value Problem with a Rough Coefficient

Consider the problem

$$\begin{aligned} Lu_0 &= -(a(x)u_0')' = f, \quad 0 < x < 1 \\ u_0(0) &= u_0(1) = 0 \end{aligned} \tag{3.1}$$

where $a(x)$ is a rather arbitrary function satisfying $0 < \alpha \le a(x) \le \beta$ and $f \in L_2(0,1)$. This simple model problem arises in the analysis of the displacements in a tapered elastic bar. f represents the load, $a(x)$ the elastic and geometric properties of the bar, and u_0 the displacement. If the bar has smoothly but rapidly or abruptly varying (as in the case of composite materials) material properties, then $a(x)$ will be a smoothly but rapidly or abruptly varying function, i.e., a rough function.

The variational formulation of (3.1) is

$$\begin{aligned} &u_0 \in \mathring{H}_1(0,1) \\ &\int_0^1 a u_0' v' dx = \int_0^1 f v dx \quad \forall v \in \mathring{H}_1(0,1). \end{aligned} \tag{3.2}$$

This has the form (2.10) with $H_1 = H_2 = \mathring{H}_1$, $B(u,v) = \int_0^1 a u' v' dx$, and $f(v) = \int_0^1 f dx$. It is known that the standard finite element method employing C^0 piecewise linear trial and test functions (i.e., the variational approximation method determined by the form B and C^0 piecewise linear trial and test spaces) does not yield accurate approximations to the solution of (3.2) (or (3.1)) when $a(x)$ is rough. We thus consider an alternate method.

Let $T_h = \{0 = x_0 < x_1 < \ldots < x_n = 1\}$ be an arbitrary mesh on $[0,1]$ and set $I_j = (x_{j-1}, x_j)$, $h_j = x_j - x_{j-1}$, and $h = \max h_j$. The points x_j are called nodes. For the trial space we choose

$$\begin{aligned} S_{1,h} = \{u:\ &\text{For each } j,\ u\big|_{I_j} = \text{a linear combination of 1 and} \\ &\int^x \frac{dt}{a(t)},\ \text{i.e., } u\big|_{I_j} \text{ is a solution of } (au')' = 0, \\ &u \text{ is continuous at the nodes, and } u(0) = u(1) = 0\} \end{aligned} \tag{3.3}$$

and for the test space

$$\begin{aligned} S_{2,h} = \{v:\ &\text{For each } j,\ v\big|_{I_j} = \text{a linear combination of 1 and } x, \\ &u \text{ is continuous at the nodes, and } u(0) = u(1) = 0\}. \end{aligned} \tag{3.4}$$

Now $S_{1,h} \subset H_1$ and $S_{2,h} \subset H_2$ and we may thus choose

$$H_{1,h} = H_1,\ H_{2,h} = H_2,\ B_h = B. \tag{3.5}$$

However, we could also choose

$$H_{1,h} = H_{2,h} = \mathring{H}_{1,h} \equiv \{u : u\Big|_{I_j} \in H^1(I_j) \quad \text{for each} \quad j,\ u(0) = u(1) = 0\},$$

$$\|u\|_{H_{1,h}} = \|u\|_{H_{2,h}} = \|u\|_{\mathring{H}_{1,h}} = [\int_0^1 u^2 dx + \sum_{j=1}^{n} \int_{I_j} (u')^2 dx]^{1/2}, \tag{3.6}$$

$$B_h(u,v) = \sum_j \int_{I_j} u'v'dx.$$

With either the choice (3.5) or (3.6), we see that (3.2) has the form (2.4) and that (2.3) is satisfied with $M = B$. We will use the choice (3.6) since it leads more naturally to the choice dictated to us in Example b below.

We next consider the variational approximation method (2.8) determined by B_h, $S_{1,h}$, and $S_{2,h}$, i.e.,

$$u_h \in S_{1,h}$$
$$B_h(u_h,v) = \int_0^1 au_h'v'dx = \int_0^1 fvdx \ \forall v \in S_{2,h}. \tag{3.7}$$

Regarding (3.7) one can prove the inf-sup condition (2.5), namely

$$\inf_{\substack{u \in S_{1,h} \\ \|u\|_{\mathring{H}_1}=1}} \ \sup_{\substack{v \in S_{2,h} \\ \|u\|_{\mathring{H}_1}=1}} \left|\int_0^1 au'v'dx\right| = \gamma(\alpha,\beta) > 0 \ \forall h \tag{3.8}$$

and the approximation result

$$\inf_{\chi \in S_{1,h}} \|u_0 - \chi\|_{\mathring{H}_1} \le N(\alpha,\beta)h\|(au_0')'\|_{L_2} = N(\alpha,\beta)h\|f\|_{L_2} \tag{3.9}$$

where $\gamma(\alpha,\beta)$ and $N(\alpha,\beta)$ depend on α and β but not otherwise on $a(x)$ nor on h. From (3.8), (3.9) and the basic estimate (2.9) we immediately have

<u>Theorem a</u>. The error $u_0 - u_h$ satisfies

$$\|u_0 - u_h\|_{\mathring{H}_1} \le C(\alpha,\beta)h\|f\|_{L_2} \tag{3.10}$$

where $C(\alpha,\beta) = [1 + \gamma^{-1}(\alpha,\beta)M(\alpha,\beta)]N(\alpha,\beta)$ depends only on α and β. Thus we have a 1st order estimate for $\|u_0 - u_h\|_{\mathring{H}_1}$ which is uniform over all $a(x)$ satisfying $\alpha \le a(x) \le \beta$. For a proof of (3.8) and (3.9) see [3]. The estimate (3.10) is the best possible estimate for $f \in L_2$ This can be seen by an application of the theory of N-widths (see, e.g., [2, 8]).

(3.9) shows that $S_{1,h}$, as defined in (3.3), yields accurate approximations to the solutions u_0 of (3.1). (3.8) shows that (2.5) holds with our choice of $S_{1,h}$ and $S_{2,h}$, as defined in (3.4). Thus our trial and test spaces have been chosen in accordance with the suggestions in Remark 2 in Section 2. We further note that if $\phi_1,\ldots,\phi_N$ is the usual basis for $S_{2,h}$ defined by $\phi_i(x_j) = \delta_{ij}$ and $\tilde{\phi}_1,\ldots,\tilde{\phi}_N$ is the basis for $S_{1,h}$ defined by $\tilde{\phi}_i(x_j) = \delta_{ij}$, then the matrix of (3.7) (the stiffness matrix) is given by

$$\int_0^1 a\tilde{\phi}_j'\phi_i'dx = \int_0^1 \underline{a}_h\phi_j'\phi_i'dx \tag{3.11}$$

where $\underline{a}_h$ is the piecewise harmonic average of $a(x)$, i.e., the step function defined by

$$a_h\Big|_{I_j} = \left(\frac{\int_{I_j} \frac{dx}{a}}{h_j}\right)^{-1}. \tag{3.12}$$

Thus the stiffness matrix is symmetric (which is not immediate since we are using different trial and test spaces) and is as easily computed as is the stiffness matrix for the standard method employing $S_{2,h}$ for both trial and test space, namely $\int_0^1 \bar{a}_h\phi_j'\phi_i'dx$, where $\bar{a}_h$ is the piecewise average of $a(x)$.

The accuracy and robustness of the approximation u_h is further shown by the estimate

$$|u(x_j) - u_h(x_j)| \le C(\alpha,\beta)V_0^1(a)h^2\|f\|_{L_\infty}, \; j=1,\ldots,n-1 \tag{3.13}$$

where $V_0^1(a)$ denotes the total variation of $a(x)$. Thus we have second order convergence at the nodes even if $a(x)$ has several jumps. The proof of (3.13), which does not follow the lines suggested by (2.9), can be found in [3].

b. <u>A Special Class of Two Dimensional Boundary Value Problems with Rough Coefficients</u>

Consider the problem

$$Lu = -(a(x)u_x)_x - (b(y)u_y)_y = f(x,y), (x,y) \in \Omega = [0,1]^2$$
$$u = 0 \quad \text{on} \quad \Gamma = \partial\Omega \tag{3.14}$$

where $0 < \alpha \leq a(x), b(y) \leq \beta$ and $f \in L_2$. This problem generalizes the model problem considered in a . The variational formulation of (3.14) is

$$u \in \mathring{H}_1(\Omega)$$
$$B(u,v) = f(v) \quad \forall v \in \mathring{H}_1(\Omega) \tag{3.15}$$

where

$$B(u,v) = \int_\Omega (au_xv_x + bu_yv_y)dxdy$$

and

$$f(v) = \int_\Omega fvdxdy.$$

This is of the form (2.10) with $H_1 = H_2 = \mathring{H}_1$.

Let T_h be a uniform triangulation or mesh on Ω with triangles of size h, as shown in Fig. 1. The vertices of the triangles $T \in T_h$ are

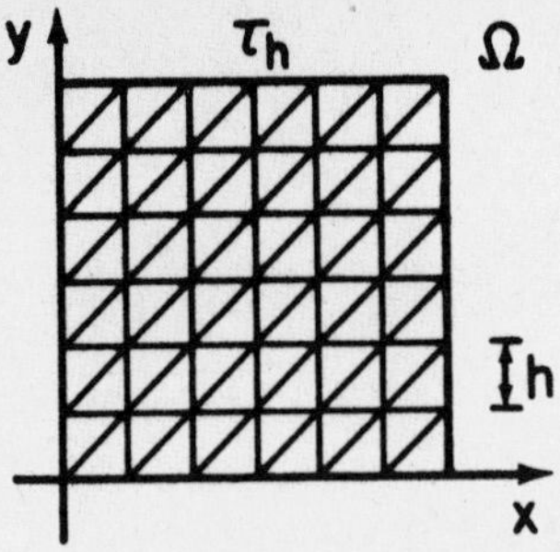

Fig. 1

called the nodes of T_h. By analogy with the definition of $S_{1,h}$ in a , we choose as trial space

$$S_{1,h} = \{u\text{: For each } T \in T_h,\ u\big|_T = \text{linear combination of } 1,$$

$\int^x \frac{dt}{a(t)}$, and $\int^y \frac{ds}{b(s)}$, u continuous at the nodes, and $u = 0$ at the boundary nodes}. (3.16)

For the test space we choose

$S_{2,h}$ = {v: For each $T \in \mathcal{T}_h$, $v\Big|_T$ = linear combination of 1, x, and y, u continuous at the nodes, and $u = 0$ at the boundary nodes}. (3.17)

In contrast to the situation in Example a, the nodal constraints imposed on the functions in $S_{1,h}$ do not imply the functions are continuous and we have $S_{1,h} \not\subset \mathcal{H}_1$, i.e., the trial space is nonforming. $S_{2,h} \subset \mathcal{H}_2$ in both examples.

We now define

$$\mathcal{H}_{1,h} = \mathcal{H}_{2,h} = H_{1,h} = \{u\colon u \in L_2(\Omega),\ u\Big|_T \in H_1(T) \quad \text{for all} \quad T \in \mathcal{T}_h\},$$

$$\|u\|_{\mathcal{H}_{1,h}} = \|u\|_{\mathcal{H}_{2,h}} = \|u\|_{H_{1,h}} = \Big[\int_\Omega u^2 dxdy + \sum_{T\in\mathcal{T}_h} \int_T |\nabla u|^2 dxdy\Big]^{1/2},$$

and

$$B_h(u,v) = \sum_{T\in Th} \int_T (ay_x v_x + bu_y v_y)dxdy.$$

B_h is defined on $H_{1,h} \times H_{1,h}$ and (2.3) holds with $M = \beta$.

The approximate solution is then defined by

$$u_h \in S_{1,h}$$
$$B_h(u_h,v) = \int_\Omega fvdxdy, \ \forall v \in S_{2,h}. \tag{3.18}$$

It is possible to show that (2.5) holds, i.e., that

$$\inf_{\substack{u\in S_{1,h} \\ \|u\|_{H_{1,h}}=1}} \ \sup_{\substack{v\in S_{2,h} \\ \|v\|_{H_{1,h}}=1}} |B_h(u,v)| \ge \gamma(\alpha,\beta) > 0 \quad \forall h \tag{3.19}$$

where $\gamma(\alpha,\beta)$ depends on α and β but not otherwise on $a(x)$ and $b(y)$. We have also shown that the functions in $S_{1,h}$ approximate the unknown solution well. In fact

$$\inf_{\chi\in S_{1,h}} \|u - \chi\|_{H_{1,h}} \le C(\alpha,\beta)h\|u\|_{H_{L,h}} \tag{3.20}$$

where

$$H_{L,h} = \{u\colon u \in L_2(\Omega),\ u \in H_1(T),\ au_x,\ bu_y \in H_1(T)\ \forall T \in \mathcal{T}_h,\ u \text{ is continuous at the nodes of } \mathcal{T}_h, \text{ and } u = 0 \text{ at the boundary nodes}\}$$

and

$$\|u\|^2_{H_{L,h}} = \|u\|^2_{H_{1,h}} + \sum_T \int_T \{|(au_x)_x|^2 + 2ab|u_{xy}|^2 + |(bu_y)_y|^2\}dxdy.$$

It is easy to see that $u,\ au_x,\ bu_y \in H_1(T)$ implies u is continuous on T and so the requirement of continuity at the nodes makes sense. Now, combining (3.19), (3.20), and (2.9) we have

$$\|u - u_h\|_{H_{1,h}} \le C(\alpha,\beta)h\|u\|_{H_{L.h}} \quad \text{for all } h. \tag{3.21}$$

We have been able to prove the following regularity result for (3.14) (or (3.15)): If $f \in L_2$, then $u \in H_L$ and

$$\|u\|_{H_L} \le C(\alpha,\beta)\|f\|_{L_2} \tag{3.22}$$

where

$$H_L = \{u\colon u \in \mathring{H}_1(\Omega),\ au_x,\ bu_y \in H_1(\Omega)\}$$

and

$$\|u\|^2_{H_L} = \|u\|^2_{H_1} + \int_\Omega \{|(au_x)_x|^2 + 2ab|u_{xy}|^2 + |(bu_y)_y|^2\}dxdy.$$

Because it is immediate that $\|u\|_{H_{L,h}} = \|u\|_{H_L}$ for any $u \in H_L$, from (3.21) and (3.22) we get

<u>Theorem b</u>. The error $u - u_h$ satisfies

$$\|u - u_h\|_{H_{1,h}} \le C(\alpha,\beta)\|f\|_{L_2} \quad \forall h. \tag{3.23}$$

(3.23) shows that we have first order convergence in the "energy norm," with an estimate that is uniform with respect to the class of coefficients satisfying $0 < \alpha \le a(x), b(y) \le \beta$.

If $\phi_1,\ldots,\phi_N$ form the usual basis for $S_{2,h}$ defined by $\phi_i(z_j) = \delta_{ij}$, for all interior nodes $z_1,\ldots,z_n$ of $\mathcal{T}_h$, and $\tilde{\phi}_1,\ldots,\tilde{\phi}_N$

form the basis for $S_{1,h}$ defined by $\phi_i(z_j) = \delta_{ij}$, for all interior nodes z_j, then the stiffness matrix of (3.18) is given by

$$B_h(\tilde{\phi}_j,\phi_i) = \int_\Omega (\underline{a}_h\phi_{j,x}\phi_{i,x} + \underline{b}_h\phi_{j,y}\phi_{i,x})\,dxdy \tag{3.24}$$

where $\underline{a}_h$ and $\underline{b}_h$ are the (one dimensional) piecewise harmonic averages of $a(x)$ and $b(y)$, respectively. The result in (3.24) is completely analogous to that expressed in (3.11) and (3.12) for Example a . We exphasize that although we are treating a two dimensional problem, the harmonic averages are one dimensional.

As we have seen, trial spaces with good approximation properties are required in an accurate method. It is thus natural to seek optimal trial functions. We have, in fact, been able to show that $S_{1,h}$, as defined in (3.16), is an optimal approximation space in the sense that

$$\sup_{\substack{u\in H_{L,h}\\ \|u\|_{H_{L,h}}=1}} \inf_{\chi\in S_{1,h}} \|u-\chi\|_{H_{1,h}} \le \sup_{\substack{u\in H_{L,h}\\ \|u\|_{H_{L,h}}=1}} \inf_{\chi\in S_N} \|u-\chi\|_{H_{1,h}} \tag{3.25}$$

for all $S_N \subset H_{L,h}$ with $\dim S_N = \dim S_{1,h} = N =$ the number of interior nodes of T_h. Therefore $S_{1,h}$ is optimal in the sence of N-widths (cf. [8]) relative to the norms $\|\cdot\|_{H_{1,h}}$ (the norm we are measuring the error in) and $H_{L,h}$ (a $2^{\underline{nd}}$ derivative norm in which u can be bounded by $\|f\|_{L_2}$). We have thus chosen an "ideal" approximation subspace for our problem. We have chosen the test space $S_{2,h}$, as defined in (3.17), so that $B_h(u,v)$ can be calculated from the input data for $v \in S_{2,h}$ (cf. (2.7)) and so that the inf-sup condition (2.5) holds. These features of the choices for the trial and test space lead to the accuracy of the method. We further note that the choice of $S_{2,h}$ led to an easily computed stiffness matrix. (3.23) is the precise statement of the accuracy and robustness of the method.

We note that (3.25) shows (in the case $a = b = 1$) that the space $S_{2,h}$ of continuous piecewise linear functions is an optimal approximating subspace in the case when the differential operator is the Laplacian. Furthermore, in an asymptotic sense, $S_{2,h}$ is optimal for problems with smooth coefficients.

c. <u>A Singular Two Point Boundary Value Problem</u>

Consider the problem

$$-(\sqrt{x}\,u')' = f(x),\ 0 < x < 1$$
$$u(0) = u(1) = 0. \tag{3.26}$$

Let

$$H_1 = H_2 = H = \{u: \int_0^1 \sqrt{x}\,(u')^2 dx < \infty,\ u(0) = u(1) = 0\}$$

and

$$\|u\|_{H_1}^2 = \|u\|_{H_2}^2 = \|u\|_H^2 = \int_0^1 \sqrt{x}\,(u')^2 dx.$$

The variational formulation of (3.26) is given by

$$u \in H$$
$$\int_0^1 \sqrt{x}\,u'v'dx = \int_0^1 fvdx\ \forall v \in H.$$

Given a mesh T_h (as in Example a), let

$$H_{1,h} = H_{2,h} = \{u: \int_{I_j} \sqrt{x}\,(u')^2 dx < \infty \quad \text{for} \quad j = 1,\ldots,n,$$
$$u(0) = u(1) = 0\}$$

and

$$\|u\|_{H_{1,h}}^2 = \|u\|_{H_{2,h}}^2 = \|u\|_{L_2}^2 + \sum_{j=1}^n \int_{I_j} \sqrt{x}\,(u')^2 dx.$$

We define the trial and test spaces by

$$S_{1,h} = \{u: \text{For each } j,\ u\big|_{I_j} = \text{a linear combination of } 1 \text{ and}$$
$$\sqrt{x},\ u \text{ continuous at the nodes, and } u(0) = u(1) = 0\}$$

and $S_{2,h} = C^\circ$ piecewise linear functions, as in Example a .
The approximate solution $u_h \in S_{1,h}$ is then characterized by

$$B_h(u_h, v) = \int_0^1 fvdx \quad \forall v \in S_{2,h}$$

where

$$B_h(u,v) = \sum_j \int_{I_j} \sqrt{x}\,u'v'dx.$$

One can show that (2.3) and (2.5) hold with M and γ independent of h. Furthermore, one can show that

$$\inf_{\chi \in S_{1,h}} \|u - \chi\|_H \le Ch \int_0^1 \frac{|[\sqrt{x}\,u')'|^2}{\sqrt{x}}\,dx.$$

Combining these facts with (2.9) leads immediately to

Theorem c. The error $u - u_h$ satisfies

$$\|u - u_h\|_H \le Ch\Big(\int_0^1 \frac{|f|^2}{\sqrt{x}}\,dx\Big)^{1/2} \quad \forall h. \tag{3.27}$$

We emphasize that (3.27) holds for an arbitrary mesh. It is not necessary to refine the mesh in the neighborhood of the singular point 0. The rate of convergence in (3.27) is the highest possible. This follows from the theory of N-widths. Problems similar to (3.26) have recently been considered in [6,11].

d. A Boundary Value Problem in a Domain with a Corner

Consider the model problem

$$\begin{aligned} -\Delta u &= f(x,y),\ (x,y) \in \Omega \\ u &= 0,\ (x,y) \in \Gamma = \partial\Omega \end{aligned} \tag{3.28}$$

where Ω is a polygonal domain with a convex angle (see Fig. 2), which

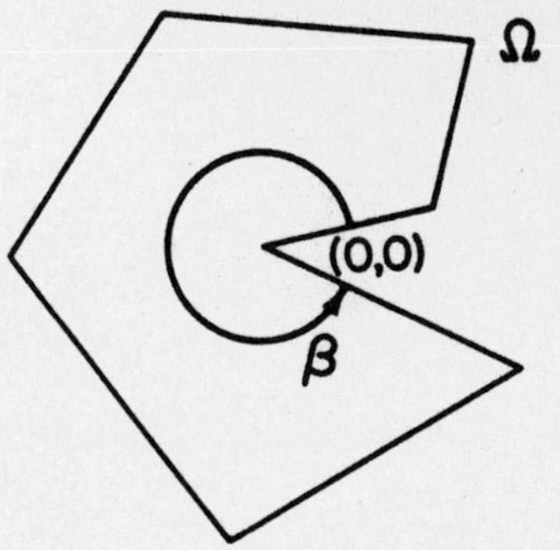

Fig. 2

we assume is placed at the origin, and $f \in C_0^\infty(\Omega)$. The solution u of (3.28) is singular at (0,0) and as a consequence the standard finite element method with piecewise linear approximating functions and a quasiuniform mesh gives an inaccurate approximate solution. In fact

$$\|u - u_h\|_{H_1(\Omega)} \geq Ch^{\rho}$$

with $\rho < 1$ depending on the angle β. (If $\beta \leq \pi$, then $\rho = 1$.) Appropriate refinement of the mesh in the neighborhood of the concave angle leads to the optimal convergence rate $O(N^{-1/2})$, where N is the number of degrees of freedom (the dimension of the trial and test space). A different way to achieve accuracy for problems with corners is to augment the trial space with singular functions (which will not have local supports). See, for example [5,7]. We will outline an approach that preserves the local nature of the finite element method by selecting special trial functions. The resulting method will have the highest possible rate of convergence, namely $O(h)$.

Toward this end suppose T_h is a quasi-uniform triangulation of Ω, let

$S_{1,h} = \{u$: For each $T \in T_h$, $u|_T$ is a linear combination of 1, $r^{\alpha} \cos \alpha\theta$, and $r^{\alpha} \sin \alpha\theta$, where $\alpha = \pi/\beta$ and (r,θ) are polar coordinates of (x,y), u is continuous at the nodes, and $u = 0$ at the boundary nodes$\}$,

and let $S_{2,h}$ be the usual C° piecewise linear functions relative to T_h that vanish on Γ. The triangulation will be required to satisfy certain technical restrictions in addition to being quasi-uniform. We describe these now. Let $0 < \tau$ and $0 < h << 1$ be given. Choose K sufficiently large, to be precise, choose $K \geq K(\beta,\gamma)$, where $K(\alpha,\tau)$ is an explicitly known expression. Triangulate the portion of Ω near the concave angle with several isoceles triangles such as shown in Fig. 3. Triangulate the rest of Ω with triangles

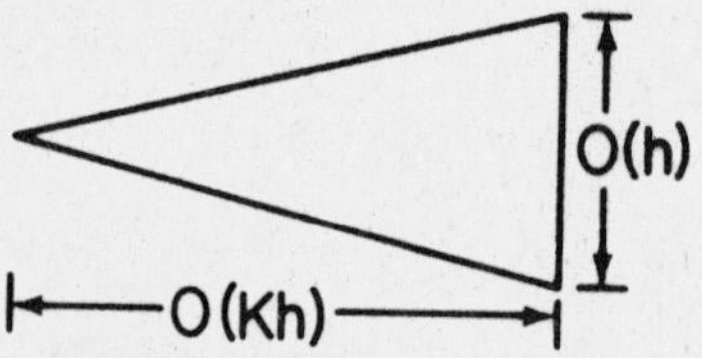

Fig. 3

of size h and minimal angle $\geq \tau$ as in Fig. 4. The bilinear form B_h, the approximate solution u_h, and the norm $\|\cdot\|_{H_{1,h}}$ are defined

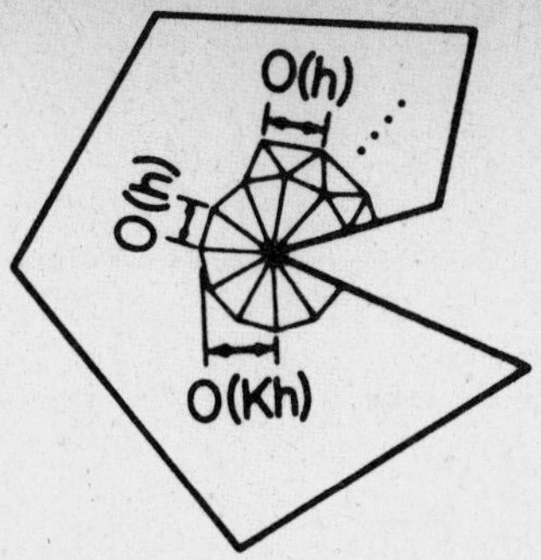

Fig. 4

in the obvious way. Then we can verify (2.3) and (2.5) and use (2.9) to prove

Theorem d. The error $u - u_h$ satisfies

$$\|u - u_h\|_{H_{1,h}} \leq \begin{cases} Ch|\log h|^{1/2}, & \beta = 2 \\ Ch, & \beta < 2 . \end{cases} \tag{3.29}$$

We thus have a first order estimate for the "energy" norm with a nearly uniform mesh.

e. An Interface Problem

Consider the problem

$$\begin{aligned} -\mathrm{div}\ (a\ \mathrm{grad}\ u) &= f(x,y),\ (x,y) \in \Omega \\ u = 0 \quad \text{on} \quad \Gamma &= \partial\Omega \end{aligned} \tag{3.30}$$

where Ω is a convex polygon and

$$a = \begin{cases} a_1 & \text{on } \Omega_1 \\ a_2 & \text{on } \Omega_2 \end{cases}$$

where a_1 and a_2 are positive constants and $\overline{\Omega}_1 \subset \Omega$, $\partial\Omega_1$ is smooth, and $\Omega_2 = \Omega - \overline{\Omega}_1$. The first derivatives of u have jumps across the interface $\partial\Omega_1$ and thus the standard finite element method is inaccurate unless the mesh is chosen so that the interface lies on edges of triangles in the trinagulation. We will now show that maximal accuracy can be achieved without aligning the mesh with the interface if we modify the trial functions properly.

Let T_h be a quasi-uniform triangulation or mesh on Ω. We describe a function $u \in S_{1,h}$.

i) For $T \in \mathcal{T}_h$,

if $T \cap \partial\Omega_1 = \phi$, $u|_T$ is linear

if $T \cap \partial\Omega_1 \neq \phi$, $u|_T$ is as follows:

Choose $Q \in \partial\Omega_1 \cap T$. Denote the tangent and normal to $\partial\Omega_1$ by t and n, respectively. Then

α) $u_1 = u|_{T\cap\Omega_1}$ and $u_2 = u|_{T\cap\Omega_2}$ are linear,

β) $u_1(Q) = u_2(Q)$,

γ) $\frac{\partial u_1}{\partial t}(Q) = \frac{\partial u_2}{\partial t}(Q)$.

δ) $a_1\frac{\partial u_1}{\partial n}(Q) = a_2\frac{\partial u_2}{\partial n}(Q)$.

ii) u is continuous at the nodes of $\mathcal{T}_h$.

Note that i)δ) is a requirement that the trial functions approximately model the interface condition $a_1 \frac{\partial u|_{\Omega_1}}{\partial n} = a_2 \frac{\partial u|_{\Omega_2}}{\partial n}$ satisfied by the exact solution. For test space we choose the usual C° piecewise linear functions. Having defined these spaces we then define the approximate solution by

$$u_h \in S_{1,h}$$

$$B_h(u,v) \equiv \sum_T \left(\int_{T\cap\Omega_1} a\nabla u \cdot \nabla v dxdy + \int_{T\cap\Omega_2} a\nabla u \cdot \nabla u dxdy\right) =$$

$$\int_\Omega f v dxdy \quad \forall v \in S_{2,h}. \tag{3.31}$$

For the method defined in (3.31) we have proved, with the aid of (2.9), the estimate

$$\|u - u_h\|_{\overline{H}_{1,h}} \le Ch\|f\|_{L_2} \quad \forall h \tag{3.32}$$

where

$$\|v\|^2_{\overline{H}_{1,h}} = \int_\Omega v^2 dxdy + \sum_T \left(\int_{T\cap\Omega_1} |\nabla v|^2 dxdy + \int_{T\cap\Omega_2} |\nabla v|^2 dxdy\right).$$

(3.32) is a first order estimate for the "energy" norm of the error. We emphasize that there is no relationship between the mesh and the interface. The method proposed here would provide an alternative to methods in which the interface is modeled by a mesh line, as in [9].

4. An Approach to the Design of Finite Element Methods for Problems with Rough Input Data

Our treatment of the Examples in Section 3 suggests an approach to the design of finite element methods for problems with rough input data. In this section we outline the steps in this approach.

a) Choose the space of right hand sides or source terms f to be considered. In all of our example we chose $f \in L_2$ except in Example 3.c where we considered f's satisfying $\int_0^1 \frac{f^2}{\sqrt{x}}\,dx < \infty$.

b) Find the space of solutions corresponding to the space of right-hand sides under consideration. This will involve a regularity result. For example, in Example 3.b , the space H_L is the space of solutions corresponding to f's in L_2. Often regularity results are available for problems with smooth data, but are not available in sufficient generality for problems with rough data.

c) Choose the mesh dependent bilinear forms and spaces (norms) B_h, $\mathcal{H}_{1,h}(\|\cdot\|_{\mathcal{H}_{1,h}})$, $\mathcal{H}_{2,h}(\|\cdot\|_{\mathcal{H}_{2,h}})$. This choice is usually very natural, following directly from the basic variational formulation, considered triangle by triangle.

d) Select trial spaces which have good (optimal) approximation properties. This is the major problem. Usually such spaces are closely related to local solutions of the equation under consideration. In many situations the proper choice leads to non-conforming functions. Inter-triangle continuity is enforces only at the nodes. The approximation properties of the trial functions is directly tied to the space of solutions (see b)). The problem of selecting optimal trial functions is often not simple. In practice one would like to find a trial space that performs almost as well as the optimal trial space but which is easily implemented.

e) Select a test space so as to ensure the inf-sup condition is satisfied with constant γ that is not too small and so that the stiffness matrix can be calculated. In contrast to the trial space, the test space is chosen to be conforming.

We have illustrated these steps on the relatively simple examples discussed in Section 3. We restricted our attention to first order methods in which the maximal rate of convergence was $\mathcal{O}(h)$. As mentioned in the introduction, the ideas in the paper can be generalized in various directions.

References

1. I. Babuška, Error-bounds for finite element method, Numer. Math., 16(1971), pp. 322-333.

2. I. Babuška and A. Aziz, Survey lectures on the mathematical foundations of the finite element method, in the Mathematical Foundations of the Finite Element Method with Applications to Partial Differential Equations, Academic Press, New York, 1973, A.K. Aziz, Editor, pp. 5-359.

3. I. Babuška and J. Osborn, Generalized finite element methods: Their performance and their relation to mixed methods, SIAM J. Numer. Anal. 10(1983), pp. 510-536.

4. P.G. Ciarlet, The Finite Element Method for Elliptic Problems, North Holland, New York, 1978.

5. M. Dobrowolski, Numerical Approximations of Elliptic Interface Problems, Habilitationsschrift, University of Bonn, 1981.

6. K. Eriksson and V. Thomée, Balerkin methods for singular boundary value problems in one space dimension, Technical Report No. 1982-11, Department of Mathematics, Chalmers University of Technology and the University of Göteborg.

7. G.J. Fix, S. Galati and T.I. Wakoff, On the use of singular functions with finite element approximation, J. Computational Phys. 13(1976), pp. 209-228.

8. A. Kolmogorov, Über die beste Annäherung vor Funktion einer gegeben Funktionenklasse, Ann. of Math. 37(1936), pp. 107-110.

9. O. McBryan, Elliptic and hyperbolic interface refinement in Boundary Layers and Interior Layers-Computational and Asymptotic Methods, Boole Press, Dublin, 1980, J. Miller, Editor.

10. A.H. Schatz and L.B. Wahlbin, Maximum norm estimates in the finite element method on plane polygonal domains. Part 1, Math. Comp. 32(1978), pp. 73=109.

11. R. Schreiber, Finite element methods of high-order accuracy for singular two-point boundary balue problems with nonsmooth solutions, SIAM Numer. Anal. 17(1980), pp. 547-566.

+ Institute for Physical Science and Technology and Department of Mathematics, University of Maryland, College Park, Maryland 20742. The work of this author was partially supported by the Office of Naval Research under contract N00014-77-C-0623.

‡ Department of Mathematics, University of Maryland, College Park, Maryland 20742. The work of this author was partially supported by the National Science Foundation under Grant MCS-78-02851.

TREATMENT OF A RE-ENTRANT VERTEX IN A THREE-DIMENSIONAL POISSON PROBLEM

A.E. Beagles and J.R. Whiteman
Institute of Computational Mathematics
and
Department of Mathematics and Statistics
Brunel University, Uxbridge, England

1. Introduction

The difficulties of obtaining accurate numerical solutions to Poisson problems containing boundary singularities are well known. In any adaptation of a numerical technique to treat a singular situation, it is advantageous to have knowledge of the form of the singularity. Although much work has been done on singularities in two-dimensional problems, with the result that for many cases the singular forms are known, three-dimensional problems have been much less studied so that the three-dimensional situation is less well understood.

Boundary singularities in three-dimensional problems can occur when the region of the problem contains a re-entrant edge or vertex, or when there is an abrupt change of boundary condition. The form of the singularity is determined by the combination of geometry and boundary conditions. For some three-dimensional configurations the exact form of the singularity has been derived analytically, see e.g. Walden and Kellogg [7], Stephan and Whiteman [6]. In this paper we consider the case of a particular vertex singularity, previously treated by Fichera [2], where analytic evaluation has not proved possible. For this we present a preliminary account of a numerical technique for estimating the singular form.

2. Three-Dimensional Problems with Re-entrant Vertices and Edges

We consider first the case where the function $u(\underline{x})$, $\underline{x} \in \mathbb{R}^3$, satisfies

$$-\Delta[u(\underline{x})] = f(\underline{x}) , \quad \underline{x} \in \Omega ,$$
$$u(\underline{x}) = 0 , \quad \underline{x} \in \partial\Omega , \tag{2.1}$$

where $f \in L_2(\Omega)$ and $\Omega \subset \mathbb{R}^3$ is a simply connected open bounded polyhedral domain with boundary $\partial\Omega$. The region $\overline{\Omega} \equiv \Omega \cup \partial\Omega$ may contain re-entrant vertices and edges as in Fig. 1, where edges result from the intersection of two planes and vertices from the intersection of three (or more) planes.

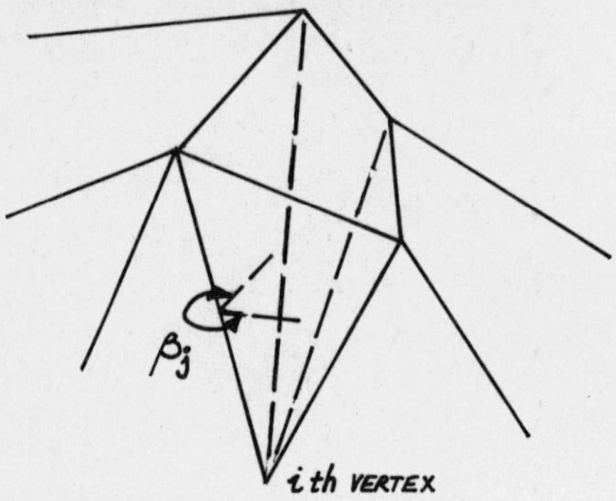

Fig. 1

The Sobolev spaces $H^1(\Omega)$ and $\overset{\circ}{H}{}^1(\Omega)$ are defined in the usual manner as

$$H^1(\Omega) \equiv \{v : v \in L_2(\Omega) \;,\; D^1 v \in L_2(\Omega)\} \;,$$
$$\overset{\circ}{H}{}^1(\Omega) \equiv \{v : v \in H^1(\Omega) \;;\; v(\underline{x}) = 0 \;,\; \underline{x} \in \partial\Omega\} \;.$$

In $\overset{\circ}{H}{}^1(\Omega)$ problem (2.1) has the well known weak form in which $u(\underline{x}) \in \overset{\circ}{H}{}^1(\Omega)$ satisfies

$$\int_\Omega \nabla u \, \nabla v \, d\underline{x} = \int_\Omega f \, v \, d\underline{x} \qquad \forall \; v(\underline{x}) \in \overset{\circ}{H}{}^1(\Omega) \quad . \tag{2.2}$$

It has been shown by Grisvard [4] that the solution $u(\underline{x})$ of (2.2) admits an expansion of the form

$$u(\underline{x}) = \sum_{\text{vertices}} a_i \chi_i(r_i,\theta_i,\phi_i) u_i(r_i,\theta_i,\phi_i)$$
$$+ \sum_{\text{edges}} b_j(z_j) \Xi_j(\rho_j,z_j,\phi_j) v_j(\rho_j,\phi_j) + \omega(\underline{x}), \tag{2.3}$$

where $\omega(\underline{x}) \in H^2(\Omega)$ and, in terms of spherical polar coordinates (r_i,θ_i,ϕ_i) and cylindrical polar coordinates (ρ_j,z_j,ϕ_j) respectively local to the ith vertex and j^{th} edge, $u_i(r_i,\theta_i,\phi_i)$ and $v_j(\rho_j,\phi_j)$ are vertex and edge *singular* functions not in $H^2(\Omega)$, whilst χ_i and Ξ_j are cut-off functions. The effect of the cut-off functions is to localise the influence of the singular functions to the neighbourhood of the vertex or edge. In (2.3) the a_i are constants and the $b_j(z_j)$ are edge functions, and each set depends both on the geometry of Ω and on the function $f(\underline{x})$ of (2.1).

In many practical situations these quantities have important physical significance.

In order to make use of the expansion (2.3) we need to know the forms of the singular functions u_i and v_j. These depend only on the geometry of the region and boundary conditions of the problem in the neighbourhood of the ith vertex or j^{th} edge and not on $f(\underline{x})$. They can thus be found by solving harmonic problems in regions chosen sufficiently local to the relevant vertex or edge for the coordinates to be separable.

For the jth edge, see Fig. 1, this entails the solution of a two-dimensional problem in a sector of the plane orthogonal to the edge through a point of the edge, as in Fig. 2, where the re-entrant angle is β_j. This two-dimensional solution will satisfy the homogeneous boundary conditions on the arms of the corner. In this case the separated

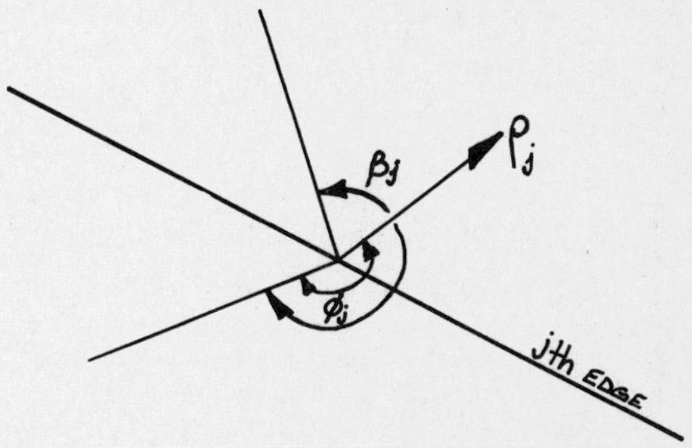

Fig. 2

equation has solutions

$$v_j^{(n)} = \rho_j^{\frac{n\pi}{\beta_j}} \sin\left(\frac{n\pi\phi_j}{\beta_j}\right) , \quad n = 1,2,... \tag{2.4}$$

where $\frac{n\pi}{\beta_j}$ is the nth eigenvalue, and $\sin \frac{n\pi\phi_j}{\beta_j}$ the corresponding eigenfunction of the problem in the angular variable for the separation constants. However, for $n \geqq 2$, $v_j^{(n)} \in H^2(\Omega)$ so that, with reference to (2.3), terms of this type are assumed to be subsumed in $\omega(\underline{x})$. Thus $v_j^{(1)}$ is the singular function appropriate to the jth edge. For the particular case of $\beta_j = \frac{3\pi}{2}$ we have from (2.4) that

$$v_j^{(1)} \equiv v_j = \rho_j^{\frac{2}{3}} \sin\left(\frac{2\phi_j}{3}\right) . \tag{2.5}$$

A similar procedure can be adopted for the ith vertex where the singular function $u_i(r_i,\theta_i,\phi_i)$ can be found by solving a three-dimensional harmonic problem in that part of Ω which is interior to the unit sphere centred on the vertex with the boundary conditions of (2.1) on those parts of the planes meeting at the vertex cut off by the sphere. It is assumed that the size of Ω is such that the unit sphere contains no vertices of $\partial\Omega$ other than the ith nor points of any faces other than those defining the ith vertex so that, inside the intersection of the unit sphere and $\Omega, u_i(r_i,\theta_i,\phi_i)$ is separable in spherical polar coordinates into the product of a function $R_i(r_i)$ and a function $Q_i(\theta_i,\phi_i)$. On separating the variables in the harmonic problem we find that

$$R_i(r_i) = r_i^{\alpha_i^{(n)}} \quad ,$$

with $\alpha_i^{(n)} > 0$, $\alpha_i^{(n)}(\alpha_i^{(n)} + 1) = \lambda_n$ and $\lambda_n (\equiv \lambda_{i,n})$ is the nth eigenvalue (with corresponding eigenfunction $Q_i^{(n)}$) of the Laplace-Beltrami eigenvalue problem, see Grisvard [4],

$$- \Delta_{\theta_i\phi_i}[Q_i(\theta_i,\phi_i)] = \lambda Q_i(\theta_i,\phi_i) \ , \quad (\theta_i,\phi_i) \in G_i \ ,$$

$$Q_i(\theta_i,\phi_i) = 0 \ , \quad (\theta_i,\phi_i) \in \partial G_i \ , \tag{2.6}$$

where G_i is that part of the surface of the unit sphere centred on the ith vertex interior to Ω, ∂G_i consists of the arcs of intersection of the surface with the planes forming the vertex, and

$$\Delta_{\theta\phi} \equiv \frac{1}{\sin\theta}\frac{\partial}{\partial\theta}\left(\sin\theta\frac{\partial}{\partial\theta}\right) + \frac{1}{\sin^2\theta}\frac{\partial^2}{\partial\phi^2} \ .$$

It is in fact possible to take $n = 1$, giving the smallest eigenvalue, since the second eigenvalue for the most singular case possible, that of a line vertex, produces a "singular function" $\tilde{u}_i \in H^2(\Omega)$. It should also be noted that, since in three-dimensions $r^\alpha \in H^2(\Omega)$, $\alpha \geq \frac{1}{2}$, only values $\lambda_1 < \frac{3}{4}$ produce an $R_i \notin H^2(\Omega)$.

3. Problem with Re-entrant Vertex

We now consider the particular case in which problem (2.1) has a re-entrant vertex at 0, formed by the intersection of three mutually orthogonal planes, see Fig. 3, and we investigate the form of the singular function relevant to 0.

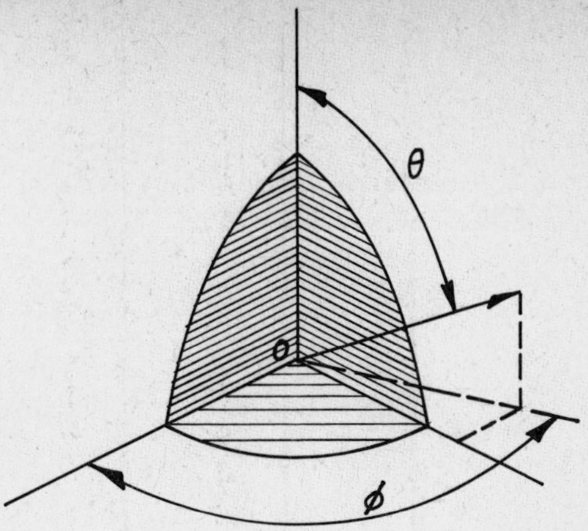

Fig. 3

When the suffix i is omitted, using the coordinates of Fig. 3, on the unit sphere

$$G \equiv \left\{ \begin{array}{ll} (\theta,\phi) \; ; & 0 < \theta < \pi/2 \; , \quad \pi/2 < \phi < 2\pi \\ & \pi/2 < \theta \leqq \pi \; , \quad 0 \leqq \phi \leqq 2\pi \end{array} \right\} \tag{3.1}$$

$$\partial G \equiv \left\{ \begin{array}{ll} (\theta,\phi) \; ; & \theta = \pi/2 \; , \; 0 \leqq \phi \leqq \pi/2 \\ & \phi = \pi/2 \; , \; 0 \leqq \theta \leqq \pi/2 \quad . \\ & \phi = 2\pi \; , \; 0 \leqq \theta \leqq \pi/2 \end{array} \right\} \tag{3.2}$$

For the problem

$$\begin{aligned} -\Delta_{\theta\phi}[Q(\theta,\phi)] &= \lambda Q(\theta,\phi) \quad \text{in} \quad G \\ Q(\theta,\phi) &= 0 \qquad\qquad \text{on } \partial G \; . \end{aligned} \tag{3.3}$$

the smallest eigenvalue λ_1 and corresponding eigenfunction $Q^{(1)}$ must be found so that

$$\alpha^{(1)} = -\frac{1}{2} + \sqrt{\frac{1}{4} + \lambda_1} \tag{3.4}$$

can be calculated, thus determining at the vertex the singular function

$$u(r,\theta,\phi) = r^{\alpha^{(1)}} Q^{(1)}(\theta,\phi) \; . \tag{3.5}$$

4. Numerical Evaluation of Singular Form

A weak form of problem (3.3) is produced by multiplying the differential equation by $v \in \overset{\circ}{H}{}^1(G)$ and integrating by parts over G, using the measure for the spherical surface $dG \equiv \sin\theta \, d\theta \, d\phi$. Thus $Q \in \overset{\circ}{H}{}^1(G)$ satisfies

$$a(Q,v) = \lambda b(Q,v) \qquad \forall \quad v \in \overset{\circ}{H}{}^1(G) \,, \tag{4.1}$$

where

$$a(Q,v) \equiv \int_G \left\{ \frac{\partial Q}{\partial \theta} \frac{\partial v}{\partial \theta} \sin\theta + \frac{1}{\sin\theta} \frac{\partial Q}{\partial \phi} \frac{\partial v}{\partial \phi} \right\} d\theta \, d\phi \,, \quad Q,v \in \overset{\circ}{H}{}^1(G) \,, \tag{4.2}$$

$$b(Q,v) \equiv \int_G Qv \sin\theta \, d\theta \, d\phi \,, \quad Q,v \in \overset{\circ}{H}{}^1(G) \,. \tag{4.3}$$

It will be noted that the second part of the integrand in (4.2) contains the term $\frac{1}{\sin\theta}$ which tends to ∞ as $\theta \to 0,\pi$. This effect occurs as a result of the coordinate system and does not cause difficulties.

Problem (4.1) is solved numerically using the finite element method. For this the spherical surface G is considered in the (θ,ϕ)-plane, as in Fig. 4a, with periodic boundary conditions $Q(\theta,0) = Q(\theta,2\pi), \pi/2 \leqq \theta \leqq \pi$,

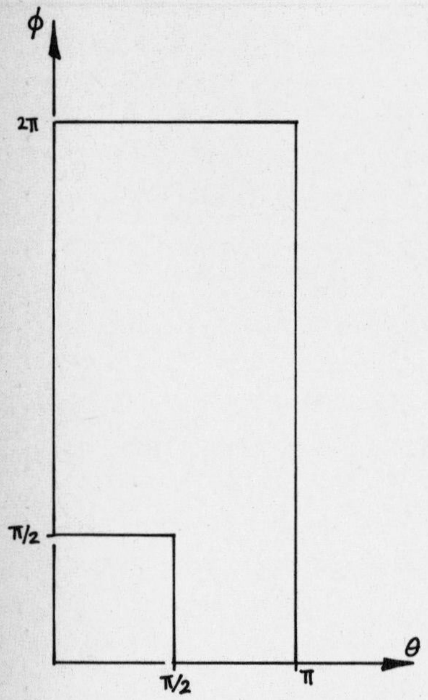

Fig. 4a

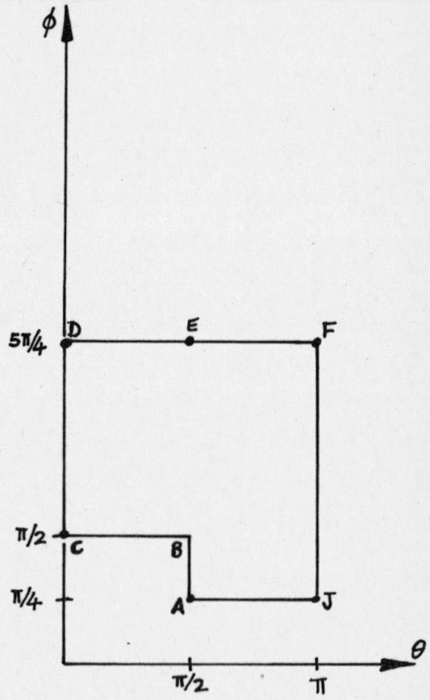

Fig. 4b

and $Q(\theta,\phi) = 0$ on the remainder of the boundary. The edge $\theta = \pi$ in Fig. 4a is in fact an interior point of the spherical surface G and so a natural condition holds there. However, problem (4.1) is symmetric about the plane $\phi = \pi/4, 5\pi/4$, with the result that the eigenfunction, corresponding to the smallest eigenvalue, also has this symmetry. Thus in the numerical method only that part of the problem in the region $\tilde{G}$ of Fig. 4b in the (θ,ϕ)-plane, where $\pi/4 \leq \phi \leq 5\pi/4$, is considered, and homogeneous Neumann boundary conditions are imposed on $\phi = \pi/4, \pi/2 \leq \theta \leq \pi$ and $\phi = 5\pi/4$, $0 \leq \theta \leq \pi$.

The region $\tilde{G}$ is partitioned using a mesh of rectangular elements and piecewise bilinear basis functions $\{B_i(\theta,\phi)\}_{i=1}^{n} \in \tilde{H}^1(\tilde{G})$ are defined in the usual manner, where

$$\tilde{H}^1(\tilde{G}) \equiv \{v : v \in H^1(\tilde{G}) , \quad v = 0 \text{ on } \overline{ABCDE}\} \quad .$$

The finite dimensional space $\tilde{S}^h \subset \tilde{H}^1(\tilde{G})$ is formed using the B_i's, giving the problem approximating (4.1) : find $Q_h \in \tilde{S}^h$ and λ^h such that

$$a(Q_h, v_h) = \lambda^h b(Q_h, v_h) \qquad \forall\, v_h \in \tilde{S}^h \ , \tag{4.4}$$

where Q_h and λ^h respectively are approximations to Q and λ based on a mesh of length h. Problem (4.4) leads to the generalised matrix eigenproblem

$$A\,\underline{Q}^h = \lambda^h B\,\underline{Q}^h \tag{4.5}$$

where the matrices A and B are symmetric and $\underline{Q}^h$ is the vector of nodal point evaluations of Q_h.

The matrices A and B are set up using the MODEL finite element code, see [1] and [3], with Gaussian quadrature employed to perform the integrations in each element. In this the quadrature points for elements involving the lines $\theta = 0,\pi$ are sufficiently far from these lines to avoid possible difficulties due to the $1/\sin\theta$ term in (4.2). System (4.5) in which of course we incorporate the essential boundary conditions of problem (4.1) in $\tilde{G}$, is then *solved* using NAG routines, [5], by being reduced to tridiagonal form using similarity transformations and then bisection to locate the first "non-zero" eigenvalue λ_1^h, followed by inverse iteration on the full system to find the corresponding eigenvector $\underline{Q}^h$.

The following values for λ_1^h were calculated using meshes of square elements of side h with bilinear test and trial functions.

h	λ_1^h
$\pi/4$	0.707
$\pi/8$	0.678
$\pi/12$	0.670
$\pi/16$	0.667

5. Discussion

The process of solving (4.1) for λ_1 is equivalent to finding the minimum of the Rayleigh quotient,

$$\lambda_1 = \min_{Q \in \overset{\circ}{H}{}^1(G)} \left\{ \frac{a(Q,Q)}{b(Q,Q)} \right\} .$$

The values given above have in effect been calculated over a subspace S^h of $\widetilde{H}^1(\widetilde{G})$. If the integrations in a(.,.) and b(.,.) had been evaluated exactly, λ_1^h would be an *upper* bound on λ_1. Fichera [2] using a modification of the Faber-Krahn principle gives the value 0.6215 as a *lower* bound for λ_1. Thus the value λ_1^h, $h = \pi/16$, together with the Fichera value indicates that $0.6215 < \lambda_1 < 0.667$, so that from (3.4) correspondingly $0.4335 < \alpha^{(1)} < 0.4576$.

In the above no attempt has been made to *optimise* the Galerkin technique (e.g. by local mesh refinement) to cater for the singularities in the solution of (4.1) due to the corners in G. Clearly this affects the accuracy and rate of convergence of the numerical method. As ultimately a method is required to treat general vertices, we have also not taken full advantage of the symmetry of this special problem.

It is clear that a theoretical error analysis for the above technique is required. Work is currently in progress to this end. Finally it must be emphasised that the motivation behind the derivation of the singular form is the construction of suitable finite element methods for treating three-dimensional problems with singularities. Whether these methods are based on adaptive mesh refinement, singular elements or augmentation, knowledge of the singular forms is required.

References

1. Akin, J.E., Application and Implementation of Finite Element Methods. Academic Press, London, 1982.
2. Fichera, G., Asymptotic behaviour of the electric field and density of the electric charge in the neighbourhood of singular points of a conducting surface. Russian Math. Surveys 30, 107-127,1975.
3. Harrison, D., Ward, T.J.W. and Whiteman, J.R., Proc. 4th Int. Seminar on Finite Element Systems. Computational Mechanics Centre, Southampton, July 1983.
4. Grisvard, P., Behaviour of the solutions of an elliptic boundary value problem in a polygonal or polyhedral domain. pp.207-274 of B. Hubbard (ed.), Numerical Solution of Partial Differential Equations III, SYNSPADE 1975. Academic Press, New York, 1976.
5. NAG Routines f01buf, f01bvf, f01bwf, f02bff and f02sdf, NAG Manual Vol.4. Numerical Algorithms Group, Oxford.

6. Stephan E. and Whiteman, J.R., Singularities of the Laplacian at corners and edges of three dimensional domains and their treatment with finite element methods. Technical Report BICOM 81/1. Institute of Computational Mathematics, Brunel University, 1981.
7. Walden, H. and Kellogg, R.B., Numerical determination of the fundamental eigenvalue of the Laplacian operator on a spherical domain. J. Eng. Math. 11, 299-318, 1977.

ON THE APPROXIMATION OF LINEAR ELLIPTIC SYSTEMS ON POLYGONAL DOMAINS

H. Blum
Institut für Angewandte Mathematik
Universität Bonn
5300 Bonn 1
Federal Republic of Germany

SUMMARY. Least squares approximations of first order systems are considered for a model problem. In the presence of corner singularities the convergence may become arbitrarily slow for increasing inner angles. For non-convex polygonal domains the method is even divergent.
For improvement, the method of dual singular functions (DSFM) is proposed. In the DSFM the approximating spaces are augmented by singular functions of the continuous problem. The so-called stress intensity factors (SIF) are approximated by means of an integral representation formula which is obtained using the singular functions of the adjoint problem. Here, the solution and the SIF are calculated with the same rates as is done by the non-modified method for the regular part of the solution. Finally, in some numerical examples the DSFM is compared with the well-known singular function method.

1. LEAST SQUARES APPROXIMATION

As a model problem we consider the first order system

$$Lv = \begin{pmatrix} \partial_x & \partial_y \\ -\partial_y & \partial_x \end{pmatrix} \begin{pmatrix} v_1 \\ v_2 \end{pmatrix} = \begin{pmatrix} f_1 \\ f_2 \end{pmatrix} = f \quad \text{in } \Omega \tag{1}$$

together with the boundary conditions

$$Bv = v \cdot t = 0 \quad \text{on } \partial\Omega. \tag{2}$$

Here, Ω denotes a polygonal domain in the plane $\mathbb{R}^2$ and t is a unit tangent vector at the boundary. For $f_2 = 0$ this is just the dual form of Poisson's equation $-\Delta u = -f_1$, $u = \text{const.}$ on $\partial\Omega$ with $v = \nabla u$.
The operator L is homogeneous elliptic and is covered by the boundary condition B. Moreover, we have a Poincaré-type inequality

$$\| v \|_o \leq c \| Lv \|_o \quad \forall v \in V, \tag{3}$$

where $V = \{v \in (L^2)^2 \mid Lv \in (L^2)^2,\ v \cdot t = 0 \text{ on } \partial\Omega\}$ is the space of admissible functions. In the interior of Ω and in subdomains Ω' with positive distance from possible singular points the usual shift theorems hold

$$\| v \|_{m+1;\Omega'} \leq c \| Lv \|_{m;\Omega}, \quad m \geq 0. \tag{4}$$

Here and in the following $\| . \|_{m;\Omega}$ denotes the norm of the usual Sobolev space $H^m(\Omega) = H^{m,2}(\Omega)$ of L^2-functions with m-th order derivatives in $L^2(\Omega)$.

First order systems like (1),(2) are frequently approximated by least squares methods, see e.g. [1]. We define the spaces of piecewise linear H^1-fields

$$V_h = \{\phi_h \in (S_h^2)^2 | \phi_h \cdot t = 0 \text{ on } \partial\Omega\} \subset V$$

with respect to a quasiregular subdivision of Ω into triangles.

We assume for the moment that (4) holds for $\Omega' = \Omega$ and $m = 1$, e.g. if Ω is smooth. In this case the H^1-norm and the L-norm $\| L. \|_o$ are equivalent and optimal convergence for the least squares method is easily obtained.

<u>Theorem 1</u>.([1]) (i) The solution $Pv \in V_h$ of

$$\| L(v-Pv) \|_o \leq \| L(v-\phi_h) \|_o \quad \forall \phi_h \in V_h$$

is uniquely determined.

(ii) The L^2-convergence rates are optimal

$$\| v-Pv \|_o + h \| L(v-Pv) \|_o \leq ch^2 \| \nabla^2 v \|_o .$$

Proof. The uniqueness of Pv follows from (3), the order O(h) for the error in the L-norm is obtained using $v \in H^2$. For the L^2-estimate the auxiliary problem

$$(Lw, L\phi) = (v-Pv, \phi) \quad \forall \phi \in V$$

has to be solved. But it is an easy consequence of the regularity assumption (4) that this problem has a unique solution $w \in H^2$.

QED

2. APPROXIMATION ON POLYGONAL DOMAINS

The preceding theorem shows that the least squares approach leads to very good results as long as the solution v is sufficiently regular. The gradient of the solution of Poisson's equation is approximated with $O(h^2)$-accuracy by linear finite elements, i.e. we gain one additional power of h compared to the discretization of the primal formulation. On polygonal domains, however, regularity can not be guaranteed and the method turns out to be inefficient in the presence of possible corner

singularities, see theorem 4 below.
The asymptotic behaviour of solutions of elliptic problems near angular corners has been investigated by Kondrat'ev [6] and many others for a single equation. The analysis was completed by Maz'ja-Plamenevskij during the seventieth. They also extended the results to general linear systems in the Douglis-Nirenberg sense, see e.g. [7].
For simplicity we only consider the case where Ω has one significant corner 0 with inner angle ω. Let (r,ϕ) denote the polar coordinates with respect to 0. We define the weighted Sobolev spaces

$$H^m_\beta(\Omega) = \{u \in (H^m_{loc}(\Omega))^2 \,|\, r^{\beta/2-m+k}\nabla^k u \in (L^2(\Omega))^2\}$$

with the natural norms

$$\|u\|^2_{m;\beta} = \Sigma^m_{k=0} \int_\Omega r^{\beta-2m+2k}|\nabla^k u|^2 dx \;.$$

Using this notation we can describe the asymptotic behaviour near 0 as follows.

<u>Theorem 2</u>.(Maz'ja-Plamenevskij) Let $k \geq 0$, $\beta \in \mathbb{R}$ such that $H^k_\beta(\Omega) \subset (L^2(\Omega))^2$, $f \in H^k_\beta(\Omega)$. Then the solution v of (1),(2) admits the representation

$$v = \Sigma_{i<n(k,\beta)}\, k_i s_i + w \;. \tag{5}$$

For the regular part w we have the a priori estimate

$$\|w\|_{k+1;\beta} \leq c\|f\|_{k;\beta} \tag{6}$$

The so-called singular functions s_i have the form

$$s_i = \tau\tilde{s}_i = \tau r^{i\alpha-1}\,(\sin(i\alpha-1)\phi \;,\; \cos(i\alpha-1)\phi)^T \;,\; \alpha = \pi/\omega$$

and $n(k,\beta)$ is chosen such that $s_{n+1} \in H^{k+1}_\beta$ but $s_n \notin H^{k+1}_\beta$. The functions $\tilde{s}_i$ satisfy the homogeneous equations (1),(2) and so do s_i in a neighbourhood of 0. τ is a cut-off function with respect to 0.

In addition to (5) Maz'ja and Plamenevskij have proved a representation formula for the coefficients k_i, the so-called stress intensity factors (SIF). We shall make use of it in the following two forms.

<u>Theorem 3</u>. For k_i we have the following identities

$$\text{(i)}\quad k_i = (f,s_{-i}) - (v,L^* s_{-i}) \tag{7}$$

$$\text{(ii)}\quad k_i = (f,\bar{s}_{-i}) := (f,s_{-i}) - (f,s'_{-i}) \;. \tag{8}$$

Here, $s_{-i} = \tau\tilde{s}_{-i}$ are the singular functions of the dual boundary value problem with

$$L^* w = \begin{pmatrix} -\partial_x & \partial_y \\ -\partial_y & -\partial_x \end{pmatrix} \begin{pmatrix} w_1 \\ w_2 \end{pmatrix} , \; B^* w = w_1 \; , \; \tilde{s}_{-i} = -(1/\omega) r^{-i\alpha} (\sin i\alpha\phi \, , \, \cos i\alpha\phi \,)^T .$$

The function s'_{-i} is obtained by solving

$$L^* s'_{-i} = L^* s_{-i} \quad \text{in } \Omega,$$
$$B^* s'_{-i} = 0 \qquad \text{on } \partial\Omega$$

for the smooth right hand side $L^* s_{-i}$. Thus, the function $\bar{s}_{-i}$ solves the homogeneous problem for L^*, B^* and has the characteristic singular behaviour of s_{-i}.

One should note that both formulae in theorem 3 do not give any information how to calculate the stress intensity factors directly from the data since neither v nor s'_{-i} are known. Both of them, however, are useful for theoretical considerations and for numerical purposes, see chapter 3.

Let us now consider how the presence of corner singularities may influence the convergence behaviour of the least squares approach. The regularity of a solution v of (1),(2) will in general be reflected in the size of the exponent of the first singular function s_1, namely α-1. Since this number is less than 1 for $\omega > \pi/2$ and thus $s_1 \notin H^2$ the assumptions of theorem 1 are not satisfied. The following theorem shows the reduced convergence rates in that case.

Theorem 4.

(i) $$\| L(v-Pv) \|_0 = \begin{cases} O(h) & , \alpha \geq 2 \\ O(h^{\alpha-1}) & , 1 < \alpha < 2 \\ O(1) & , \alpha < 1 \end{cases}$$

(ii) $$\| (v-Pv) \|_0 = \begin{cases} O(h^2) & , \alpha \geq 2 \\ O(h^{2(\alpha-1)}) & , 1 < \alpha < 2 \\ O(1) & , \alpha < 1 \end{cases}$$

(iii) All results are sharp.

Remark. The last statement especially means that the least squares method does not converge on non-convex domains, if the solution contains the first singular function.

Proof. Part (i) and (ii) of the theorem are easily obtained by estimation of the interpolation error and duality arguments. Furthermore, for $\alpha \geq 2$ there is nothing to show in part (iii).

In the case $1 < \alpha < 2$ the L-norm and the H^1-norm are equivalent. Thus, we get the estimate from below

$$\| L(s_1 - Ps_1) \|_o \geq ch^{\alpha-1}$$

by analysis of the error in radial direction. The main point is the divergence result for $\alpha < 1$. To see this, we simply use (8) for the SIF of s_1, namely

$$1 = (Ls_1, \bar{s}_{-1}) .$$

On the other hand, an H^1-spline Ps_1 can not contain s_1 for $\alpha < 1$, i.e.

$$0 = (LPs_1, \bar{s}_{-1}).$$

Finally, we note that $\bar{s}_{-1} \in L^2$, since $\bar{s}_{-1} \sim r^{-\alpha}$ and $\alpha < 1$. Thus, we may use Hölder's inequality

$$1 = (L(s_1 - Ps_1), \bar{s}_{-1}) \leq \| L(s_1 - Ps_1) \|_o \| \bar{s}_{-1} \|_o \tag{9}$$

which proves divergence in the L-norm.
There remains to show the sharpness of the L^2-estimates for $\alpha < 2$. We observe that $L^* Ls_1 \in L^2$ and $(Ls_1)_1 = 0$ on $\partial\Omega$. Therefore,

$$\| s_1 - Ps_1 \|_o \geq \frac{(L^* Ls_1, s_1 - Ps_1)}{\| L^* Ls_1 \|_o} = \frac{\| L(s_1 - Ps_1) \|_o^2}{\| L^* Ls_1 \|_o} . \tag{10}$$

This shows that duality is sharp for our problem.

QED

Remark. The lower bounds in the estimates (9) and (10) are independent of the mesh. Therefore, the well-known refinement techniques will not lead to better results in the non-convex case.

3. THE METHOD OF DUAL SINGULAR FUNCTIONS (DSFM)

In this chapter we want to define an improved approximation procedure for (1),(2) which is based on the singular structure of the solution. Here, the trial spaces V_h are augmented by several singular functions s_i

$$V_h^n := V_h \cup \{s_1, \ldots, s_n\} ,$$

while the stress intensity factors k_i are calculated using formula (7), i.e. by means of the the dual singular functions s_{-i}. This method, the DSFM, has first been proposed in Blum-Dobrowolski [3],[4] in connection with single equations and has proved to be very efficient also for forth order problems [2].

The method is defined as follows.

0. Determine $a_{ij}^{-} := (Ls_j, s_{-i}) - (Ps_j, L^* s_{-i})$.

1. Calculate $Pv \in V_h = V_h^o$.

2. Set $k_i^o := (Lv, s_{-i}) - (Pv, L^* s_{-i})$.

3. Define k_i^{-n} by solving $\Sigma_{j=1}^n \; k_j^{-n} a_{ij}^{-} = k_i^o$.

4. Set $P_{-n}v = Pv + \Sigma_{i=1}^n \; k_i^{-n}(s_i - Ps_i)$.

Remarks. (i) Step 0 has to be passed only once for a given domain Ω and a given subdivision.

(ii) The numbers a_{ij}^{-} are the approximate coefficients of the singular functions s_i, i.e. $a_{ij}^{-} \to \delta_{ij}$ $(h \to 0)$, except for a_{1j}^{-} in the non-convex case.

(iii) The main step is the correction step 3 which eliminates the bad convergence behaviour of the original method for the singular functions, i.e. $P_{-n}s_i = s_i$ $(1 \le i \le n)$.

(iv) $P_{-n}v$ can be calculated in two different ways. Either the values of Ps_i are stored from step 0 or one has to solve an additional linear system for the right hand side $f - \Sigma_{i=1}^n \; k_i^{-n} Ls_i$ (and starting values Pv).

In order to derive error estimates for the DSFM we make use of some identities which are obtained very easily from the definition of the method.

<u>Lemma</u>.

(i) $\quad v - P_{-n}v = \Sigma_{i=1}^n \; (k_i - k_i^{-n})(s_i - Ps_i) + (w - Pw)$

(ii) $\quad k_i - k_i^{-n} = -(v - P_{-n}v, L^* s_{-i})$

(iii) $\quad \Sigma_{j=1}^n \; (k_j - k_j^{-n}) a_{ij}^{-} = -(w - Pw, L^* s_{-i})$

Proof. We only want to show indentity (iii). From the definition of a_{ij}^{-} we see

$$\Sigma_{j=1}^n \; (k_j - k_j^{-n}) a_{ij}^{-} =$$
$$\Sigma_{j=1}^n \; k_j((Ls_j, s_{-i}) - (Ps_j, L^* s_{-i})) - ((Lv, s_{-i}) - (Pv, L^* s_{-i})) =$$
$$-(Lw, s_{-i}) + (Pw, L^* s_{-i}) =$$
$$-(w, L^* s_{-i}) + (Pw, L^* s_{-i})$$

The last identity follows from the fact that w does not contain the functions s_i $(1 \le i \le n)$.

QED

Let us now state our main theorem.

Theorem 5. Let $\| \cdot \|_a$ be an arbitrary norm with $\| s_1 - Ps_1 \|_a < C$. Then for the DSFM the following estimates hold

$$\| v - P_{-n} v \|_a + \Sigma_{i=1}^{n} |k_i - k_i^{-n}| \leq c \| w - Pw \|_a .$$

Remarks. (i) The question of convergence rates is completely reduced to estimates for the regular part in a given norm.
(ii) The error order for all SIF is determined by the highest possible order for w in an arbitrary weak norm.

Proof. The theorem is a direct consequence of the preceding lemma. From (iii) we have

$$|k_i - k_i^{-n}| \leq c \| w - Pw \|_a$$

since the matrix $(\bar{a}_{ij})$ does not degenerate. From (i) the estimate for $v - P_{-n} v$ follows.

QED

Error bounds for the DSFM are now obtained from the corresponding rates for the regular part w. First, let us consider a convex polygon where no optimal convergence is guaranteed by theorem 4.

Theorem 6. If $1 < \alpha < 2$ we get the following error estimates for w.

(i) $w \in H^2(\Omega) \Rightarrow \| w - Pw \|_o + h^{\alpha-1} \| L(w - Pw) \|_o = O(h^\alpha)$

(ii) $w \in H^2_\gamma(\Omega)$ for some $\gamma \in \mathbb{R}$ with $-2 < \gamma < 2\alpha - 4 \Rightarrow$

$$\| w - Pw \|_{(\gamma)} + h \| L(w - Pw) \|_{(\gamma)} = O(h^2)$$

Here, $\| u \|^2_{(\gamma)} = \int_\Omega \sigma^\gamma |u|^2 dx$ with $\sigma^2 = r^2 + \kappa^2 h^2$, $\kappa > 0$, are the weighted norms introduced by Nitsche [9] and Natterer [8] in order to derive L^∞-estimates for the Finite Element Method.

The estimates stated in the theorem are formally the same as the results for single second order equations and also the proof carries over from this case [5]. It is always possible to get $O(h^2)$-accuracy if sufficiently many singular functions are added to the trial space. We do not know whether estimate (i) is optimal for the L^2-norm. In the duality argument we only get an additional power $h^{\alpha-1}$ since it is by no means clear that the solution of the auxiliary problem is regular.
The situation is even worse in the non-convex case $1/2 \leq \alpha < 1$. Here, no optimal convergence can be shown since it is not possible to use a weight

$\gamma \leq -2$ in part (ii). The numerical results in chapter 4, however, indicate that these estimates are not optimal.

Theorem 7. Let Ω be non-convex, $1/2 \leq \alpha < 1$. Then for the regular part w the following estimates are valid.

(i) $w \in H^2(\Omega) \Rightarrow \| w-Pw \|_o + \| L(w-Pw) \|_o = O(h)$

(ii) $w \in H^2_{-2+\varepsilon}(\Omega) \Rightarrow \| w-Pw \|_{(-2+\varepsilon)} = O(h^{1+\alpha-\varepsilon})$

4. NUMERICAL RESULTS

We want to illustrate the theoretical results of the preceding chapters by some numerical examples. First, we consider the non-convex case. On an L-shaped domain ($\omega = 1.5\pi$) the first three singular functions with exponents -1/3, 1/3, 1 are approximated by linear finite elements. Table 1 shows the divergence for s_1 as predicted by theorem 4 even for arbitrary weak norms (see last column). In fact, the functions Ps_1 converge to a wrong limit.

The L-norm behaviour in table 2 is in accordance with the theory whereas the error rates for the solution itself are too high. This indicates that the L^2-results in theorems 6 and 7 may be not optimal. The computed and predicted powers of h are given in every column.

h^{-1}	$L(s_1-Ps_1)$	s_1-Ps_1	$k_1-k_1^o$
8	2.19	0.89	0.62
16	2.04	0.83	0.56
24	2.0	0.81	0.55
32	2.0	0.80	0.54
40	1.97	0.80	0.55
48	1.97	0.79	0.54
56	1.96	0.80	0.53
64	1.96	0.79	0.54
	$h^o(h^o)$	$h^o(h^o)$	$h^o(h^o)$

Table 1

h^{-1}	$L(s_2-Ps_2)$	s_2-Ps_2	$L(s_3-Ps_3)$	s_3-Ps_3
8	0.66	0.34	0.19	0.38(-1)
16	0.47	0.18	0.10	0.14(-1)
24	0.38	0.13	0.68(-1)	0.36(-2)
32	0.34	0.11	0.50(-1)	0.19(-2)
40	0.31	0.91(-1)	0.41(-1)	0.14(-2)
48	0.29	0.81(-1)	0.34(-1)	0.89(-3)
56	0.27	0.72(-1)	0.30(-1)	0.82(-3)
64	0.26	0.66(-1)	0.26(-1)	0.60(-3)
	$h^{0.4}(h^{1/3})$	$h^{0.7}(h^{1/3})$	$h(h)$	$h^2(h)$

Table 2

In table 3 stress intensity factors for the DSFM using 1 and 2 singular functions and the corresponding results for the well-known singular function method (SFM) are computed for $f_1=-1$ ($y\geq 0$), $f_1=-1-y$ ($y<0$), $f_2=0$ in an L-shaped domain. In the SFM the trial and test spaces are augmented by singular functions s_i and the approximate SIF k_i^n are implicitly calculated. Here, the convergence for k_i in general is determined by the difference between the rates for subsequent singular functions in the L-norm, not by the regularity of w (see e.g. [5]).

h^{-1}	$k_1-k_1^{\pm 1}$	$k_2-k_2^{-2}$	$k_2-k_2^{+2}$
8	0.55(-1)	0.36(-1)	
16	0.25(-1)	0.16(-1)	
24	0.22(-1)	0.62(-2)	diver-
32	0.14(-1)	0.35(-2)	
40	0.14(-1)	0.22(-2)	gence
48	0.12(-1)	0.15(-2)	
56	0.10(-1)	0.12(-2)	
64	0.85(-2)	0.89(-3)	
	$\sim h(h^{1/3})$	$h^2(h)$	$(h^{2/3})$

Table 3

There is no difference between k_1^1 and k_1^{-1} since Ps_1 does not converge. For k_2 not even the theoretical error order in the SFM is obtained, k_2^2 converges to a wrong limit. This behaviour for the higher SIF is often observed, compare also [2], [5].
If $\| s_1-Ps_1 \|_o \to 0$ also for the first stress intensity factor bad results are obtained by the SFM compared with the DSFM. Here, instead of our problem (1),(2) on a convex polygon we consider Poisson's equation $-\Delta u = 1$ on a slit domain with mixed boundary conditions. Again, for k_2 the errors of the SFM are unacceptably high whereas in the DSFM all coefficients are convergent with the same order. The results in table 4 were obtained in [5].

h^{-1}	$k_1-k_1^{-2}$	$k_1-k_1^{2}$	$k_2-k_2^{-2}$	$k_2-k_2^{2}$
10	0.11	0.58	0.34(-1)	1.0
20	0.43(-1)	0.23	0.12(-1)	0.78
30	0.20(-1)	0.14	0.53(-2)	0.74
40	0.11(-1)	0.96(-1)	0.30(-2)	0.71
60	0.53(-2)	0.54(-1)	0.14(-2)	0.67
80	0.30(-2)	0.37(-1)	0.74(-3)	0.65
	$h^2(h^{3/2})$	$h^{1.3}(h)$	$h^2(h^{3/2})$	$h^{0.3}(h^{1/2})$

Table 4

REFERENCES.

1. Aziz, A.K.-Leventhal, S.: Finite element approximation for first order systems, SIAM J. Numer. Anal. 15(6), 1103-1111 (1978)
2. Blum, H.: Der Einfluß von Eckensingularitäten bei der numerischen Behandlung der biharmonischen Gleichung, Dissertation, Bonn 1981 (=Bonner Math. Schr. 140 (1982))
3. Blum, H.-Dobrowolski, M.: Une méthode d'éléments finis pour la résolution des problêmes elliptiques dans des ouverts avec coins, C.R. Acad. Sc. Paris, t.293, ser.I, 99-101 (1981)
4. Blum, H.-Dobrowolski, M.: On finite element methods for elliptic equations on domains with corners, Computing 28, 53-63 (1982)
5. Dobrowolski, M.: Numerical approximation of elliptic interface and corner problems, Habilitationsschrift, Univ. Bonn 1981
6. Kondrat'ev, V.A.: Boundary value problems for elliptic equations in domains with conical or angular points, Trans. Mosc. Math. Soc. 16, 227-313 (1967)
7. Maz'ja, V.G.-Plamenevskij, B.A.: Coefficients in the asymptotics of the solutions of elliptic boundary-value problems in a cone, J. Sov. Math. 9, 750-764 (1978)
8. Natterer, F.: Über die punktweise Konvergenz finiter Elemente, Numer. Math. 25, 67-77 (1975)
9. Nitsche, J.A.: L^∞-convergence of finite element approximation, 2. Conference on Finite Elements, Rennes 1975

The Regularity of Interface-Problems on Corner-Regions

Matthias Blumenfeld
Freie Universität Berlin
Arnimallee 2-6, 1000 Berlin 33

The singularity functions are determined for plane interface-problems with piecewise smooth boundary- and interface-curves and with coefficients, that have bounded first derivatives within subregions and are allowed to be discontinuous across the interfaces.

1. Introduction

We consider interface-problems in their ***variational form***: let a sourcefunction $f \in L_2(\Omega)$ be given, where Ω is an open bounded domain in $\mathbb{R}^2$ with piecewise smooth boundary $\partial\Omega$. Let $H_E^1(\Omega)$ contain all functions from $H^1(\Omega)$, that vanish on some part E of the boundary $\partial\Omega$.

A function $u \in H_E^1(\Omega)$ is a solution of the ***interface-problem***, if for all testfunctions $v \in H_E^1(\Omega)$

$$\iint_\Omega (p\,\nabla u \cdot \nabla v)\,dx\,dy = \iint_\Omega (f v)\,dx\,dy. \tag{1.1}$$

The coefficient $p \in L_2(\Omega)$ with $p \geq p_{\min} \in \mathbb{R}_+$ is assumed to have bounded first derivatives in all n_Ω subregions Ω_i of Ω, while p is allowed to be discontinuous on the piecewise smooth interfaces $\partial\Omega_i \cap \partial\Omega_j$.

Such interface-problems arise, when diffusion is studied in composite media consisting of different materials, as e.g. in nuclear reactors or waveguides.

We are interested in the regularity of solutions u of (1.1) since this determines the asymptotic convergence of the finite element approximations of u. In particular we are interested to determine the singularity functions of (1.1), which have by definition the following property: ***for any $f \in L_2(\Omega)$ there are coefficients $\sigma_j \in \mathbb{R}$, such that the solution u has the form***

$$u = \sum_{j \in J} \sigma_j \cdot s_j + \bar{u}, \qquad \text{where } \bar{u} \text{ is regular, e.g. } \bar{u}|_{\Omega_i} \in H^2(\Omega_i). \tag{1.2}$$

It is well known, that the solution of (1.1) is regular outside the neighborhoods of points Q_k , where the boundary $\partial\Omega$ has a corner, where the Dirichlet boundary conditions on E and the Neumann boundary conditions on $\partial\Omega\setminus E$ meet or where interfaces have corners. These points Q_k we will call ***cornerpoints*** and we ***assume***, that the number n_Q of these cornerpoints is finite.

We also ***assume***, that at any cornerpoint Q_k the interior angle towards an adjacent subregion Ω_i is ***positive*** , that is, we exclude any cusp regions.

Regularity results for (1.1) have been published in the following special cases:

- **Kondratiev** [7] derives the singularity functions for $n_\Omega = 1$, that is in the absence of interfaces; see also **Grisvard** [2] and [3];
- **Kellog** [4] and [6] studies coefficients of the form $p = a \cdot \kappa$, where a is continuous and κ is piecewise constant for problems with Dirichlet boundary conditions;
- **Kellog** [5] studies piecewise constant coefficients on polygonal domains with mixed boundary conditions;

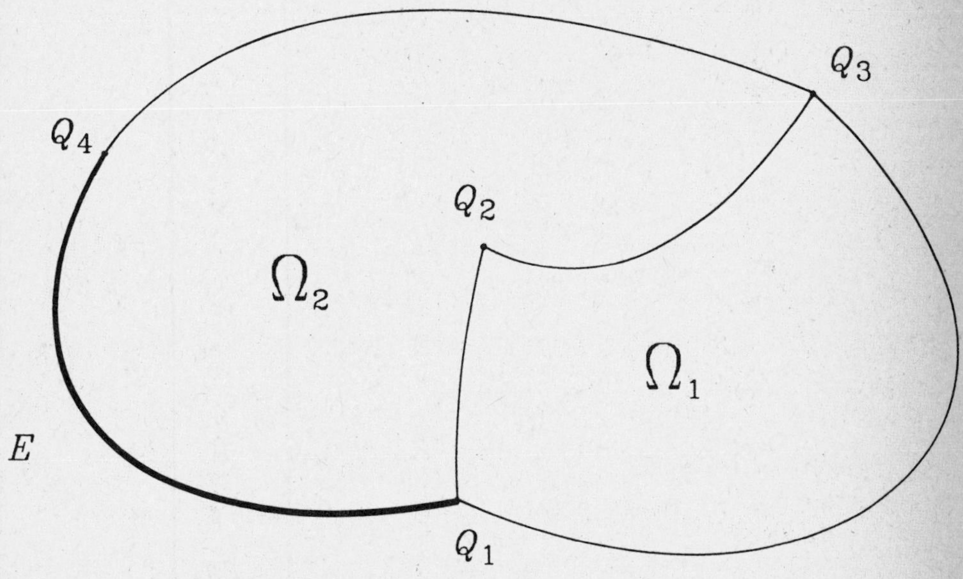

Figure (1.3): example of Ω, with subregions Ω_1 and Ω_2

- **Lemrabet** [9] analyses one intersecting interface ($n_\Omega = 2$ and $n_Q = 2$) for variable coefficients p and Dirichlet boundary conditions using the techniques of Grisvard.

We will use a different approach that allows us to construct the singularity functions ***explicitly*** for problem (1.1). As we have shown in [1] this approach also covers more general problems with third boundary conditions.

2. Local Problems

Let Q be any cornerpoint. For convenience we *assume* Q to be shifted into the origin and that the adjacent subregions Ω_i have been renumbered in anticlockwise manner to $0 \le i \le n$; here Ω_0 denotes the outside-region $\mathbb{R}^2 \setminus \bar{\Omega}$, if Q is a boundary cornerpoint.

Let $B(\rho) := \{ (r \cos\vartheta, r \sin\vartheta) \in \mathbb{R}^2 \mid r < \rho \}$ be a ball around Q with radius ρ. We ***define***

- a ***local domain*** $G := \Omega \cap B(\rho)$,
- with ***local regions*** $G_i := \Omega_i \cap B(\rho)$ for $0 \le i \le n$,
- and their ***interface- or boundary-arcs***
  $\Gamma_i := \partial\Omega_i \cap \partial\Omega_j$ for $0 \le i \le n$ and $j := (i+1) \bmod (n+1)$.

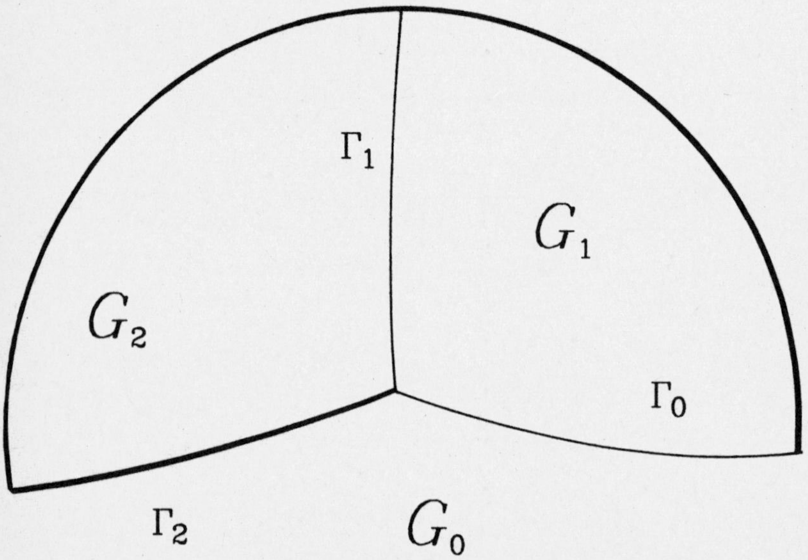

Figure (2.1): example of a local domain G

We *assume*, that the radius ρ was chosen *small enough* to insure, that no other cornerpoints are in $B(\rho)$, so that Γ_i can be *parameterized* as

$$\Gamma_i := \{(r\cos\vartheta, r\sin\vartheta) \in B(\rho) \mid \vartheta = \alpha_i(r)\} \tag{2.2}$$

for functions $\alpha_i \in C^\infty([0,\rho])$ and $0 \le i \le n$. We have

$$G_i \subset G \quad \text{for } n \ge i \ge n_0 := \begin{cases} 0, & \text{if } Q \text{ is an interface-corner,} \\ 1, & \text{if } Q \text{ is an boundary-corner} \end{cases}$$

and, since positive interior angles were assumed, we also have

$$\alpha_{i-1}(r) < \alpha_i(r) \quad \text{for } r \in [0,\rho] \text{ and } n_0 \le i \le n, \tag{2.3}$$

where $\alpha_{-1}(r) := \alpha_n(r) - 2\pi$.

Let $\chi \in C^\infty([0,\rho])$ with $\chi(r) := \begin{cases} 1, & \text{for } r < \rho/3 \\ 0, & \text{for } r > 2\rho/3 \end{cases}$

be a *cutoff-function* and let $EG := \partial G \setminus (\partial\Omega \setminus E)$. Then $u \cdot \chi \in H^1_{EG}(G)$ is the solution of the following *local problem with respect to Q*

$$\begin{aligned} &\iint_G p\nabla(u\chi)\cdot\nabla v \;\; r dr\, d\vartheta = \\ &\iint_G (f\chi - p\,\nabla u\cdot\nabla\chi)v + (p\,u\,\nabla\chi)\cdot\nabla v \;\; r dr\, d\vartheta, \quad \forall\; v \in H^1_{EG}(G). \end{aligned} \tag{2.4}$$

In order to simplify the task, we seek a transformation that *flattens* the boundary- and interface-arcs Γ_i and *keeps the angles* $(\alpha_i(0) - \alpha_{i-1}(0))$ of G_i at the origin. This can be achieved by mapping $(r, \alpha_i(r))$ onto $(r, \alpha_i(0))$ or more precisely, by mapping (r,ϑ) for $\vartheta \in [\alpha_{i-1}(r), \alpha_i(r))$ via interpolation onto $(r,\hat\vartheta)$, where $\hat\vartheta$ is defined as

$$\hat\vartheta(r,\vartheta) := \alpha_{i-1}(0) - \frac{\alpha_i(0) - \alpha_{i-1}(0)}{\alpha_i(r) - \alpha_{i-1}(r)}(\vartheta - \alpha_{i-1}(r)). \tag{2.5}$$

This defines a *transformation*

$$\hat{} : \begin{cases} G & \to & B(\rho) \\ (r,\vartheta) & \to & (r,\hat\vartheta) \end{cases} \tag{2.6}$$

which is well defined, since for $\delta := \det \dfrac{\partial(r,\hat\vartheta)}{\partial(r,\vartheta)}$ follows from (2.3)

$$\delta|_{G_i} = \delta_i(r) := \frac{\alpha_i(0) - \alpha_{i-1}(0)}{\alpha_i(r) - \alpha_{i-1}(r)} > 0.$$

Any subregion G_i is mapped by $\hat{}$ onto the *sector* $S_i := \{(r,\vartheta) \in B(\rho) \mid \alpha_{i-1}(0) < \vartheta < \alpha_i(0)\}$, G onto the *sector-domain* S, arcs Γ_i onto *straight lines* $L_i := \{ (r,\vartheta) \in B(\rho) \mid \vartheta = \alpha_i(0) \}$ and the Dirichlet-boundary EG is mapped onto ES.

LEMMA 2.1: **(isomorphism)**

The mapping

$$\iota : \left\{ \begin{array}{ccl} L_2(G) & \to & L_2(S) \\ u & \to & \hat{u} \\ & & \text{for } \hat{u}(r,\hat{\vartheta}) := u(r,\vartheta) \end{array} \right. \tag{2.7}$$

is an isomorphism of

$$L_2(G) \to L_2(S) \quad , \quad H^1_{EG}(G) \to H^1_{ES}(S) \quad \text{and}$$

$$H^2(G_i) \to H^2(S_i) \quad \text{for } n_0 \le i \le n.$$

PROOF: The L_2- and H^1- seminorms are equivalent respectively, since

$$\iint_G u^2\, r dr\, d\vartheta = \iint_S \hat{u}^2 \frac{r}{\delta} dr\, d\vartheta \quad \text{and}$$

$$\iint_G \nabla u \cdot \nabla u r dr\, d\vartheta = \iint_S \nabla\hat{u} \cdot D \cdot \nabla\hat{u}\, r dr\, d\hat{\vartheta}, \tag{2.8}$$

where $D \in L_2(S)^{2\times 2}$ is defined as $\begin{bmatrix} 1 & \eta \\ \eta & \delta^2+\eta^2 \end{bmatrix}$

for $\delta|_{G_i} := \delta_i$ and $\eta := r\hat{\vartheta}_r$,

because D as well as $D^{-1} \in L_2(S)^{2\times 2}$ are positive definite.

$$\iint_{G_i} \left[u_{rr}^2 + 2\left[\frac{u_\vartheta}{r}\right]_r^2 + \left[\frac{u_{\vartheta\vartheta}}{r^2} + \frac{u_r}{r}\right]^2 \right] r dr\, d\vartheta$$

is a seminorm on $H^2(G_i)$, see [1;p. 70], and it can be shown to be *equivalent* to the corresponding seminorm on $H^2(S_i)$, see [1; p.108]. ■

REMARK:

To prove Lemma 2.1 the boundary- and interface-arcs are required only to be in $W^{2,\infty}$ in local cooordinates, see [1].

The mapping ι can only be an isomorphism from $H^3(G_i)$ to $H^3(S_i)$, if $\partial_r\, \alpha_i(0) = \partial_r \alpha_{i-1}(0) = 0$. Only in this situation we may obtain higher order regularity results using our transformation ^.

Lemma 2.1 insures, that the regularity of u at Q may be obtained by studying $w := \hat{u}\cdot\chi \in H^1_{ES}(S)$, which is the solution of the ***transformed problem:***

$$\iint_S \hat{p}\ \nabla w\cdot D\cdot\nabla v\ r dr\, d\vartheta = \iint_S (\, h\, v + g\cdot\nabla v)\ r dr\, d\vartheta \qquad (2.9)$$

for all $v \in H^1_{ES}(S)$, where $h := \hat{f}\ \chi/\delta - \hat{p}\ \nabla\hat{u}\cdot D\cdot\nabla\chi$ and $g^T := \hat{p}\ \hat{u}\ \nabla\chi\cdot D$ with $g|_{S_i} \in H^1(S_i)\times H^1(S_i)$.

In abbreviated form we write the transformed problem (2.9) as

$$P(\, w\, ,v\,) = l(\, v\,), \qquad \forall\ v \in H^1_{ES}(S). \qquad (P)$$

3. Piecewise constant coefficients

Before we can treat (2.9) we need some results on problems with piecewise constant coefficients.

DEFINITION 3.1:

Let $\kappa \in L_\infty(S)$ be the piecewise constant function, that represents the discontinuities of p at Q :

$$\kappa|_{S_i} := \kappa_i := \hat{p}|_{G_i}(Q) \quad \text{for } n_0 \le i \le n. \qquad (3.1)$$

The corresponding bilinear form on $H^1_{ES}(S)$ is

$$K(u,v) := \iint_S \kappa\nabla u\cdot\nabla v\ r dr\, d\vartheta\ ,$$

and we define the following *piecewise constant Eigenvalue-problem :*

$$K(\, u\, ,v\,) = \mu^2\cdot\iint_S \kappa\, u\, v\ r dr\, d\vartheta\ , \quad \forall\ v \in H^1_{ES}(S) \qquad (K)$$

Let $I := \partial S \cap \partial B(\rho)$ and $EI := \{\, (\rho,\alpha_i(0))\ |\ \Gamma_i \subset EG\}$ be the Dirichlet points.

The *Sturm-Liouville Eigenvalueproblem* that corresponds to the the Eigenvalue-problem (K) is :

determine eigensolutions ($\lambda^2 \in \mathbb{R}$, $\varphi \in H^1_{EI}(I)$) of

$$\int_I \kappa\, \varphi_\vartheta\, \psi_\vartheta\, d\vartheta = \lambda^2\cdot\int_I \kappa\, \varphi\, \psi\, d\vartheta\ , \quad \forall\ \psi \in H^1_{EI}(I)\,. \qquad (KI)$$

THEOREM 3.2:

Let $(\lambda_j,\varphi_j)_{j\geq 1}$ be the complete eigensystem of (KI). Let $J^{(\lambda_j)}$ be the Bessel-function of fractional order λ_j . Let $(\mu_{jk}\rho)$ be its k'th positive zero and let

$$J_{jk}(r) := J^{(\lambda_j)}(\mu_{jk}\, r).$$

Then the complete eigensystem of (K) is given by

$$(\,\mu_{jk}^2\,, J_{jk}\varphi_j\,)_{j,k\geq 1}\,. \tag{3.2}$$

PROOF: (see [1 ; p.65])
$(J_{jk}\,\varphi_j)_{j,k\geq 1}$ are eigenfunctions, since $J_{jk}\,\varphi_j \in H^1_{ES}(S)$ and since $(\mu_{jk}^2, J_{jk}\,\varphi_j)$ solves the variational equation (K) for any $v \in H^1_{ES}(S) \cap C^\infty(S)$:

$$K(J_{jk}\varphi_j,v) = \int_0^\rho \left[-\frac{1}{r}(r\,(J_{jk})_r)_r + \lambda_j^2 r^{-2}\, J_{jk} \right] \int_I \kappa\,\varphi_j\, v\, d\vartheta \;\; r dr$$

$$= \int_0^\rho \mu_{jk}^2\; J_{jk} \int_I \kappa\,\varphi_j\, v\, d\vartheta \;\; r dr$$

$$= \mu_{jk}^2 \iint_S \kappa\,(J_{jk}\;\varphi_j)\; v\; d\vartheta\, r dr.$$

The eigensystem is complete, since $(\,J_{jk}\varphi_j)_{j,k\geq 1}$ can shown to be dense in $L_2(S)$. ■

This theorem allows us to *expand* $v \in H^1_{ES}(S)$ *into eigenfunctions*

$$v = \sum_{j\geq 1} v_j\;\varphi_j = \sum_{j,k\geq 1} b_{jk}\; u_{jk} \quad \text{for} \quad u_{jk} := J_{jk}\varphi_j \;/\; \|J_{jk}\varphi_j\|_{L_2(S)}\,.$$

Then the following forms separate as well:

$$\begin{aligned}
(v,v)_{T^0(S)} &:= (\kappa\, v, v)_{L_2(S)} &&:= \sum_{j,k\geq 1} b_{jk}^2 \\
&= \sum_{j\geq 1} \int_0^\rho v_j^2\; r dr &&= \sum_{j\geq 1} (v_j, v_j)_{BH^0} \\
(v,v)_{T^1(S)} &:= K(v,v) &&:= \sum_{j\geq 1} \mu_{jk}^2 \cdot b_{jk}^2 \\
&= \sum_{j\geq 1} \int_0^\rho [(v_j)_r^2 + (\frac{\lambda_j}{r} v_j)^2\,]\; r dr &&= \sum_{j\geq 1} (v_j, v_j)_{BH^1(\lambda_j)}
\end{aligned} \tag{3.3}$$

LEMMA 3.3:

For a given $f \in L_2(S)$ let $u \in H^1_{ES}(S)$ be the solution of the variational equation

$$K(u,v) = (\kappa f, v) \quad , \quad \forall v \in H^1_{ES}(S) \ . \tag{3.4}$$

Then

$$\begin{aligned} \| u \|^2_{T^2(S)} &:= \left(\frac{1}{\kappa} \nabla(\kappa \nabla u), \nabla(\kappa \nabla u) \right)_{L_2(S)} \\ &= (\kappa f \, , f)_{L_2(S)} =: \| f \|^2_{T^0(S)} \ . \end{aligned} \tag{3.5}$$

PROOF: For $u = \sum_{j \geq 1} u_j \cdot \varphi_j = \sum_{j,k \geq 1} c_{jk} \cdot u_{jk}$ we conclude from (3.4), that

$$K(u, u_{jk}) = \mu^2_{jk} \cdot c_{jk} = (\kappa f, u_{jk}) := f_{jk} \ , \quad \text{so that}$$

$$\| u \|^2_{T^2(S)} = \sum_{j \geq 1} \int_0^\rho [-\frac{1}{r}(r \ (u_j)_r)_r + \lambda_j^2 r^{-2} \, u_j \ ]^2 \, r dr$$

$$:= \sum_{j \geq 1} \| -\Delta_{\lambda_j} u_j \|^2_{BH^0} = \sum_{j,k \geq 1} \mu^4_{jk} \cdot c^2_{jk}$$

$$= \sum_{j,k \geq 1} (f_{jk})^2 = \| f \|^2_{T^0(S)} \ .$$

■

We have defined on the sector S the following spaces :

$$\begin{array}{ccccc} T^0(S) \equiv L_2(S) & \supset & T^1(S) \equiv H^1_{ES}(S) & \supset & T^2(S) \quad \text{and} \\ BH^0 & \supset & BH^1(\lambda_j) & \supset & BT^2(\lambda_j) \end{array}$$

are the corresponding spaces on $(0, \rho)$ with

$$\| u_j \|_{BT^2(\lambda_j)} := \| -\Delta_{\lambda_j} u_j \|_{BH^0} \ . \tag{3.6}$$

The remaining problem is to find out, when a function $u \in T^2(S)$ is piecewise in $H^2(S_i)$ or , more precisely, is bounded in

$$(u,u)_{H^{2,\kappa}(S_i)} := \iint_S \left[u_{rr}^2 + 2\left(\frac{u_\vartheta}{r}\right)_r^2 + \left(\frac{(\kappa u_\vartheta)_\vartheta}{\kappa r^2} + \frac{u_r}{r}\right)^2 \right] \kappa r dr\, d\vartheta$$

$$= \sum_{j\geq 1} \int_0^\rho \left[(u_j)_{rr}^2 + 2\left(\frac{\lambda_j u_j}{r}\right)_r^2 + \left(\frac{\lambda_j^2 u_j)}{r^2} - \frac{u_r}{r}\right)^2 \right] r dr\, d\vartheta$$

$$=: \sum_{j\geq 1} (u_j, u_j)_{BH^2(\lambda_j)} .$$

LEMMA 3.4:

Let w *be in* $C_\lambda^\infty := \{ r^\lambda z \mid z \in C^\infty([0,\rho])$ and $z(\rho) = 0 \}$ *for* $\lambda \geq 0$.
Iff $\lambda \notin (0,1)$ *or* $z(0) = 0$, *then*

$$\|w\|_{BH^2(\lambda)} \leq \|w\|_{BT^2(\lambda)} . \qquad (3.7)$$

PROOF: In [1; p. 72] we show

$$(w,w)_{BT^2(\lambda)} - (w,w)_{BH^2(\lambda)} = \qquad (3.8)$$

$$\left[(w_r - \lambda^2 \frac{w}{r})^2\right]_0^\rho + \lambda^2(1-\lambda)^2 \left[(\frac{w}{r})^2\right]_0^\rho = w_r(\rho)^2 \geq 0 ,$$

if $\lambda = 0$ or if $\lambda \geq 1$ or if $\lambda \in (0,1)$ and $z(0) = 0$, since

$$\lim_{r\to 0} \frac{w}{r} = \lim_{r\to 0} w_r = \lim_{r\to 0} \frac{r^\lambda}{(1-\lambda)} z_r = 0 .$$

If on the other hand $\lambda \in (0,1)$ and $z(0) \neq 0$, then

$$\|w\|_{BH^2(\lambda)}^2 \geq \frac{1}{2} z(0)^2 \int_0^\varepsilon (r^\lambda)_{rr}^2\, r dr \geq \infty > \|w\|_{BT^2(\lambda)}^2 .$$

■

From this Lemma we may conclude at once, that eigenfunctions J_{jk} are *singular* $\forall\ k \geq 1$. , iff $\lambda_j \in (0,1)$, since $J_{jk} \in C_{\lambda_j}^\infty$. Therefore an eigenfunction-expansion of the solution u of (3.4) does not directly lead to a singular-function-expansion.
But we may regroup this expansion:

THEOREM 3.5: **(singular-function-expansion)**

Given $f \in L_2(S)$ *let* $u \in H^1_{ES}(S)$ *be the solution of*

$$K(u,v) = (\kappa f, v) \quad , \quad \forall\, v \in H^1_{ES}(S) \ . \tag{3.4}$$

Then the following singular-function-expansion holds :

$$u = \sum_{j \in J} \sigma_j\, r^{\lambda_j} \chi \varphi_j + \bar{u} \ , \ \text{where}$$

$$\sum_{j \in J} \sigma_j^2 + (\bar{u},\bar{u})_{H^{2,\kappa}(S_i)} \le C\, \|f\|_{L_2(S)} \quad \text{for} \tag{3.9}$$

$$\sigma_j := \frac{1}{\lambda_j}(r^{-\lambda_j}\chi\varphi_j \, , u\,)_{H^1_{ES}(S)} \quad \text{and} \quad J := \{\, j \in \mathbb{N} \mid \lambda_j \in (0,1)\,\} \ .$$

PROOF: From Lemma 3.3 we have $u = \sum_{j \ge 1} u_j\, \varphi_j$ with

$$\sum_{j \ge 1} (u_j, u_j)_{BT^2(\lambda_j)} = (\kappa f, f\,)_{L_2(S)} \ . \tag{3.10a}$$

From Lemma 3.4 follows for $\lambda_j \notin (0,1)$

$$(u_j, u_j)_{BH^2(\lambda_j)} \le (u_j, u_j)_{BT^2(\lambda_j)} \ , \tag{3.10b}$$

since $\operatorname{span}_{1 \le k < \infty} J_{jk} \subset C^\infty_{\lambda_j}$ is dense in $BT^2(\lambda_j)$.

Let $w \in C^\infty_\lambda$ for $\lambda \in (0,1)$, that is $w = r^\lambda z$ with $z \in C^\infty([0,\rho])$ and $z(\rho) = 0$. Then

$$z(0) = -\int_0^\rho (r^{-\lambda} w)_r \, dr = (\frac{1}{\lambda} r^{-\lambda} \, , w\,)_{BH^1(\lambda)}$$

$$= (\frac{1}{\lambda} r^{-\lambda}\chi \, , -\Delta_\lambda w\,)_{BH^0} + (\frac{1}{\lambda} r^{-\lambda}(1-\chi) \, , w)_{BH^1(\lambda)}$$

$$\le C(\lambda)\Big[\, \|w\|_{BT^2(\lambda)} + \|w\|_{BH^1(\lambda)} \Big] .$$

Since $C^\infty_{\lambda_j}$ is dense in $BT^2(\lambda)$, we have a bounded

$$\sigma_j := \frac{1}{\lambda_j}(r^{-\lambda_j}\chi \, , u_j\,)_{BH^1(\lambda_j)} = \frac{1}{\lambda_j}(r^{-\lambda_j}\chi\varphi_j \, , u\,)_{H^1_{ES}(S)} \tag{3.10c}$$

and we may define $\bar{u}_j := u_j - \sigma_j\, r^{\lambda_j}\chi$ with $\bar{u}_j(0) = 0$.

From Lemma 3.4 we therefore have again

$$(\bar{u}_j, \bar{u}_j)_{BH^2(\lambda_j)} \le (\bar{u}_j, \bar{u}_j)_{BT^2(\lambda_j)} \tag{3.10d}$$

and (3.9) follows from (3.10a,b,c,d). ■

4. Variable coefficients

We are now ready to tackle the transformed problem with variable coefficients, see (2.9): determine a solution $w \in H^1_{ES}(S)$, such that

$$P(w,v) = l(v), \quad \forall\, v \in H^1_{ES}(S) \ . \tag{P}$$

Since it will play a crucial role, whether the constants depend on the radius ρ, we will denote those constants, that do ***not depend on*** ρ by $C_k \in \mathbb{R}_+$ for $k \geq 1$.

To achieve this we will use the transformation

$$\sim : \begin{cases} B(\rho) & \rightarrow \quad B(1) \\ (r,\vartheta) & \rightarrow \quad (\frac{r}{\rho},\vartheta) \end{cases} \tag{4.1}$$

The following seminorms are ***invariant*** under this transformation $\sim$:

$$\frac{1}{\rho}\cdot\| v \|_{L_2(S)} \quad \text{and} \quad \| v \|_{H^1_{ES}(S)} \quad \text{and}$$

$$\rho\cdot\| v \|_{T^2(S)} \quad \text{and} \quad \rho\cdot\| v \|_{H^{2,\kappa}(S_i)} \ .$$

DEFINITION 4.1: (spaces $Z^2(S)$ and $SZ^2(S)$)

A function $v \in H^1_{ES}(S)$ is in $Z^2(S)$, iff

$$\| v \|^2_{Z^2(S)} := | v |^2_{H^1_{ES}(S)} + \rho^2 \cdot \sum_{i=n_0}^{n} | v |^2_{H^2(S_i)} < \infty \ . \tag{4.2}$$

A function $w \in H^1_{ES}(S)$ is in $SZ^2(S)$, iff it may be decomposed into

$$\begin{aligned} w &= \sum_{j\in J} \sigma_j r^{\lambda_j} \chi \varphi_j + \hat{w} \quad \text{such that} \\ \| v \|^2_{SZ^2(S)} &:= \sigma_j^2 \cdot \rho^{2\lambda_j} + \| v \|^2_{Z^2(S)} \ . \end{aligned} \tag{4.3}$$

This space of regular functions $Z^2(S)$ and the augmented space $SZ^2(S)$ have norms that are by construction invariant under the transformation $\sim$.

THEOREM 4.2: **(abstract regularity)**

Assume there exists a solution $w_0 \in SZ^2(S)$ *of*

$$K(w_0,v) = l(v)\,, \quad \forall\, v \in H^1_{ES}(S) \qquad (\mathrm{P}_0)$$

and assume there exist solutions $w_m \in SZ^2(S)$ *for* $m \geq 1$ *of*

$$K(w_m,v) = K(w_{m-1},v) - P(w_{m-1},v)\,, \quad \forall\, v \in H^1_{ES}(S) \qquad (\mathrm{P}_m)$$

such that there is a constant C_1 *not depending on* ρ *with*

$$\|w_m\|_{SZ^2(S)} \leq C_1 \cdot \rho \cdot \|w_{m-1}\|_{SZ^2(S)} \,. \qquad (4.4)$$

Then if the radius $\rho < C_1$

$$w := \sum_{m \geq 0} w_m \in SZ^2(S)$$

and w *solves*

$$P(w,v) = l(v), \quad \forall\, v \in H^1_{ES}(S)\,. \qquad (\mathrm{P})$$

PROOF: The sum converges since

$$\|w^{(n)}\|_{SZ^2(S)} := \|\sum_{m=0}^{n} w_m\|_{SZ^2(S)} \leq$$
$$\sum_{m=0}^{n} (\rho C_1)^m \|w_0\|_{SZ^2(S)} \leq \frac{1}{1-\rho C_1}\|w_0\|_{SZ^2(S)} \,. \qquad (4.5)$$

Adding problems (P_m) for $0 \leq m \leq n$ we obtain

$$|\, P(w^{(n)},v) - l(v)\,| = |\, K(w_n,v) - l(v)\,|$$

$$\leq \max_i \kappa_i \;\|\, w_{m+1}\,\|_{SZ^2(S)} \;|\, v\,|_{H^1_{ES}(S)} = (\rho C_1)^{m+1}\, C \to 0$$

for fixed $v \in H^1_{ES}(S)$ and $m \to \infty$. ∎

In order to show that we may fulfill the assumptions of this theorem, we first transform problem (P_m) *into standart form.*

LEMMA 4.3: **(standart form)**

For $y \in SZ^2(S)$ and $n_0 \le i \le n$ there exist functions $f_i \in L_2(S_i)$ and $g_i \in H^1(S_i)$, such that for all $v \in H^1_{ES}(S)$

$$K(y,v) - P(y,v) = \sum_{i=n_0}^{n} \left[(\kappa \cdot f_i, v)_{L_2(S_i)} + \int_{L_i} g_i \, v \, dr - \int_{L_{i-1}} g_i \, v \, dr \right], \text{with} \tag{4.6}$$

$$\sum_{i=n_0}^{n} \left[\| f_i \|^2_{L_2(S_i)} + | g_i |^2_{H^1(S_i)} \right] \le C_2 \cdot \| y \|^2_{SZ^2(S)} ,$$

where $L_i := \{ (r,\vartheta) \in \overline{S} \mid \vartheta = \alpha_i(0) \}$.

PROOF: For $E := \kappa \cdot I - \widehat{p} \cdot D$ we have

$$K(y,v) - P(y,v) = \iint_S \nabla y \cdot E \cdot \nabla v \; r dr \, d\vartheta \ . \tag{4.6a}$$

Since D_r and $\frac{D_\vartheta}{r} \in L^\infty(S_i)^{2\times 2}$ for $n_0 \le i \le n$ and from (3.1) we find

$$\frac{E}{r} , \; E_r , \; \frac{E_\vartheta}{r} \in L^\infty(S_i)^{2\times 2} . \tag{4.7}$$

Let $\begin{bmatrix} h_i \\ g_i \end{bmatrix} := E \cdot \nabla y$ on S_i . Then

$$\begin{aligned} &\iint_{S_i} \left[E \cdot \nabla y \right] \cdot \nabla v \; r dr \, d\vartheta = \\ &\iint_{S_i} f_i \cdot v \; r dr \, d\vartheta + \int_{L_i} g_i \, v \, dr - \int_{L_{i-1}} g_i \, v \, dr \end{aligned} \tag{4.6b}$$

for $f_i := -\nabla(E \cdot \nabla y) = -\frac{1}{r}(r\, h_i)_r - \frac{1}{r}(g_i)_\vartheta$.
Adding (4.6a) and (4.6b) for all i gives (4.6).
Using (4.7) we can bound

$$\iint_{S_i} (g_i)^2_\vartheta \frac{1}{r} r dr \, d\vartheta \tag{4.8}$$

$$\le C_{31} \, | y |^2_{H^1(S_i)} + \iint_{S_i} r^2 \left[\frac{y_\vartheta}{r} \right]^2 + r^2 \left[\frac{y_{\vartheta\vartheta}}{r^2} + \frac{y_r}{r} \right]^2 \; r dr \, d\vartheta$$

$$\le C_3 \left[\sum_{j \in J} \sigma_j^2 \cdot \rho^{2\lambda_j} + | \overline{y} |^2_{H^1(S_i)} + \rho^2 \cdot | \overline{y} |^2_{H^2(S_i)} \right]$$

$$= C_3 \cdot \| y \|^2_{SZ^2(S)} \quad \text{for } y = \sum_{j \in J} \sigma_j \, r^{\lambda_j} \chi \varphi_j + \overline{y} \ .$$

Similar bounds can be obtained for $(g_i)_r$ and $\frac{(r\, h_i)}{r}$. ■

We are now ready to fulfill the assumption (4.4).

THEOREM 4.4: (Regularity of iterated solutions)
The solution $w \in H^1_{ES}(S)$ of

$$K(w,v) = K(y,v) - P(y,v)\,, \quad \forall\, v \in H^1_{ES}(S) \qquad (P_m)$$

for a given $y \in SZ^2(S)$ is again in $SZ^2(S)$ and

$$\|w\|_{SZ^2(S)} \le C_1 \cdot \rho \cdot \|y\|_{SZ^2(S)}\,. \qquad (4.9)$$

PROOF: From Lemma 4.2 we find

$$K(w,v) = \sum_{i=n_0}^{n} \Big[(\kappa f_i, v)_{L_2(S_i)} - \int_{\partial S_i} g_i v\, dr \Big]$$

and after using the transformation $\tilde{}$ in (4.2) :

$$\iint_{\tilde{S}} \kappa \nabla\tilde{w} \cdot \nabla\tilde{v}\; \tilde{r}\, d\tilde{r}\, d\vartheta = \sum_{i=n_0}^{n} \Big[(\kappa\rho^2 \tilde{f}_i, \tilde{v})_{L_2(\tilde{S})} - \int_{\partial\tilde{S}_i} \rho\tilde{g}_i\, \tilde{v}\, d\tilde{r} \Big]\,.$$

From *inverse trace theorems*, see [3,Th.1.5.2.4] we know, that there exist functions $\tilde{z}_i \in H^1_0(\tilde{S}_i) \cap H^2(\tilde{S}_i)$ with

$$\partial_n(\kappa_i \tilde{z}_i) = -\rho\, \tilde{g}_i \quad \text{on } \partial\tilde{S}_i \quad \text{and}$$
$$|\tilde{z}_i|_{H^1(\tilde{S}_i)} + |\tilde{z}_i|_{H^2(\tilde{S}_i)} \le C_{11} |\rho\, \tilde{g}_i|_{H^1(\tilde{S}_i)}\,. \qquad (4.10)$$

Subtracting $\tilde{z}$ with $\tilde{z}|_{\tilde{S}_i} := \tilde{z}_i$ leads to

$$\iint_{\tilde{S}} \kappa \nabla(\tilde{w}-\tilde{z}) \cdot \nabla\tilde{v}\; \tilde{r}\, d\tilde{r}\, d\vartheta = \iint_{\tilde{S}} \kappa \tilde{F}\, \tilde{v}\; \tilde{r}\, d\tilde{r}\, d\vartheta$$
$$\text{for} \quad \tilde{F}|_{\tilde{S}_i} := \rho^2 \tilde{f}_i + \frac{1}{\kappa} \nabla(\kappa \nabla \tilde{z}_i)\,.$$

Application of Theorem (3.5) gives

$$\tilde{w} - \tilde{z} = \sum_{j \in J} \tilde{\sigma}_j \tilde{r}^{\lambda_j} \tilde{\chi} \varphi_j + (\tilde{\bar{w}} - \tilde{z}) \quad \text{with}$$
$$\sum_{j \in J} \tilde{\sigma}_j^2 + |\tilde{\bar{w}} - \tilde{z}|^2_{H^1_{ES}(\tilde{S})} + \sum_{i=n_0}^{n} |\tilde{\bar{w}} - \tilde{z}|^2_{H^2_{ES}(\tilde{S})} \le C_{12} \|\tilde{F}\|^2_{L_2(\tilde{S})} \qquad (4.11)$$

Transforming back to S leads to $w = \sum_{j \in J} \sigma_j r^{\lambda_j} \chi \varphi_j + \bar{w}$ for $\sigma_j := \dfrac{\tilde{\sigma}_j}{\rho^{\lambda_j}}$ and

$$
\begin{aligned}
\|w\|^2_{SZ^2(S)} &\le \left(\|w-z\|_{SZ^2(S)} + \|z\|_{SZ^2(S)}\right)^2 \\
&\le 3\left(\rho^{-2} C_{12}\|F\|^2_{L_2(S)} + \|z\|^2_{Z^2(S)}\right) \quad \text{from (4.11)} \\
&\le 9\rho^{-2} C_{12}\|\rho^2 f\|^2_{L_2(S)} + (9C_{12}+1)\|z\|^2_{Z^2(S)} \\
&\le (9C_{12}+1)\cdot \sum_{i=n_0}^{n} \left(\rho^2 \|f_i\|^2_{L_2(S_i)} + C_{11}\,|\rho g_i|^2_{H^1(S_i)}\right) \\
&\le (9C_{12}+1)\,(C_{11}+1)\,C_2\,\rho^2\,\|y\|^2_{SZ^2(S)} \quad \text{from (4.7)}\ .
\end{aligned}
\tag{4.9'}
$$

■

Following the same lines as in the previous theorem we can proove, see [1], that the solution $w_0 \in H^1_{ES}(S)$ of

$$K(w_0,v) = l(v)\ , \quad \forall\, v \in H^1_{ES}(S) \tag{P_0}$$

is again in $SZ^2(S)$ and

$$\|w_0\|^2_{SZ^2(S)} \le C\left(\|\hat{f}\|^2_{L_2(S)} + \|\hat{u}\|^2_{H^1_{ES}(S)}\right).$$

Therefore the assumptions of Theorem 4.1 are fulfilled and for $\rho < C_1$ we obtain the ***regularity result for the transformed problem***

$$P(w,v) = l(v)\ , \quad \forall\, v \in H^1_{ES}(S) \ \text{with} \tag{P}$$

$$\|w\|^2_{SZ^2(S)} \le \frac{C}{1-\rho C_1}\left(\|\hat{f}\|^2_{L_2(S)} + \|\hat{u}\|^2_{H^1_{ES}(S)}\right). \tag{4.12}$$

A backtransformation from S to G of $w = \hat{u}\,\chi$ supplies us with the ***regularity result for the local problem***

$$
\begin{aligned}
&\iint_G p\,\nabla(u\chi)\cdot\nabla v \;\; r dr\, d\vartheta = \\
&\iint_G (f\chi - p\;\nabla u\cdot\nabla\chi) + (p\,u\;\nabla\chi)\cdot\nabla v\;\, r dr\, d\vartheta\ , \quad \forall\; v \in H^1_{EG}(G)\ ,
\end{aligned}
\tag{2.4}
$$

where $u\,\chi$ may be decomposed into

$$u\,\chi = \sum_{j\in J} \sigma_j\cdot s_j + \bar{u}\,\chi \tag{4.13}$$

for singularity functions $\hat{s}_j := r^{\lambda_j}\chi\varphi_j$ where $j \in J$ such that

$$\sum_{j\in J}\sigma_j^2 + |\bar{u}\,\chi|^2_{H^1_{EG}(G)} + \sum_{i=n_0}^{n} |\bar{u}\,\chi|^2_{H^2(G_i)} \tag{4.14}$$

$$\leq C \left[\|f\|^2_{L_2(G)} + \|u\|^2_{H^1_{EG}(G)} \right] .$$

5. Conclusion

Let us denote the singular functions with respect to a singular point Q_k by

$$s_j^{(k)} \text{ for } j \in J^{(k)} .$$

Since we may extend $s_j^{(k)}$ on $\Omega \setminus G^{(k)}$ by zero, we have $s_j^{(k)} \in H^1_E(\Omega)$.

Our *main regularity result* is

THEOREM 5.1: (regularity on general domains)
The solution $u \in H^1_E(\Omega)$ of

$$\iint_\Omega (p \nabla u \cdot \nabla v) dx dy = \iint_\Omega (f v) dx dy , \quad \forall\, v \in H^1_E(\Omega) \tag{1.1}$$

may be decomposed into

$$u = \sum_{k=1}^{n_Q} \sum_{j \in J^{(k)}} \sigma_j^{(k)} s_j^{(k)} + \bar{u} \quad \text{with}$$

$$\sum_{k=1}^{n_Q} \left[\sum_{j \in J^{(k)}} (\sigma_j^{(k)})^2 + \|\bar{u}\|^2_{H^2(\Omega_i)} \right] \leq C \|f\|^2_{L_2(\Omega)} . \tag{5.1}$$

PROOF: First we construct a partition of unity for Ω :
Around any singular point Q_k we construct a circle $B_k(\rho_k)$ of radius ρ_k , such that

$$\rho_k < C_1^{(k)} ,$$

where $C_1^{(k)}$ is the constant from (4.9). Then the remaining points on the boundary and interfaces are covered and finally the interior. This gives the required partition of unity.

Application of theorem 4.2 leads to equation (4.14) for $u \chi^{(k)}$ on $G^{(k)} := \Omega \cap B_k(\rho_k)$ and since $\chi^{(k)} \equiv 1$ on $B_k(\frac{\rho_k}{3})$, we obtain (4.14) for u on $\Omega \cap B_k(\frac{\rho_k}{3})$.

From the a priori estimate

$$\|u\|_{H^1_E(\Omega)} \leq C \|f\|_{L_2(\Omega)}$$

and the well-established regularity of u away from the singular points we conclude (5.1). ∎

REFERENCES

[1] **Blumenfeld,M.**, *Eigenlösungen von gemischten Interface-Problemen auf Gebieten mit Ecken — ihre Regularität und Approximation mittels Finiter Elemente;* Dissertation, Freie Universität BERLIN, 1983.

[2] **Grisvard,P.**, *Behaviour of the solutions of an elliptic boundary value problem in polygonal or polyhedral domains;* SYNSPADE III (Hubbard,B. ed.), pp. 207 - 274, 1976.

[3] **Grisvard,P.**, *Boundary value problems in nonsmooth domains;* Lecture Notes 19, Dept. Math., University of Maryland, 1980.

[4] **Kellog,R.B.**, *Singularities in interface problems;* SYNSPADE II (Hubbard,B. ed.), pp. 351 - 400, 1971.

[5] **Kellog,R.B.**, *Higher order singularities for interface problems;* the mathematical foundations of ... (Aziz,A.K., ed.), Academic, pp. 589 - 602, 1972.

[6] **Kellog,R.B.**, *On the Poisson equation with intersecting interfaces;* Applicable Analysis, *4*, pp. 101 - 129, 1975.

[7] **Kondrat'iev,V.A.**, *Boundary problems for elliptic equations in domains with conical or angular points;* Trans. of the Moscow Math. Soc. , *16*, pp. 227 - 313, 1967.

[8] **Ladyzhenskaya,O.A., Rivkin,V.J., Ural'ceva,N.N.**, *the classical solvability of diffraction problems;* Trudy Math. Inst. Steklova, *92*, pp. 132 - 166, 1966.

[9] **Lemrabet,K.**, *Régularité de la solution d'un probleme de transmission;* J. Math. pures et appl., *56*, pp. 1 - 38, 1977.

[10] **Schechter,M.**, *A generalisation of the problem of transmission;* Ann. Scuola Norm. Sup. Pisa, *14*, 207 - 236, 1960.

SPECTRAL ANALYSIS OF SINGULAR STURM-LIOUVILLE PROBLEMS WITH OPERATOR COEFFICIENTS

Jochen Brüning

Institut für Mathematik
der Universität Augsburg
Memminger Str. 6
D-8900 Augsburg
BRD

We give a report on work in progress which has been done largely in collaboration with Bob Seeley.

1. Motivating examples

a) Consider a surface of revolution M in $\mathbb{R}^3$ homeomorphic to S^2, generated by a smooth curve

$$c(t) = (c_1(t), 0, c_3(t)), \quad t \in [0,L],$$

parametrized by arc length. In the natural coordinate system

$$(0,L) \times (0,2\pi) \ni (t,\varphi) \mapsto (c_1(t)\cos\varphi, c_1(t)\sin\varphi, c_3(t)) \in M$$

the metric of M takes the form

$$g(t,\varphi) = \begin{pmatrix} 1 & 0 \\ 0 & c_1(t)^2 \end{pmatrix}$$

and the Laplacian Δ (which we define to be positive) becomes

$$= -\frac{1}{c_1(t)} \frac{\partial}{\partial t} \left(c_1(t) \frac{\partial}{\partial t}\right) - \frac{1}{c_1(t)^2} \frac{\partial^2}{\partial \varphi^2} .$$

The isometric S^1 action on M induces a unitary S^1 action on all eigenspaces of Δ and hence subdivides the spectrum according to the irreducible unitary representations of S^1. Let us denote the spectrum by

σ and by σ_κ the subset belonging to the representation κ. It turns out that the invariant spectrum σ_1 is determined by σ and contains already interesting geometric information on M (see [4]). Moreover, the invariant spectrum is precisely the spectrum of the Friedrichs extension T of the operator

$$-d_t^2 + \frac{2c_1(t)c_1''(t) - c_1'(t)^2}{4c_1(t)^2} =: -d_t^2 + q(t)$$

in $L^2([0,L])$ with domain $C_o^\infty((0,L))$ (see [4]). By construction we must have

$$c_1(0) = c_1(L) = 0, \; c_1'(0) = -c_1'(L) = 1.$$

Thus the potential q can be written

$$q(t) =: \frac{a(t)}{t^2(L-t)^2}$$

where $a \in C^\infty([0,L])$ and $a(0), a(L) = -1/4$ i.e. the analysis of the invariant spectrum reduces to the spectral analysis of a singular Sturm-Liouville problem. One may ask whether the method of heat invariants also generalizes to the spectra σ_κ, i.e. whether there is an asymptotic expansion of

$$\sum_{\lambda \in \sigma_\kappa} e^{-\lambda s} \text{ as } s \to 0 \quad .$$

This is in fact so (see [3]) and we have an asymptotic expansion of the form

$$\sum_{\lambda \in \sigma_\kappa} e^{-\lambda s} \sim (4\pi s)^{-1/2} \sum_{j \geq o} a_j^\kappa s^{j/2} \quad ,$$

where $a_o^\kappa = L$. For $\kappa = 1$ this implies the existence of an asymptotic expansion for tr e^{-sT} as $s \to 0$ and we are lead to ask whether this remains true for more general singular Sturm-Liouville operators, which cannot be handled by the group action approach.

b) It seems desirable to extend the spectral theory of the Laplacian on compact smooth manifolds to more general spaces allowing singularities. The most simple type of singularity is a cone over a smooth compact manifold N: we put

$$C_r(N) := (0,r) \times N$$

equipped with the metric

$$dx \otimes dx + x^2 g, \; g \text{ a smooth Riemannian metric on } N,$$

and call it a metric cone over N. As a special case we obtain $C_r(S)$, the n+1-ball of radius r. Cheeger ([7]) has developed a precise functional calculus for the Laplacian on $C_r(N)$ in terms of functions of the Laplacian on N, based on separation of variables and the Hankel transform. It is, however, possible to attack this problem without using separation of variables. Let us denote by $A^p(M)$ the smooth p-forms on a differentiable manifold M and by $L^2_p(M)$ the square integrable p-forms. Then we have a bijective linear map

$$\psi_p : C^\infty((0,r),A^p(N)) \times C^\infty((0,r),A^{p-1}(N)) \to C^\infty(C_r(N))$$

given by

$$\psi_p(\phi_1,\phi_2) = \pi^*\phi_1 + \pi^*\phi_2 \wedge dx$$

where $\pi : C_r(N) \to N$ is the natural projection. By a cumbersome but straightforward computation one finds that

$$\psi_p^{-1} \circ \Delta \circ \psi_p \; (\phi_1,\phi_2) \tag{1}$$
$$= \Big(-x^{2p-n} \frac{\partial}{\partial x} \big(x^{n-2p} \frac{\partial \phi_1}{\partial x}\big) + x^{-2}\Delta_N \phi_1 + (-1)^p \frac{2}{x} d_N \phi_2 \; ,$$
$$-\frac{\partial}{\partial x} \big(x^{2p-n-2} \frac{\partial}{\partial x} (x^{n+2-2p} \phi_2)\big) + x^{-2}\Delta_N \phi_2 + (-1)^p \frac{2}{x^3} \partial_N \phi_1\Big) \; ,$$

where $d_N, \partial_N, \Delta_N$ denote the intrinsic operations on N. Regarding Δ as a symmetric operator in $L^2_p(C_r(N))$ with domain the smooth functions with compact support we see that Δ is unitarily equivalent to the operator

(1) with domain $C_o^\infty((0,r),A^p(N)) \times C_o^\infty((0,r),A^{p-1}(N)$ in the Hilbert space $L^2((0,r),L_p^2(N),x^{n-2p}dx) \oplus L^2((0,r),L_{p-1}^2(N),x^{n-2p+2}dx)$. The obvious transformation brings us to the operator

$$(-\frac{\partial^2}{\partial x^2}\phi_1 + (\frac{n}{2}-p)(\frac{n}{2}-p-1)\ x^{-2}\phi_1 + x^{-2}\Delta_N\phi_1 + (-1)^p\frac{2}{x^2}d_N\phi_2\ ,$$

$$-\frac{\partial^2}{\partial x^2}\phi_2 + (\frac{n}{2}+2-p)(\frac{n}{2}+1-p)\ x^{-2}\phi_2 + x^{-2}\phi_2 + x^{-2}\Delta_N\phi_2 + (-1)^p\frac{2}{x^2}\partial_N\phi_1)$$

in the Hilbert space $L^2((0,r),L_p^2(N),dx) \oplus L^2((0,r),L_{p-1}^2(N),dx)$ $\simeq L^2((0,r),L_p^2(N) \oplus L_{p-1}^2(N),dx)$ with domain $C_o^\infty((0,r),A^p(N) \times A^{p-1}(N))$. Setting

$$A := \begin{pmatrix} \Delta_N + (\frac{n}{2}-p)(\frac{n}{2}-p-1) & (-1)^p\ 2d_N \\ (-1)^p\ 2\partial_N & \Delta_N + (\frac{n}{2}+2-p)(\frac{n}{2}+1-p) \end{pmatrix}$$

we see that A is an elliptic formally selfadjoint operator on $A^p(N) \times A^{p-1}(N)$ hence is essentially selfadjoint in $L_p^2(N) \oplus L_{p-1}^2(N)$ and the unique selfadjoint extension has a pure point spectrum. It is also not hard to see that $A \geq -1/4$, and that $-1/4$ is in fact an eigenvalue if there are harmonic $(n+1)/2$ forms on N. Therefore, the selfadjoint extension of Δ on forms with compact support on $C_r(N)$ are unitarily equivalent to selfadjoint extensions of

$$-d_x^2 + \frac{A}{x^2} \tag{2}$$

in $L^2((0,r),L_p^2(N) \oplus L_{p-1}^2(N),dx)$ with domain $C_o^\infty((0,r),A^p(N) \times A^{p-1}(N))$. We will see that the operator (2) is bounded below so its Friedrichs extension T exists. As pointed out by Cheeger there is a choice of boundary conditions if the middle cohomology $H^{(n+1)/2}(N)$ does not vanish but we will restrict attention to T in what follows. Our problem is now to express spectral invariants of T by spectral invariants of A. Away from the singular point the analysis of Δ is classical so this will amount to expressing spectral invariants of Δ by spectral invariants of A. In particular, we are interested in the asymptotic expansion of tr e^{-sT} as $s \to 0$. A general calculus would also allow us to replace A by a more general operator, assuming for example that N already has cone-like singularities i.e. we could attack the analysis of iterated cone singularities as done in [8].

c) The calculus developed by Cheeger is special in the sense that it does not allow pertubations of the metric. A very natural extension of the metric cone would be a metric on $(0,r) \times N$ of the form

$$dx \otimes dx + x^2 g_x \tag{3}$$

where g_x is a smooth family of metrics on $[0,r]$ (this includes for example all normal geodesic balls); a metric of this type is called asymptotically cone-like. In the approach described above this leads to a singular operator Sturm-Liouville problem of the form

$$-d_x^2 + \frac{A(x)}{x^2} \tag{4}$$

where $A(x)$ is a smooth family of selfadjoint elliptic operators with $A(0) \geq -1/4$. If $A(x)$ is scalar this is essentially the problem discussed in a).

2. The heat equation: scalar case

We start with the problem discussed in 1a): consider an intervall $I = (0,L)$, $a \in C^{\infty}(\bar{I})$ with $a(0), a(L) \geq -1/4$, and the singular Sturm-Liouville operator

$$-d_x^2 + \frac{a(x)}{x^2(L-x)^2}, \; x \in I. \tag{5}$$

This operator is symmetric in $L^2(I)$ with domain $C_o^{\infty}(I)$, and by Hardy's inequality it is bounded below (here we need the $-1/4$ condition). Hence the Friedrichs extension T does exist and we want to determine the asymptotic expansion of tr e^{-sT} as $s \to 0$ (if it exists). This has been done in [6], [5], and [1] with three different methods. We give a short description of the approach in [1]. If $a(0), a(L) > -1/4$ then the domain $\mathcal{D}(T)$ of T is contained in $H^1(I)$ so the Sobolev inequality implies the existence of the heat kernel,

$$e^{-sT} u(x) = \int_I \Gamma_s(x,y)\, u(y)\, dy, \quad u \in L^2(I) .$$

Using the fact that (5) is a differential equation with regular singularities this follows in general and we also obtain good estimates of $\Gamma_s(x,y)$ near the singular points. In particular, e^{-sT} is trace class and

$$\operatorname{tr} e^{-sT} = \int_I \Gamma_s(x,x)\, dx .$$

By a simple reflection argument one may restrict attention to the left endpoint i.e. it suffices to expand

$$I(s) := \int_0^{L/2} \Gamma_s(x,x)\, dx .$$

Now choose $0 < \varepsilon < L/2$ and split the integral at $x = \varepsilon$,

$$I(s) = \int_0^{\varepsilon} + \int_{\varepsilon}^{L/2} \Gamma_s(x,x)\, dx =: I_<(s,\varepsilon) + I_>(s,\varepsilon) .$$

Away from 0 we obtain an expansion of $\Gamma_s(x,x)$ by the classical method of Minakshisundaram-Pleijel,

$$\Gamma_s(x,x) \sim (4\pi s)^{-1/2} \sum_{j\geq 0} s^j \frac{w_j(x)}{x^{2j}(L-x)^{2j}} ,$$

where w_j is a universal polynomial in the variables $a^{(k)}$, $0 \leq k \leq 2j$, and with coefficients in $C^\infty(\bar{I})$. This gives an expansion for $I_>(s,\varepsilon)$ as $s \to 0$ but we are not allowed to let $\varepsilon \to 0$ because the resulting integrals are divergent beyond the first one. The arbitrariness of ε suggests that we treat ε as an additional variable and the homogeneity of the equation forces us to introduce $\sigma := s/\varepsilon^2$ as second variable. Then it is easy to see that $I_>(\varepsilon^2\sigma,\varepsilon)$ has an asymptotic expansion with respect to the system of functions $(\sigma^{i/2}\varepsilon^j \log^k \varepsilon)_{\substack{i\geq -1\\ j\geq 0\\ 0\leq k\leq 1}}$ as $\varepsilon^2 + \sigma^2 \to 0$.

To achieve a similar expansion for $I_<$ near 0 we compare Γ_s with the heat kernel $\bar{\Gamma}_s$ of the Friedrichs extension of

$$-d_x^2 + \frac{a(0)}{x^2} \tag{6}$$

in $L^2(\mathbb{R}_+)$, using Duhamel's principle. Thus we obtain an asymptotic representation of $\Gamma_s(x,x)$ near 0 by a Neumann series built from $\bar{\Gamma}_s$. Since ε is now a variable we can use Taylor expansion on the coefficients rendering the terms in the series universal expressions in the variables $a^{(k)}(0)$, $k \geq 1$. It remains to expand certain convolution integrals in $\bar{\Gamma}_s$ the simplest one being

$$\int_0^\varepsilon \bar{\Gamma}_s(x,x)\,dx .$$

Now the homogeneity of (6) causes the following homogeneity of $\bar{\Gamma}_s$:

$$\bar{\Gamma}_s(x,y) = \alpha\,\bar{\Gamma}_{\alpha^2 s}(\alpha x,\alpha y),\quad \alpha,s,x,y > 0 . \tag{7}$$

Substituting $x = \varepsilon u$ and using (7) with $\alpha = \varepsilon$ we find

$$\int_0^\varepsilon \bar{\Gamma}_s(x,x)\,dx = \int_0^1 \varepsilon\,\bar{\Gamma}_{\varepsilon^2\sigma}(\varepsilon u,\varepsilon u)\,du = \int_0^1 \bar{\Gamma}_\sigma(u,u)\,du .$$

Again with (7) and $\alpha = 1/u$ and with the substitution $x = \sigma/u$ we obtain

$$\int_0^1 \bar{\Gamma}_\sigma(u,u)\,du = \int_0^1 \frac{1}{u}\,\bar{\Gamma}_{\sigma/u^2}(1,1)\,du = 2\int_\sigma^\infty \frac{1}{x}\,\bar{\Gamma}_x(1,1)\,dx . \tag{8}$$

So all we need is the asymptotic expansion of $\bar{\Gamma}_s$ at $(1,1)$ as $s \to 0$ which is classical. The other terms in the Neumann series can be treated in a similar way leading to an expansion of $I_<$ in terms of the functions $(\sigma^{i/2}\varepsilon^j \log^k \sigma)_{\substack{i\geq -1\\ j\geq 0\\ 0\leq k\leq 1}}$ (the appearance of logarithmic terms is no surprise in view of the $\frac{1}{x}$ factor in (8)).

Summing the two expansions leads to an expansion of I(s) of the following form:

$$I(s) \sim (4\pi s)^{-1/2} \sum_{j \geq o} s^{j/2}(A_j + B_j + C_j \log s) \quad . \tag{9}$$

Here A_j is the "interior term",

$$A_{2j} = \text{constant term in the expansion of} \tag{10}$$

$$\int_{\varepsilon}^{L-\varepsilon} \frac{w_j(x)}{x^{2j}(L-x)^{2j}} dx \quad \text{as } \varepsilon \to 0 \ ,$$

$$A_{2j+1} = 0, \ j \geq 0 \ .$$

B_j and C_j are "boundary terms" i.e. universal expressions in $a^{(k)}(0)$, $k \geq 0$. In particular,

$$C_o = C_1 = 0, \ C_2 = -\frac{b_0(0) + b_L(0)}{8\sqrt{\pi}} \ ,$$

where we have written $\dfrac{a(x)}{x^2(L-x)^2} = \dfrac{a_i}{(i-x)^2} + \dfrac{b_i(i-x)}{(i-x)}$ as $x \to i$, $i = 0, L$.

It is interesting to note that all C_j vanish if b_0 and b_L are odd at 0 which is the case in example 1a) above.

3. The heat equation: operator coefficients

We now consider the following situation. Let H be a Hilbertspace, H_1 a dense subspace, and A(x) a family of semibounded selfadjoint operators with common domain H_1, $x \in \bar{I}$. We assume that each A(x) has pure point spectrum and that $A(0) \geq -1/4$. Moreover, we require that the map $\bar{I} \ni x \mapsto A(x) \in L(H_1, H)$, the space of bounded linear operators from H_1 with the graph norm of A(0) to H, is smooth. Then

$$-d_x^2 + \frac{A(x)}{x^2} \tag{11}$$

is a symmetric operator in $L^2(I,H)$ with (dense) domain $C_o^\infty(I,H_1)$, and

Hardy's inequality shows again that this operator is bounded from below. So the Friedrichs extension T exists and we ask for an asymptotic expansion of tr e^{-sT} as $s \to 0$ if it exists. A natural condition for this to hold is that $e^{-sA(x)}$ is trace class for each $x \in \bar{I}$ and that expansions

$$\operatorname{tr} e^{-sA(x)} \sim s^{-\alpha} \sum_{\substack{j \geq 0 \\ 0 \leq k \leq k_0}} s^{\mu_j} \log^k s \, a_{jk}(x)$$

do exist for some $\alpha > 0$ and with $\mu_j \to \infty$. In fact, we pose the stronger condition that expansions of the form

$$\operatorname{tr} p(A, A', \ldots, A^{(k)})(x) \, e^{-sA(x)} \sim s^{-\alpha} \sum_{\substack{j \geq 0 \\ 0 \leq k \leq k_0}} s^{\mu_j} \log^k s \, a_{jk}(p,x) \quad (12)$$

do exist for $x \in \bar{I}$ and any (noncommutative) polynomial p in A and its derivatives. In this general setting one can prove the existence of an expansion

$$\operatorname{tr} e^{-sT} \sim s^{-\alpha-1/2} \sum_{\substack{j \geq 0 \\ 0 \leq k \leq k_0+1}} s^{\gamma_j}(A_j + B_j + C_{jk} \log^k s) \; . \quad (13)$$

Here A_j is again an "interior contribution" built similar to (10) from the functions $a_{jk}(p,x)$ in (12) for suitable p. B_j and C_{jk} are "boundary terms" depending only on the derivatives of A at 0 and L.

The proof of (13) can be done following essentially the pattern of the scalar case. The technical difficulties are, however, considerable. The first step is to prove the existence of an operator valued heat kernel for T i.e.

$$e^{-sT} u(x) = \int_I \Gamma_s(x,y)(u(y)) \, dy$$

for $x, y \in I$, $s > 0$, and $u \in L^2(I,H)$. Thus $\Gamma_s(x,y)$ is a bounded linear operator in H and the arguments for the scalar case can be generalized to yield good estimates for the operator norm. If Γ_s were trace class in H we would get

$$\operatorname{tr} e^{-sT} = \int_I \operatorname{tr}_H \Gamma_s(x,x)\, dx$$

as expected. The crucial step to prove this is a modification of the Minakshisundaram-Pleijel expansion:

$$\Gamma_s(x,y) \sim (4\pi s)^{-1/2} e^{-\frac{(x-y)^2}{4s}} \sum_{j\geq 0} s^j U_j(x,y)\, e^{-s\frac{A(y)}{y^2}} . \qquad (14)$$

Here the U_j are polynomials in A(x), A(y), and their derivatives, and the expansion is uniform with respect to the trace norm in compact subsets of I × I. In a neighborhood of 0 we compare with the heat kernel $\bar{\Gamma}_s$ of the Friedrichs extension of

$$-d_x^2 + \frac{A(0)}{x^2}$$

which again enjoys the scaling property (7). The treatment of the Neumann series is more complicated due to the unboundedness of the operators A(x); here we need regularity theorems for weak solutions of (11). The situation is much simpler if A is a constant function as in example 1b). Then it is enough to derive the expansion for

$$\int_0^{L/2} \operatorname{tr} \Gamma_s(x,x)\, dx .$$

Using (7) and the analogous properties of $\bar{\Gamma}_s$ we find as in the scalar case

$$\int_0^{L/2} \operatorname{tr} \Gamma_s(x,x)\, dx \sim \int_0^{L/2} \operatorname{tr} \bar{\Gamma}_s(x,x)\, dx = \int_0^{L/2} \frac{1}{x} \operatorname{tr} \bar{\Gamma}_{s/x^2}(1,1)\, dx$$

$$= 2 \int_{4s/L^2}^{\infty} \frac{1}{u} \operatorname{tr} \bar{\Gamma}_u(1,1)\, du .$$

Thus the expansion follows from (14) and the expansions for

$$\operatorname{tr} A^k e^{-sA}, \; k \geq 0 ,$$

whose coefficients are linear combinations of the coefficients in the expansion with $k = 0$. The contributions to the constant term in the expansion are (nonlocal) spectral invariants of A leading to Cheeger's formulas in the case of cone-like singularities.

4. The resolvent

For a semibounded selfadjoint operator T we can also study the resolvent $(T + z)^{-1}$, $\operatorname{Im} z \neq 0$. If γ is a suitably chosen path around the spectrum of T and if the resolvent has modest growth at infinity we have

$$e^{-sT} = \frac{1}{2\pi i} \int_\gamma e^{2z} (T + z)^{-1} \, dz .$$

Thus an asymptotic expansion of $\operatorname{tr} (T + z)^{-1}$ for large z implies an asymptotic expansion of $\operatorname{tr} e^{-sT}$ for small s. For nonsingular elliptic operators this expansion is well known ([9]). In the simplest singular case, the Friedrichs extension $\bar{T}$ in $L^2(\mathbb{R}_+)$ of the operator

$$-d_x^2 + \frac{a}{x^2} , \; a \geq -1/4 ,$$

the kernel of the resolvent $(\bar{T} + z^2)^{-1}$ can be determined explicitly,

$$(\bar{T} + z^2)^{-1}(x,y) = (xy)^{1/2} \, I_\nu(xz) \, K_\nu(yz) ,$$

where $0 < x \leq y$, $\nu := (a + 1/4)^{1/2}$, $\operatorname{Im} z \neq 0$, and I_ν, K_ν are modified Bessel functions (cf. [6]). This operator is not trace class but we can define a "distributional trace" by

$$\int_0^\infty \varphi(x) \, x \, I_\nu(xz) \, K_\nu(xz) \, dx, \; \varphi \in C_0^\infty(\mathbb{R}) .$$

The well known asymptotic expansion of Bessel functions for large arguments suggests the following generalization. Consider a "symbol"

$\sigma(x,\zeta)$ where $x \in \mathbb{R}$ and $\zeta \in C := \{z \in \mathbb{C} \,|\, |\arg z| < \pi-\varepsilon\}$ for some $\varepsilon > 0$. We require σ to be smooth in x and rapidly decreasing and to have an asymptotic expansion as $\zeta \to \infty$ in C,

$$\sigma(x,\zeta) \sim \sum_{\alpha,m} \sigma_{\alpha m}(x)\, \zeta^{\alpha} \log^{m} \zeta \quad ,$$

where α runs through a discrete set of complex numbers, Re $\sigma \to -\infty$, each m is a nonnegative integer, and there are only finitely many $\sigma_{\alpha m} \neq 0$ for fixed α. Then we may ask whether there is an asymptotic expansion of

$$\int_0^\infty \sigma(x,xz)\, dx \tag{15}$$

as $z \to \infty$ in C. This expansion has been determined in [5] and has the following form:

$$\int_0^\infty \sigma(x,xz)\, dx \sim \sum_{k \geq 0} z^{-k-1} \int_0^\infty \frac{\zeta^k}{k!}\, \sigma^{(k)}(0,\zeta)\, d\zeta \tag{16}$$

$$+ \sum_{\alpha,m} \int_0^\infty \sigma_{\alpha m}(x)\, (xz)^{\alpha} \log^{m}(xz)\, dx$$

$$+ \sum_{\substack{\alpha,m \\ -\alpha \in \mathbb{N}}} z^{\alpha} \log^{m+1} z\, \sigma_{\alpha m}^{(-\alpha-1)}(0) \Big/ (m+1)(-\alpha-1)! \ .$$

The first and third sum contains "boundary" terms and the middle sum "interior" terms all of which can be viewed as "moments" of σ. The integrals, however, may be divergent and have to be defined suitably. For example, the integral

$$\int_0^\infty \varphi(x)\, x^{\alpha} \log^{m} x\, dx =: I(\alpha)$$

with $\varphi \in C_0^\infty(\mathbb{R})$ is analytic in α if Re $\alpha > -1$. Integrating by parts we see that $I(\alpha)$ extends meromorphically to the whole complex plane. At a pole, we substract the singular part and call the resulting value the "regular analytic extension". In this way all the integrals in (16) are defined.

The more general case of the operator

$$-d_x^2 + \frac{a(x)}{x^2}$$

can also be handled by this approach. To do so we use the resolvent of $\bar{T}$ as parametrix and obtain a convergent Neumann series. Each term in the series turns out ot be of the form (15) and the expansion follows from (16).

The method can also be extended to handle (constant) operator coefficients. We choose $m \in \mathbb{N}$ such that for a given selfadjoint operator $A \geqq -1/4$ with discrete spectrum

$$\sum_{\nu^2-1/4 \in \operatorname{spec} A} \nu^{-2m} < \infty .$$

Denoting by $k_\nu^m(x,y,z)$ the kernel of

$$(-d_x^2 + \frac{\lambda}{x^2} + z^2)^{-m}, \quad \nu = \sqrt{\lambda+1/4} ,$$

we find that the distributional trace of

$(-d_x^2 + \frac{A}{x^2} + z^2)^{-m}$ is given by

$$\int_0^\infty \varphi(x)\; x^{2m-1} \sum_\nu k_\nu^m(1,1,xz)\, dx .$$

To apply (16) we have to show that

$$\sigma(x,\zeta) := x^{2m-1} \sum_\nu k_\nu^m(1,1,\zeta)$$

has an asymptotic expansion as $\zeta \to \infty$. This is done by means of an expansion for $(-d_x^2 + \frac{A}{x^2} + z^2)^{-m}$ in terms of $(A + \mu)^{-\ell}$. This argument parallels the expansion (14) and uses the calculus of pseudodifferential operators with operator coefficients.

REFERENCES

1. Brüning, J.: Heat equation asymptotics for singular Sturm-Liouville operators, to appear in Math. Ann.

2. Brüning, J., and E. Heintze: Representations of compact Lie groups and elliptic operators, Inventiones math. 50 (1979), 169 - 203

3. Brüning, J., and E. Heintze: The asymptotic expansion of Minakshisundaram-Pleijel in the equivariant case, preprint 1983

4. Brüning, J., and E. Heintze: Zur spektralen Starrheit gewisser Drehflächen, preprint 1984

5. Brüning, J., and R.T. Seeley: Regular singular asymptotics, preprint 1984

6. Callias, C.: The heat equation with singular coefficients I, Comm. Math. Phys. 88 (1983), 357 - 385

7. Cheeger, J.: On the spectral geometry of spaces with conelike singularities, Proc. Nat. Acad. Sci. USA 76 (1979), 2103 - 2106

8. Cheeger, J.: Spectral geometry of singular Riemannian spaces, preprint 1982

9. Seeley, R.T.: Complex powers of elliptic operators, Proc. Symposia in Pure Math. Vol 10, 288 - 307, Providence: AMS 1967.

ON A NUMERICAL METHOD FOR FRACTURE MECHANICS

P.DESTUYNDER
Ecole Centrales des Arts et Manufactures
Grande Voie des Vignes
92290 CHATENAY MALABRY
FRANCE

M.DJAOUA
Ecole Nationale d'Ingénieurs de Tunis
B.P.37, Belvedere TUNIS
TUNISIA

S.LESCURE
EDF - DER
92160 Clamart
FRANCE

ABSTRACT.- The energy release rate, G, is the derivative of the energy with respect to the crack length. Using Lagrangian coordinates, we can derive an expression of G as a surface integral, which is mathematically equivalent to other well known expressions as the J integral of Rice or the expression in terms of stress intensity factors. Using interior error estimates, we derive an error estimate of $O(h^{1-\varepsilon})$ on G by using piecewise linear elements. For sake of simplicity, we present these estimates for the Laplace operator. The numerical trials, which show a very good stability of the method, were performed for the elasticity system.

1 - INTRODUCTION

Let Ω be a planar slit domain, with boundary Γ, corresponding to a body with a crack in it. We consider the following problem :

$$\left\{ \begin{array}{l} \text{Find } u \in H^1_0(\Omega) \text{ such that} \\ -\Delta u = f \text{ in } \Omega \end{array} \right. \tag{1.1}$$

where f is a given function of $L^2(\Omega)$, vanishing in some neighbourhood v of the crack tip.

It is well known that the problem (1.1) can be set in a variational formulation :

$$\begin{cases} \text{Find } u \in H^1_0(\Omega) \text{ such that} \\ A(u,v) = \int_\Omega \left(\frac{\partial u}{\partial x1}\frac{\partial v}{\partial x1} + \frac{\partial v}{\partial x2}\frac{\partial v}{\partial x2}\right) dx = \int_\Omega f\, v\, dx \text{ for all } v \in H^1_0(\Omega) \end{cases} \quad (1.2)$$

and that the problem (1.2) has a unique solution in $H^1_0(\Omega)$. Furthermore, this solution can be expanded near the crack tip S, into a singular part and a regular part ([10]);

$$u(M) = K\, r^{\frac{1}{2}} \sin\frac{\beta}{2} + u^R(M) \quad (1.3)$$

where (r,β) are the polar coordinates of M with respect to the origin S, and u^R is an element of $H^2(\Omega)$.

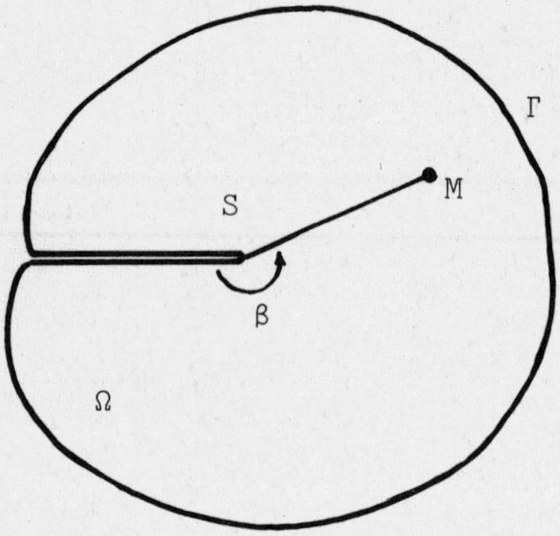

Figure 1

The calculation of the factor K, which is called in mechanics the stress intensity factor, is of great importance : In aerodynamics, it caracterizes the lift of a flow around a corner shaped body [9]. In fracture mechanics, most of the crack propagation criteria are expressed in terms of it [4].

The numerical methods used for the calculation of the coefficients K are of two kinds :

- Those which explicitely use the asymptotic expansion of the solution near the crack tip : one can introduce the singular function as a basis function

([6], [9]), or use the two first terms of the asymptotic expansion for an extrapolation ([13], or some other treatment using a very precise knowledge of the solution at the crack tip [2], ...

- The second kind of methods use implicitely the asymptotic expansion : Mesh refinements ([12], [14] or special elements such as the quarter point one [1].

Both kinds of methods need some special treatment, and cannot give an approximation of the factor K from a solution calculated by a standard finite element method.

We hereafter describe a new method, using standard meshes and standard (polynomial) approximation spaces, and leading to satisfactory error estimates.

2 - SOME PRELIMINARY RESULTS

Let θ be a vectorial function, defined on $\bar{\Omega}$, each of its components belonging to $W^{1,\infty}(\bar{\Omega})$. Furthermore, we suppose θ is as follows :

$$\theta = \begin{cases} \theta_1 \ (x_1, x_2) \\ 0 \end{cases} \tag{2.1}$$

were

$$\theta_1 = 1 \text{ in } v_1 \ ; \ \theta_1 = 0 \text{ in } v_2^c \tag{2.2}$$

were v_1 and v_2 are two neighbourhoods of S, $v_1 \subset v_2 \subset v \subset \Omega$.

θ describes a possible propagation of the crack, in the direction tangent to the crack.

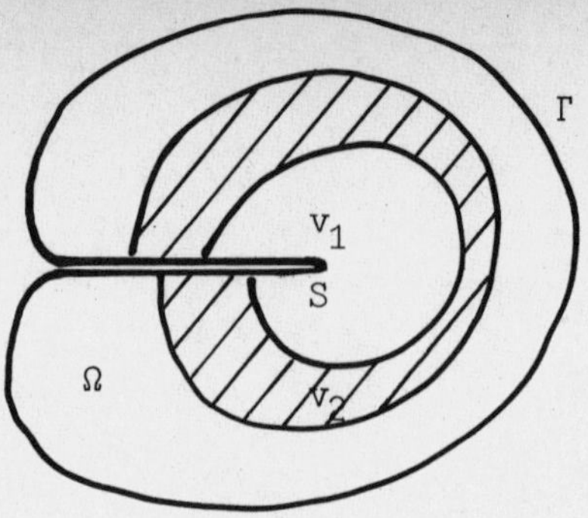

Figure 2

The following theorem gives an expression of K, in terms of u and θ.

Theorem 2.1

Let u *be the solution of problem (1.2) and* K *be the stress intensity factor defined by (1.3). With θ being any vectorial function of* $(W^{1,\infty}(\bar{\Omega}))^2$, *satisfying (2.1) and (2.2), we get*

$$K^2 = \frac{4}{\pi}\left\{\frac{1}{2}\int_\Omega\left(\left(\frac{\partial u}{\partial x_1}\right)^2 + \left(\frac{\partial u}{\partial x_2}\right)^2\right)\frac{\partial\theta_1}{\partial x_1} - \int_\Omega\left(\frac{\partial\theta}{\partial x_1}\left(\frac{\partial u}{\partial x_1}\right)^2 + \frac{\partial\theta_1}{\partial x_2}\frac{\partial u}{\partial x_1}\cdot\frac{\partial u}{\partial x_2}\right)\right\}. \quad (2.3)$$

PROOF

The proof of this theorem is given, for the elasticity system, in [7]. Let us denote by G the right hand side of (2.3), by C_r the circle of radius r and center S, and et Ω_r be

$$\Omega_r = \{x \in \Omega \; ; \; \text{dist}\,(x,S) > r\} \quad (2.4)$$

Then, we have

$$G = \lim_{r\to 0}\left\{\frac{4}{\pi}\left\{\frac{1}{2}\int_{\Omega_r}|\nabla u|^2\frac{\partial\theta_1}{\partial x1} - \int_{\Omega_r}\left(\frac{\partial\theta_1}{\partial x_1}\left(\frac{\partial u}{\partial x_1}\right)^2 + \frac{\partial\theta_1}{\partial x2}\frac{\partial u}{\partial x_1}\frac{\partial u}{\partial x_1}\right)\right\}\right\}$$

Since u is regular (i.e., in $H^2(\Omega_r)$), we can use Green's formulas and then obtain

$$G = \frac{4}{\pi}\lim_{r\to 0}\left\{\frac{1}{2}\int_{C_r}|\nabla u|^2\,\theta_1\,n_1 - \int_{C_r}\frac{\partial u}{\partial n}\cdot\frac{\partial u}{\partial x_1}\cdot\theta_1\right\} \quad (2.5)$$

For r small enough, $\theta_1 \equiv 1$ on C_r. Using expression (1.3), we can see that the only contributions which remain when $r \to 0$ are those of the singular part of u. Thus, a simple calculation gives

$$G = K^2.$$

REMARK 2.1

When f does not vanish near the crack tip, we have to modify (2.3) in the following way :

$$K^2 = \frac{4}{\pi}\left\{\frac{1}{2}\int_\Omega\left(\left(\frac{\partial u}{\partial x1}\right)^2 + \left(\frac{\partial u}{\partial x2}\right)^2\right)\theta_{1,1} - \int_\Omega\left(\frac{\partial\theta}{\partial x1}\left(\frac{\partial u}{\partial x1}\right)^2 + \frac{\partial\theta}{\partial x2}\frac{\partial u}{\partial x1}\frac{\partial u}{\partial x2}\right)\right.$$
$$\left. - \int_\Omega f.u.\theta_{1,1} - \int_\Omega\frac{\partial f}{\partial x1}\cdot\theta_1\cdot u\right\} \quad (2.6)$$

Provided that f is smooth enough (say, $f \in H^1(\Omega)$)

Expression (2.3) involves ∇u in the region where $\frac{\partial\theta}{\partial x1}$ does not vanish. This region is "far from" the crack tip. It is known that the approximation by finite elements of the solution of an elliptic problem gives better results in regions far from the crack tip than in the near regions. We recall in the following theorem, due to Nitsche and Schatz [11], some interior error estimates that we shall use to get our own estimates.

Before that, we have to recall some definitions used by these authors.

With Ω_1 and Ω_2 being two subdomains of Ω, with $\Omega_1 \quad \Omega_2$, we define

$$\operatorname{dist}_{<}(\Omega_1, \Omega_2) = \inf_{x \in \partial\Omega_1 \setminus (\partial\Omega_1 \cap \partial\Omega)} \operatorname{dist}\left(x, \partial\Omega_2 \setminus (\partial\Omega_2 \cap \partial\Omega)\right). \tag{2.7}$$

Then, $\Omega_1 < \Omega_2$ signifies that $\Omega_1 \subset \Omega_2$, and that $\operatorname{dist}_{<}(\Omega_1 \, \Omega_2) \geqslant 0$.

For any subdomain D of Ω, we also define the following spaces, for $s > \frac{1}{2}$:

$$\overset{<}{H}{}^s(D) = (v \in H^s(D) \; ; \; v = 0 \text{ on } \partial D \cap \partial\Omega\}$$

Let us denote by $V_h(\Omega)$ a space of regular finite elements ; we then define, for any subdomain D of Ω

$$V_h(D) = \{v_h|_D \; ; \; v_h \in V_h(\Omega)\} \tag{2.8}$$

$$\overset{<}{V}_h(D) = \{v_h \in V_h(D) \; ; \; v_h \equiv 0 \text{ in some neighborhood of } \partial D \setminus (\partial D \cap \partial\Omega)\} \tag{2.9}$$

The interior error estimates are as follows.

<u>*THEOREM 2.2*</u> [11]

Let p be a non negative integer and $\Omega_1 < \Omega_2 < \Omega$. There exist constants C and h_1 depending on p and $\operatorname{dist}_{<}(\Omega_1, \Omega_2)$, such that for $0 < h < h_1$, the following holds :

Let $u \in \overset{<}{H}{}^1(\Omega_2)$, and $u_h \in V_h(\Omega_2)$ satisfy

$$A(u - u_h, \phi) = \int_\Omega \nabla (u - u_h) . \nabla\phi = 0 \text{ for all } \phi \in \overset{<}{V}_h(\Omega_2). \tag{2.10}$$

Then, for any $\chi \in V_h(\Omega)$,

$$||u - u_h||_{1,\Omega_1} < C\,(||u - \chi||_{1,\Omega_2} + ||u - u_h||_{-p,\Omega_2}). \tag{2.11}$$

3 - THE FINITE ELEMENT METHOD AND THE ERROR ESTIMATES

Let B_1, B_2, B_3 and B_4 be four polygonal domains containing S, and satisfying $B_1 \subset B_2 \subset B_3 \subset B_4 \subset v$. We define

$$\Omega_1 = B_3 \setminus B_2$$

$$\Omega_2 = B_4 \setminus B_1$$

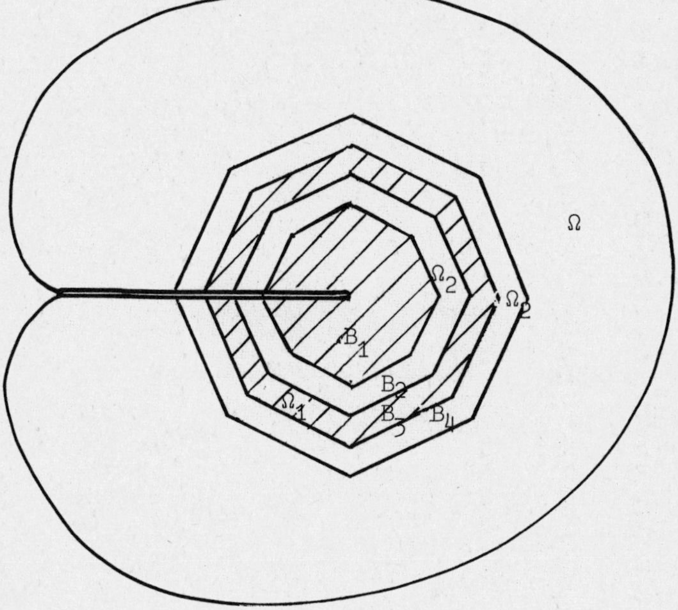

Figure 3

Let T_h be a regular triangulation of Ω, such that no triangle crosses the boundary of Ω_1 or Ω_2. Let us define

$$V_h(\Omega) = \{v_h \in C^0(\overline{\Omega}) ; v_h|_T \in P_1(T) \; \forall \, T \in T_h, \text{ and } v_h|_\Gamma = 0\} . \qquad (3.1)$$

Let u_h be the solution of the finite element discretization of problem (1.2) we calculate then an approximate value of K by the formula.

$$K_h^2 = \frac{4}{\Pi} \{\frac{1}{2} \int_\Omega |\nabla u_h|^2 \frac{\partial \theta_1}{\partial x_1} - \int_\Omega \frac{\partial \theta_1}{\partial x1}\left(\frac{\partial u_h}{\partial x1}\right)^2 + \frac{\partial \theta_1}{\partial x2} \frac{\partial u_h}{\partial x1} \frac{\partial u_h}{\partial x2}\} \quad (3.2)$$

Where θ_1 is a regular function satisfying

$$\theta_1 = \begin{cases} 1 & \text{in } B_2 \\ 0 & \text{in } \Omega \setminus B_3 \end{cases} \quad (3.3)$$

Expression (3.2) has then to be calculated only on Ω_1, since $\nabla\theta_1 = 0$ outside of this domain.

THEOREM 3.1

Let K_h be calculated by (3.2), and let K be the stress intensity factor defined by (1.3). The following error estimate then holds for any $\varepsilon > 0$.

$$|K^2 - K_h^2| \leqslant C\, h^{1-\varepsilon} \{|u|_{2,\Omega_2} + ||u||_{H^{\frac{3}{2}-\varepsilon}(\Omega)}\} \cdot \quad (3.4)$$

Where C is a constant depending on u and ε

Proof

We have, by (2.3) and (3.2)

$$K^2 - K_h^2 = \frac{4}{\Pi} \{\frac{1}{2}\int_{\Omega_1} (|\nabla u|^2 - |\nabla u_h|^2) \frac{\partial \theta 1}{\partial x_1}$$

$$- \int_{\Omega_1} \frac{\partial \theta 1}{\partial x1} \{ (\frac{\partial u}{\partial x1})^2 - (\frac{\partial u_h}{\partial x1})^2\} + \frac{\partial \theta_1}{\partial x_2} (\frac{\partial u}{\partial x1} \cdot \frac{\partial u}{\partial x2} - \frac{\partial u_h}{\partial x1} \cdot \frac{\partial u_h}{\partial x2}) \}$$

Thus, by using the Cauchy-Schwarz inequality

$$|K^2 - K_h^2| \leqslant \frac{4}{\Pi} \{ \frac{1}{2} ||u - u_h||_{H^1(\Omega_1)} ||u + u_h||_{H^1(\Omega_1)} ||\theta_1||_{W^{1,\infty}(\Omega)}$$

$$+ ||u - u_h||_{H^1(\Omega_1)} ||u + u_h||_{H^1(\Omega_1)} ||\theta_1||_{W^{1,\infty}(\Omega_1)}$$

$$+ ||\theta_1||_{W^{1,\infty}(\Omega_1)} \{ ||u||_{H^1(\Omega_1)} ||u - u_h||_{H^1(\Omega_1)} + ||u_h||_{H^1(\Omega_1)} ||u - u_h||_{H^1(\Omega_1)} \}\}$$

Since we have global error estimates, we get

$$|K^2 - K_h^2| \leqslant C ||u - u_h||_{H^1(\Omega_1)} . \tag{3.5}$$

Let χ be some regular function equal to one on Ω_2, and vanishing on $\partial\Omega \setminus (\partial\Omega \cap \partial\Omega_2)$, and let us define

$$\begin{cases} U = \chi u \\ U_h = \chi u_h \end{cases} \tag{3.6}$$

Then, $U \in \overset{<}{H}{}^1 (\Omega_2)$ and for any $\varepsilon \overset{<}{V}_h (\Omega_2)$, we have, since $\phi = 0$ outside of Ω_2

$$A(U - U_h, \phi) = \int_\Omega \nabla(\chi(u - u_h)). \nabla\phi$$
$$= \int_{\Omega_2} \nabla(u - u_h). \nabla \phi = 0.$$

By application of theorem 2.2, we have, for any $\chi \in V_h (\Omega_2)$

$$||u - u_h||_{1,\Omega_1} = ||U - U_h||_{1,\Omega_1} \leqslant C \{||u - \chi||_{1,\Omega_2} + ||u - u_h||_{-p,\Omega_2}\}.$$

For $p = 0$, we have

$$||u - u_h||_{1,\Omega_1} \leqslant C \{ \inf_{\chi \in V_h(\Omega_2)} ||u - \chi||_{1, \Omega_2} + ||u - u_h||_{0,\Omega_2} \}. \tag{3.7}$$

But, since $u \in H^{\frac{3}{2}-\varepsilon}(\Omega)$, we have

$$||u - u_h||_{0,\Omega_2} \leqslant ||u - u_h||_{0,\Omega} \leqslant C\, h^{1-\varepsilon}\, |u|_{H^{\frac{3}{2}-\varepsilon}(\Omega)}. \qquad (3.8)$$

Then, by choosing for χ the interpolate of u, we get ([5])

$$\inf_{\chi \in V_h(\Omega_2)} ||u - \chi||_{1,\Omega_2} \leqslant ||u - \Pi_h u||_{1,\Omega_2} < C\, h\, |u|_{2,\Omega_2}.$$

So $|K^2 - K_h^2| < C\, h^{1-\varepsilon}\, \{|u|_{2,\Omega_2} + ||u||_{H^{\frac{3}{2}-\varepsilon}(\Omega)}\}.$

and (3.4) is proved.

<u>REMARK 3.1</u>

When f does not vanish near the crack tip, it is sufficient to replace expression (3.2) by the following

$$K_h^2 = \frac{4}{\Pi} \{ \int_\Omega |\nabla u|^2 \theta_{1,1} - \int_\Omega \theta_{1,1} \left(\frac{\partial u_h}{\partial x1}\right)^2 + \theta_{1,2} \frac{\partial u_h}{\partial x1} \cdot \frac{\partial u_h}{\partial x2}$$

$$- \int f.u_h\, \theta_{1,1} - \int \frac{\partial f}{\partial x1} \cdot \theta_1\, u_h \} \qquad (3.9)$$

to obtain the same estimate .

<u>REMARK 3.2</u>

In order to improve estimate (3.4), we have to improve the global estimate $||u - u_h||_{0,\Omega}$. Th can be performed by using some mesh refinement [12]. But we also have to improve the local estimate, which can be done by using P_2 finite elements. We then obtain

$$|K^2 - K_h^2| = 0\,(h^2) \qquad (3.10)$$

4 - NUMERICAL RESULTS

These results were obtained at E.D.F, by using the ASKA finite element code (University of Stuttgart). The body was a plate of 60 mm x 60 mm x 1 mm, with a crack of 6 mm length. The plate was symmetric, and so were the traction efforts on a part of its boundary. The mechanical constants were :

$$E = 2\ 10^5\ \ N/mm^2$$

$$\nu = 0.3$$

Numerical trials were run for four meshes, and three cases of loading. (see figures 4 and 5)

1) $d = 10^4$ N/mm on the whole length
2) d = 4670 N/mm on the half of the length
3) d = 1000 N/mm on the crack length

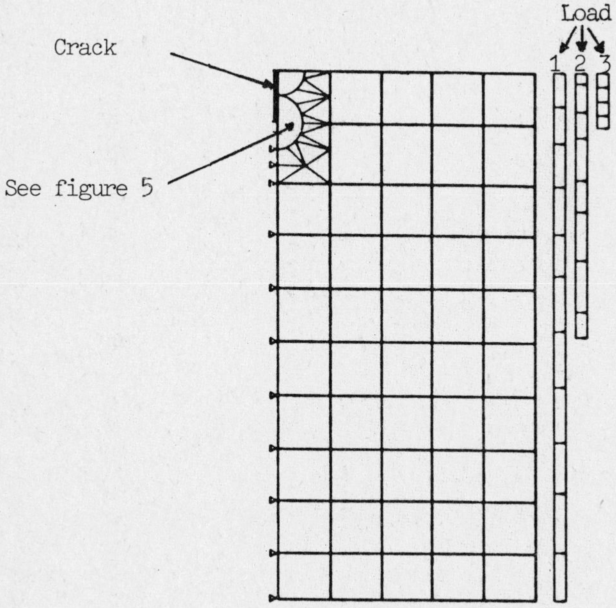

Figure 4

Mesh n°1

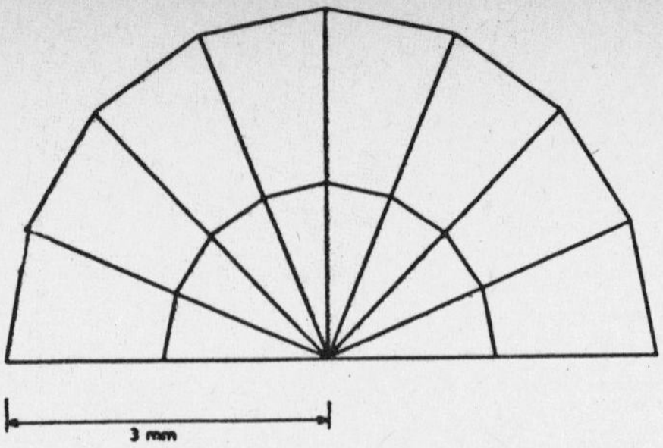

Mesh n°2

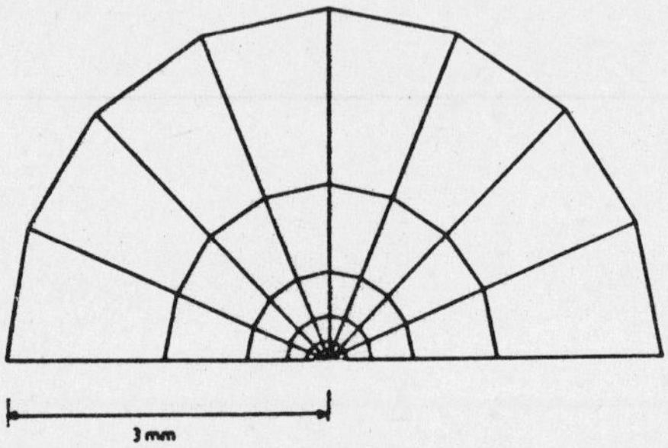

Mesh n°3

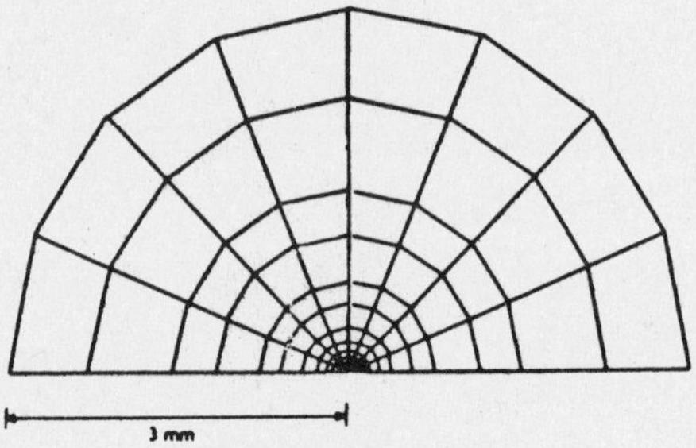

Mesh n°4

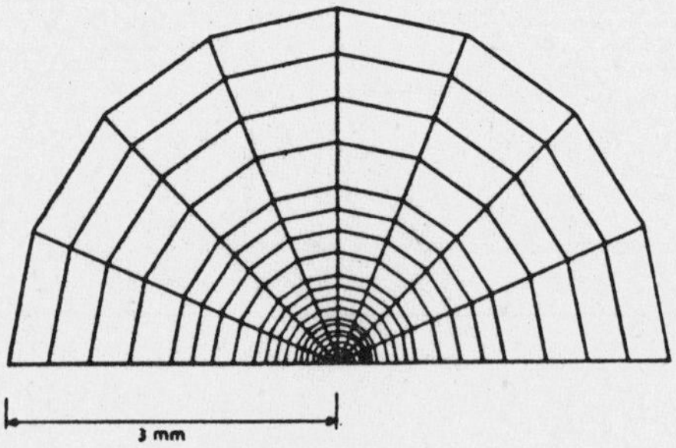

Figure 5

The results were compared to Bowie's [3], and the difference was less than 1%.

	Bowie's (N/mm)	Our results (piecewise qudratic elements, mesh n°4)
1st case	14310	14445
2rd case	18300	18410
3rd case	2998	3008

Table 1

The convergence was compared to the J integral (calculated at a distance of 1.5 mm of the crack tip) and the expression in terms of the stress intensity factors (here : $G = \frac{1}{E} K_1^2$, K_1 calculated by identification.

Mesh	1	2	3	4
J	6560	10862	11220	12600
G	14950	14432	14441,5	14445
$\frac{1}{E} K_1^2$	13520	13567	13781	13886

Table 2

All these results were obtained with piecewise quadratic elements. It was expected that the J intgral should not behave well, since the integration path is quite near the crack tip.

The choice of the θ function is quite arbitrary, provided that $\theta \varepsilon (W^{1,\infty}(\overline{\Omega}))^2$ and that $\theta = \{ {1 \atop 0}$ in some neighbourhood of the crack tip. The following results show a very good stability of G with respect to the choice of θ . It is possible then to choose θ with a small support near the crack tip.

support of $\nabla\theta$	2.625<R<3mm	0.656<R<3	0.46<R<0.94	0.023<R<0.047	0.023mm<R<3mm
J (piecewise quadratic, 8 nodes)	14028	11518	11598	11652	12200
G (piecewise quadratic, 8 nodes)	14445.44	14445.4	14445.19	14344.89	14344.66
G (piecewise linear 4 nodes)	13096	13106	13131	13079	13116

Table 3

As expected, the J integral is much better far from the crack tip. Piecewise quadratic elements lead to much better results, as announced by estimate (3.10). Many other cases were treated by this method, some of then from both numerical and theoretical points of vue : thermoelasticity, unilateral contact, crack bifurcation, etc... More details abount thes can be found in [8].

5 - CONCLUSION

It is important to notice that the estimate (3.4) is obtained without any special treatment of the space V_h.

The error estimate (3.4) is obtained on K^2, and not on K. In fact K has no interest, by itself. The very quantity to be calculated, for the prediction of crack growth, is the energy release rate, which is

$$G = \frac{\Pi}{4} K^2.$$

The numerical results show that the method is very accurate and very stable with respect to the choice of the domains Ω_1, Ω_2. They also show that the use of piecewise quadratic elements leads to much better results than the use of piecewise linear ones.

REFERENCES

[1] BARSOUM, R.S., *On the use of the isoparametric finite elements in linear fracture mechanics*, Int. J. for Num. Meth. in Eng., 10, 25-37 (1976).

[2] BELLOUT, R. and MOUSSAOUI, M., *Calculations on singularities coefficients these proceedings).*

[3] BOWIE, O.L., *Methods of analysis and solutions of crack problems, G.L.SIH Ed, Noordhorf International Publishing, Leyden (1973).*

[4] BUI, H.D., *Mécanique de la rupture fragile, Masson, (1978).*

[5] CIARLET, P.G. and RAVIART,P.A., *Lagrange and Hermite interpolation in R^n, with applications to finite element methods, Arch. Rat. Mech. Anal., 46, 177-199 (1972).*

[6] DESTUYNDER, Ph. and DJAOUA, M., *Estimation de l'erreur sur le coefficient de la singularité de la solution d'un problème elliptique sur un ouvert avec coin, R.A.I.R.O., Analyse Numérique, 14, N° 3, 239-248, (1980).*

[7] DESTUYNDER, Ph. and DJAOUA, M., *Sur une interprétation mathématique de l'intégrale de Rice en théorie de la rupture fragile, Math. Meth. in the Appl. Sci., 3, 70-87 (1981).*

[8] DESTUYNDER, Ph., LEFEBRE.J.P., LESCURE.S. and MIALON.P : *in mécanique de la rupture, Proceedings of the CEA-EDF.INRIA conference held at Savigny sur Clairis, October (1982).*

[9] DJAOUA, M., *A method of calculation of lifting flows around two dimensional carner shaped bodies, Math. of Comp., 36, N° 154, 405-425, (1981).*

[10] GRISVARD, P., *Behavior of the solutions of an elliptic boundary value problem in a polygonal or polyhedral domain, Numerical solution of Partial Differential Equations III, Synspade 1975 (Bert Hubbard, Ed.),*

[11] NITSCHE,J.A. and SCHATZ,AH., *Interior estimates for Ritz-Galerkin methods, Math. of Comp., 28, 937-958 (1974).*

[12] RAUGEL, G., *Resolution numérique de problèmes elliptiques dans les domaines avec coins, Thèse de 3ème cycle, Université de Rennes, (1978).*

[13] SCHATZ, A.H. and WAHLBIN, L.B., *Maximum norm estimates in the finite element method on plane polygonal domains, Part 1, Math. of Comp., 32, N°141, 73-109 (1978).*

[14] SCHATZ, A.H. and WAHLBIN, L.B., *Maximum norm estimates in teh finite element mthod on plane polygonal domains, Part 2, Refinements, Math. of Comp., 33, N° 146, 465-492 (1979).*

ON FINITE ELEMENT METHODS FOR NONLINEAR ELLIPTIC PROBLEMS

ON DOMAINS WITH CORNERS

Manfred Dobrowolski
Institut für Angewandte Mathematik
der Universität Bonn
Wegelerstraße 10
5300 BONN / Fed.Rep.Ger.

Abstract: For the variational problem $\int_\Omega \left\{\frac{1}{p}(K+|\nabla u|^2)^{p/2} - fu\right\} dx \to \mathrm{Min}$ in $H^{1,p}(\Omega)$, $\Omega \subset \mathbb{R}^2$, $p \geq 2$, $K \geq 0$, it is shown that in the near of corners a singular expansion $u = ks + w$ holds, where $k \in \mathbb{R}$, $s = r^\alpha t(\varphi)$, and w satisfies $|w| \leq cr^{\alpha+\varepsilon}$, $|\nabla w| \leq cr^{\alpha+\varepsilon-1}$ etc. with a small $\varepsilon = \varepsilon(\alpha,p)$. The pair $(\alpha, t(\varphi))$ is obtained by a nonlinear eigenvalue problem. It is proved that the eigenvalue α is given by a root of a quadratic polynomial with known coefficients. The theoretical results are used for the investigation of the ordinary Finite Element Method and the Dual Singular Function Method already known from the linear case. Some numerical computations illustrate the theoretical results.

1. Introduction

Consider the variational problem

$$\int_\Omega \left\{ \frac{1}{p} (K + |\nabla u|^2)^{p/2} - fu \right\} dx \to \mathrm{Min} \quad , 1 < p < \infty , K \geq 0 \tag{1.1}$$

where the Min is taken over all functions $u \in \mathring{H}^{1,p}(\Omega)$, $\Omega \subset \mathbb{R}^2$. Here and in the following, $L^p(\Omega)$ and $H^{m,p}(\Omega)$, $m \in \mathbb{N}$, $1 \leq p < \infty$, denote the usual Lebesgue and Sobolev spaces with norms

$$\|v\|_p := \|v\|_{p;\Omega} := \left(\int_\Omega |v|^p \, dx \right)^{1/p} ,$$

$$\|v\|_{m,p} := \|v\|_{m,p;\Omega} := \left(\sum_{|\alpha| \leq m} \|D^\alpha v\|_p^p \right)^{1/p} .$$

The corresponding Euler equation of the variational problem (1.1) is given by

$$\begin{aligned} Au := - \mathrm{div} \; (K + |\nabla u|^2)^{(p-2)/2} \nabla u &= f \quad \text{in } \Omega \\ u &= 0 \quad \text{on } \partial\Omega \end{aligned} \tag{1.2}$$

which has a unique weak solution $u \in \mathring{H}^{1,p}(\Omega)$ since $A : \mathring{H}^{1,p} \to (\mathring{H}^{1,p})^*$ is a monotone and coercitive operator. Moreover, the solution of (1.2) with $K > 0$ is known to be regular if the data of the problem, $\partial\Omega$ and f, are smooth.

In this paper, we are concerned with the situation in which a corner of the domain destroys the smoothness of the solution u. In order to isolate the influence of a corner we assume that Ω contains one corner with interior angle ω, $0 < \omega \leq 2\pi$, and that the other boundary and the right hand side f are sufficiently smooth for all our purposes. Hence it is sufficient to analyze problem (1.2) in the unit cone $C = \{ (r,\varphi) \in (0,1) \times (0,\omega) \}$ where (r,φ) denote the polar coordinates of the corner. Using the notaion S for the sides and B for the bottom of C we have to study the problem

$$Au = f \quad \text{in } C , \quad u = 0 \quad \text{on } S , \quad u = g \quad \text{on } B \tag{1.3}$$

where, by local regularity, g is a smooth function on B.

In order to get more familar with the problem we first review the linear case $p = 2$, $A = -\Delta$. Here the problem is completely solved by Kondrat'ev [10] who has proved that

$$u = \sum_{i=1}^{I} k_i s_i + w \quad , \; k_i \in \mathbb{R} , \tag{1.4}$$

where $s_i = r^{i\alpha} \sin i\alpha\varphi$, $\alpha := \pi/\omega$, and w satisfies $|w| \leq cr^{I\alpha+\varepsilon}$, $|\nabla w| \leq cr^{I\alpha-1+\varepsilon}$ etc. The possible choices of the integer I in (1.4) depend on the behaviour of the right hand side f in a neighbourhood of 0. In fact, for $f = 0$ we immediately obtain that

$$u = \sum_{i=1}^{\infty} k_i s_i \quad ,$$

where k_i are the Fourier-coefficients of the function g. From the linear case we learn that the singular functions are kernel functions of the operator which seperate in the polar coordinates r and φ.

By the linear case it is motivated to study solutions of the Pseudo-Laplace equation $K = 0$ which have the form $u = r^\alpha t(\varphi)$. Writing A in polar coordinates the problem $A(r^\alpha t(\varphi)) = 0$ in C, $r^\alpha t(\varphi) = 0$ on S, reduces to the nonlinear eigenvalue problem

$$\begin{aligned} &- \partial_\varphi \left\{ (\alpha^2 t^2 + |\partial_\varphi t|^2)^{(p-2)/2} \partial_\varphi t \right\} - \\ &- \alpha(\alpha(p-1) + 2 - p)(\alpha^2 t^2 + |\partial_\varphi t|^2)^{(p-2)/2} t = 0 \qquad (1.5) \\ &t(0) = t(\omega) = 0 \quad . \end{aligned}$$

From [13] and [7] we know that there exists a unique solution (α,t) of (1.5) with $\alpha > 0$ and $t(\varphi) > 0$ in $(0,\omega)$. Moreover, the eigenvalue α is given as a root of a quadratic polynomial. These results are reviewed in section 2. For the following it is sufficient to know that

$$\begin{array}{ll} 0 < \alpha < 1 & \text{for } \omega > \pi \\ 1 < \alpha & \text{for } 0 < \omega < \pi \end{array} \quad \text{for } 1 < p < \infty \,. \tag{1.6}$$

The eigenvalue problem (1.5) has no solution for $p = 1$ so that there are no results for the minimal surface equation.

The knowledge of a positive singular function $s := r^\alpha t(\varphi)$ leads to first estimates of the solution of problem (1.3) by using the following "weak comparison principle":

<u>Lemma</u>: Let A be the elliptic operator in (1.2) with $K \geq 0$ and $1<p<\infty$ and assume that for $u,v \in H^{1,p}(\Omega)$

$$Au \leq Av$$

in the weak sense. Then $u \leq v$ on $\partial\Omega$ implies that

$$u \leq v \quad \text{in } \Omega \,.$$

The <u>proof</u> of this comparison principle is very easy. We insert the function $\psi = \max\{u-v,0\} \geq 0$ and obtain that

$$\int_{u>v} (Au - Av)(u-v)\, dx \leq 0$$

and hence $u \leq v$ in Ω.

For A being the Pseudo-Laplace operator $(K = 0)$ and $f = 0$ we get for the solution u of (1.3)

$$k_1 s \leq u \leq k_2 s \quad , \quad k_i \in \mathbb{R} \,. \tag{1.7}$$

A known function (here $v = s$)) for estimating the solution by the comparison principle is called a barrier. In our case we obtain the estimate $|u| \leq cr^\alpha$ by using the barrier s. In section 3 a barrier is constructed which estimates the solution for more general problems than $Au = 0$ even we are not able to treat the problem (1.1) in full generality. Now higher regularity of u can be proved by using the local regularity results on the domains $C_{a,b} = \{(r,\varphi) \in (a,b) \times (o,\omega)\}$ and transforming them to $C_{at,bt}$, $t > 0$. For the Pseudo-Laplace equation we get the estimate $|\nabla u| \leq cr^{\alpha-1}$ and for the regular variational problem we can proceed further $|\nabla^2 u| \leq cr^{\alpha-2}$ etc. Note that the solution of the

Pseudo-Laplace equation is not regular even if the data of the problem are smooth, see section 2 for some examples.

Let us say a word about the differences between the Pseudo-Laplace equation and the regular variational problem $K > 0$. In the latter case the constant K dominates the term $|\nabla u|^2$ in

$$- \operatorname{div} \left\{ (K + |\nabla u|^2)^{(p-2)/2} \nabla u \right\} = f \tag{1.8}$$

if $|\nabla u|$ tends to zero for $r \to 0$. We conclude that u behaves like

$$r^{\pi/\omega} \sin \frac{\pi}{\omega}\varphi \quad \text{for } 0 < \omega < \pi , \quad r^{\alpha} t(\varphi) \quad \text{for } \pi < \omega ,$$

where $(\alpha, t(\varphi))$ is the positive solution of the eigenvalue problem (1.5). Since our analysis of the regular variational problem (1.8) reduces to the investigation of Laplace's equation for konvex angles we assume in the following that

$$\pi < \omega \quad \text{if } K > 0 \tag{1.9}$$

For proving a singular expansion $u = ks + w$ we have to ensure a priori that the stress intensity factor k does not vanish. This can be done by the condition $k_i > 0$ (or < 0) in (1.7) which for example, is satisfied if $f, g > 0$. Now the following "stretching" idea of Tolksdorf [13] applies. Consider the stretching operator

$$T_R u(x) = c_R u(Rx)$$

where the **normalization factor** c_R is chosen such that

$$\sup_{x \in C} |T_R u| = 1 .$$

For this operator one can prove that

$$T_R u \to s \quad \text{in } C^{1,\beta}(C) \text{ for } R \to 0 ,$$

which, by (1.6), implies that

$$u = ks + w \quad \text{with } w = o(s) .$$

In the case $p > 2$ we can conclude further by linearizing the operator at ks and applying the linear theory. This improves the regularity of the remainder term to $|w| \leq r^{\alpha+\varepsilon}$ where the maximum ε depends on p and ω . All these arguments are carried out in section 3. Moreover, we derive a representation formula for the stress intensity factor k , which is suitable for numerical purposes but may be interesting in its own.

It should be noted that a real corner theory for nonlinear elliptic problems does not exist. At the time there are some techniques which lead to relevant results only in certain special cases. On the other hand, there is no doubt that a singular expansion holds in all cases.

Consider now a general nonlinear elliptic problem $Au = f$ in a domain which contains a corner at the point 0. The following sequence of theoretical results seems to be natural:

1. Existence of a weak solution u.
2. Estimates of the Hölder coefficient of the solution: $|u| \leq cr^{\alpha}$, $|\nabla u| \leq cr^{\alpha-1}$ etc.
3. Existence of a decomposition $u = ks + w$ with a singular function which does not depend on f. The remainder part should be more regular than s.
4. A representation formula $k = g(k,u)$ for the stress intensity factor $k \in \mathbb{R}$ in $u = ks + w$.

The following numerical methods correspond to the above stated theoretical results:

1. The ordinary Finite Element Method (FEM)
2. The ordinary Finite Element Method on a refined mesh near the corner (RFEM)
3. The Singular Function Method (SFM)
4. The Dual Singular Function Method (DSFM)

For defining the FEM approximation Pu to u we divide the domain Ω into triangles of size h. The finite element space $S^{m,h}$ consists of continuous functions on Ω which are polynomials of degree $\leq m$ on each triangle of the subdivision. $Pu \in S^{m,h}$ is given by the finite dimensional nonlinear equation

$$(APu, v_h) = (f, v_h) \qquad \forall v_h \in S^{m,h} .$$

In our case (1.1) the analysis of the FEM can be carried out by using the monotonicity of the operator A. It turns out that the FEM is quasi-optimal in $H^{1,p}(\Omega)$. The related results will be proved in section 4.

A well known idea for improving the bad convergence properties of the FEM is the use of a refined mesh near the corner. For problem (1.1) one can compute the optimal design of the refinement depending on the Hölder exponent of the solution and the order of the Spline space m. Blum [3] has given a detailed analysis of the RFEM both theoretically and numerically. The results are similar to the linear case treated in [1].

A second method for improving the FEM is the use of an augmented spline space $S_1^{m,h} := S^{m,h} \oplus s$ in the FEM which is called the Singular Function Method first analyzed in [9]. The method is obviously based on the existence of a singular expansion $u = ks + w$. The investigation of the SFM following the lines of section 4 shows that there only is a slight improvement in comparison to the FEM. Since the solution of the SFM can be written in the form $k^h s + w_h$, $w_h \in S^{m,h}$, one obtains an approximate k^h of the stress intensity factor k in a natural way. From the numerical calculations in the linear case we know that this method of stress intensity factor computation fails in many important cases (see [5]).

The DSFM is a modification of the FEM which also uses the singular function in the trial space but the test space only consists in the usual splines $S^{m,h}$. Hence one further equation can be defined by using the representation formula $k = g(k,u)$. In the linear case one obtains a method with optimal convergence properties for the error $u - P_{-1}u$ as well as for $k - k_h$ (see [4]). A variant of the DSFM which uses a stress intensity factor computation of least squares type (see [12]) is proposed in [8]. The DSFM for problem (1.1) is studied in section 5, some numerical results are presented in section 6.

Finally, let us remark that most of our results carry over to the more general equation

$$- \operatorname{div} \left\{ a(|\nabla u|^2) \nabla u \right\} = f$$

under the conditions

$$c_1 (K + t)^{p-2} \leq a(t^2) \leq c_2 (K + t)^{p-2} , \quad K \geq 0 ,$$

$$c_1 a(t) \leq a'(t)\, t \leq c_2 a(t) ,$$

$$\lim_{t \to \infty} \frac{a'(t)\, t}{a(t)} = (p-2)/2 .$$

2. A nonlinear eigenvalue problem

Here we consider the Pseudo-Laplace equation

$$Au := - \operatorname{div} \left\{ |\nabla u|^{p-2} \nabla u \right\} = 0 \quad , \; 1 < p < \infty \tag{2.1}$$

in the unit cone $C = \{(r,\varphi) \in (0,1) \times (0,\omega)\} \subset \mathbb{R}^2$ where (r,φ) denote the polar coordinates of the plane. On the sides S of the cone we assume the boundary condition

$$u = 0 \quad \text{on } S \tag{2.2}$$

Writing (2.1) in polar coordinates yields

$$-\frac{1}{r}\partial_r\left\{\left(|\partial_r u|^2 + |\tfrac{1}{r}\partial_\varphi u|^2\right)^{(p-2)/2} r\,\partial_r u\right\} -$$

$$-\frac{1}{r^2}\partial_\varphi\left\{\left(|\partial_r u|^2 + |\tfrac{1}{r}\partial_\varphi u|^2\right)^{(p-2)/2}\partial_\varphi u\right\} = 0$$

Hence, the solutions of (2.1), (2.2) which have the form $r^\alpha t(\varphi)$ satisfy the nonlinear eigenvalue problem

$$\partial_\varphi\left\{\left(\alpha^2 t^2 + |\partial_\varphi t|^2\right)^{(p-2)/2}\partial_\varphi t\right\} +$$

$$+\,\alpha(\alpha(p-1)+2-p)\left(\alpha^2 t^2 + |\partial_\varphi t|^2\right)^{(p-2)/2} t = 0\,, \tag{2.3}$$

$$t(0) = t(\omega) = 0\ .$$

The first positive eigenvalue α of (2.3) is given by the following

Theorem 1: Let $\gamma(\omega) = (\frac{\omega}{\pi}-1)^2 - 1$ and $s(\gamma,p) = \frac{(\gamma-1)p-2\gamma}{2\gamma(p-1)}$. For

$$\alpha = \begin{cases} s(\gamma,p) + \sqrt{s^2(\gamma,p)+\gamma^{-1}} & ,\ \omega \le \pi \\ s(\gamma,p) - \sqrt{s^2(\gamma,p)+\gamma^{-1}} & ,\ \pi \le \omega < 2\pi \\ (p-1)/p & ,\ \omega = 2\pi \end{cases}$$

there is a positive solution $t(\varphi)$ of (2.3).

Remarks: (i) A simple computation shows that $r^\alpha t(\varphi) \in H^{1,p}(C)$. $(\alpha, t(\varphi))$ is unique in the class of positive solutions $\alpha > 0$, $t > 0$ in $(0,\omega)$, which lie in the space $H^{1,p}(C)$. Assuming the converse we have for two solutions $(\alpha_i, t_i(\varphi))$ in that class by the weak comparison principle (see section 1)

$$c_1 r^{\alpha_1} t_1 \le r^{\alpha_2} t_2 \le c_2 r^{\alpha_1} t_1 \quad \text{in } C$$

and thus $\alpha_1 = \alpha_2$. The uniqueness of t follows from the proof of the Theorem.

(ii) The eigenvalue problem (2.3) has infinitely many solutions (α_i, t_i) with $0 < \alpha = \alpha_1 < \alpha_2 < \dots$. The i-th eigenvalue can be obtained by considering the problem (2.3) on the intervall $(0,\omega/i)$ and using a reflection argument.

(iii) The same situation occurs in higher space dimensions. The solutions of $Au = f$ with $u = r^{\alpha} t(\varphi_1, .., \varphi_{n-1})$ satisfy a similar eigenvalue problem in a domain of S^{n-1}. Tolksdorf [13] has shown that this problem also has a positive solution which builds up the first singular function of $Au = 0$. On the other hand, there is no hope to determine the first eigenvalue exactly (even for $p = 2$!).

(iv) By using the result of the Theorem for $\omega = \frac{\pi}{2}$ and by reflecting the corresponding singular function three times one obtains an example of an irregular solution of $Au = 0$ in the interior of the domain. Thus the Hölder exponent of any solution of $Au = 0$ is estimated by

$$\alpha = \frac{7p - 6 + \sqrt{p^2 + 12p - 12}}{6p - 6} .$$

It is not known whether this example is sharp, but seems to be better than the inhomogeneous case. For instance, a solution of $Au = 1$ is $cr^{p/(p-1)}$.

(v) For some numerical methods for approximating $Au = f$ the knowledge of the eigenfunction $t(\varphi)$ is necessary. This can be determined by solving the eigenvalue problem numerically or by inverting the inverse function of $t(\varphi)$ which is known by the proof of the Theorem.

(vi) Since the formula for determing α is rather complicated we state some properties of α:

$$0 < \omega < 1 \Leftrightarrow \omega > \pi \quad , \quad \lim_{p\to\infty} \alpha = \begin{cases} \infty & \text{for } \omega < \pi \\ 0 & \text{for } \omega > \pi \end{cases} \quad , \quad \lim_{p\to\infty} \alpha = 1 .$$

Proof of Theorem 1: We consider (2.3) in the intervall $(-\frac{\omega}{2}, \frac{\omega}{2})$ with boundary conditions $t(-\frac{\omega}{2}) = t(\frac{\omega}{2}) = 0$ and seek a positive solution whose first derivative vanishes in the point $\varphi = 0$. (2.3) can be written in the form ($t' := \partial_{\varphi} t$)

$$(p-1)|t'|^2 t'' + \alpha^2 t^2 t'' + \left\{\alpha^2(2p-3) + \alpha(2-p)\right\} |t'|^2 t + \\ + \left\{\alpha^4(p-1) + \alpha^3(2-p)\right\} t^3 = 0 .$$

Setting $t' = ut$ gives $u(0) = 0$, $u(\frac{\omega}{2}) = -\infty$, $t'' = t(u + u^2)$, and, after dividing through t^3,

$$-\frac{1}{u'} = (a + u^2)/\left\{(u^2 + f)(u^2 + g)\right\} , \tag{2.4}$$

where $a = \alpha^2/(p-1)$, $f = \alpha^2$, $g = \alpha^2 + \alpha(2-p)/(p-1)$. From the assumptions on u it follows that

$$\frac{\omega}{2} = \int_{-\infty}^{0} - \frac{1}{u'}\,du = \frac{\pi}{2}\left\{ \frac{a-f}{h\sqrt{f}} + \frac{g-a}{h\sqrt{g}} \right\} ,$$

where $h = \alpha(2-p)/(p-1)$, and thus

$$\omega = \pi \left\{ 1 + (1-\alpha)\left(\alpha^2 + \alpha\frac{2-p}{p-1}\right)^{-1/2} \right\} .$$

The solution of this equation is given in the assertion. Now our approach can be justified. In view of $\alpha > 0$ for $1 < p \leq 2$ and $\alpha > (p-2)/(p-1)$ for $p \geq 2$ we have $g > 0$ in (2.4). Hence (2.4) defines a strictly monotone inverse function of u , such that u with the boundary conditions stated above is uniquely determined. Now $t = \exp(\int u)$.

3. Singular expansion

In the first part of this section we consider the homogeneous Pseudo-Laplace equation for $2 \leq p < \infty$,

$$\begin{aligned} Au &= -\operatorname{div}\left\{ |\nabla u|^{p-2}\,\nabla u \right\} = 0 \quad \text{in } C \\ u &= 0 \quad \text{on } S \ , \quad u = g \quad \text{on } B \ , \end{aligned} \tag{3.1}$$

where we have adopted the notation of the introduction.
seen as a part of a general domain $\Omega \subset \mathbb{R}^2$ which contains a corner of angle ω .

Theorem 2: Assume that $g \in C^1(B)$ with $g(\partial B) = 0$ and $g > 0$ in B . Then the solution u of (3.1) admits the singular expansion

$$u = ks + w \quad , \quad k \in \mathbb{R}^+ \ ,$$

where $s = r^\alpha t(\varphi)$ and $(\alpha, t(\varphi))$ is the solution of the eigenvalue problem from Theorem 1. The regular part w is C^∞ in a neighbourhood $U(0)$ of 0 and satisfies the estimates

$$|\nabla^i w| \leq c r^{\alpha+\varepsilon-i} \qquad \forall i \in \mathbb{N} \ ,$$

where the maximum $\varepsilon > 0$ depends on p and ω .

Proof: The proof will be given in three steps.

step 1: A priori estimate $k_1 s \leq u \leq k_2 s$ in C with $k_i > 0$.

Step 1 is trivial for problem (3.1) but may be crucial for more general equations. Since g is assumed to be positive in B we can find positive constants k_1, k_2 with

$$k_1 t(\varphi) \le g \le k_2 t(\varphi)$$

(in the case that $g' = 0$ on ∂B one has to use Hopf's maximum principle). By the weak comparison principle the a priori estimate is proved.

<u>step 2</u>: $u = ks + w$ with $k > 0$ and $w = \mathcal{O}(s)$.

Since this result is due to Tolksdorf [13] we will only give the key idea of the proof. Consider for $R > 0$ the stretching operator

$$T_R u(x) = c_R\, u(Rx) \quad ,$$

where the normalization factor c_R is chosen such that

$$\max_{x \in C} |T_R u(x)| = 1 \quad .$$

Now we observe that the Pseudo-Laplace equation is not affected by the stretching operator: $A(T_R u) = 0 \;\; \forall R > 0$. Hence the theory of the homogeneous Pseudo-Laplace equation can be applied to analyze the behaviour of $T_R u$ and to obtain that

$$T_R u \to s \quad \text{in} \quad C^{1,\beta}(C) \quad \text{for a} \quad \beta > 0 \; .$$

Thus step 2 is finished by using the a priori estimate of step 1.

<u>step 3</u>: $u = ks + w$ with $|w| \le cr^{\alpha+\varepsilon}$ for an $\varepsilon > 0$.

Here the corner theory of linear elliptic operators is applied to the linear part L of $A(ks + w)$, i.e.

$$Lw := -\operatorname{div}\left\{ |\nabla s|^{p-2}\, \nabla w \right\} - (p-2)\operatorname{div}\left\{ |\nabla s|^{p-4} \nabla s \nabla w\, \nabla s \right\} \; . \tag{3.2}$$

Since $c_1 r^{\alpha-1} \le |\nabla s| \le c_2 r^{\alpha-1}$ and $(Lv,v) > 0$ for all $v \in \mathring{C}^{\infty}(C)$, L is a linear elliptic operator with degenerating coefficients at the point 0 which satisfies a maximum principle. Moreover, the coefficients l_{ij} have the form

$$l_{ij} = r^{(\alpha-1)(p-2)} s_{ij}(\varphi) \quad ,$$

so that the results of Maz'ja-Plamenevskij [12] can be applied. These refer to the boundary value problem ($\vec{f} = (\vec{f}_1, \vec{f}_2)$)

$$Lv = \operatorname{div} \vec{f} \;\text{ in }\; C \;,\quad v = 0 \;\text{ on }\; S \;,\quad v = \vec{g} \;\text{ on }\; B \tag{3.3}$$

for "weak" solutions v satisfying

$$\int_C r^{(\alpha-1)(p-2)} |\nabla v|^2 \, dx < \infty \quad .$$

The singular functions $\tilde{s}_i = r^{\alpha_i} t_i(\varphi)$, $i \in Z/\{0\}$, of (3.3) are given by the eigenvalue problem

$$L r^{\alpha_i} t_i(\varphi) = 0 \quad \text{in } C \ , \quad r^{\alpha_i} t_i(\varphi) = 0 \quad \text{on } S \ , \tag{3.4}$$

which has infinitely many solutions. These are ordered by the conditions $\alpha_i < \alpha_j$ for $i < j$ and

$\tilde{s}_1$ is the first weak solution of $L\tilde{s}_i = 0$ in the sequence of singular functions $\tilde{s}_i$.

By the maximum principle, there is only one positive weak solution of (3.4). Thus

$$\tilde{s}_1 = s \ .$$

To each singular function $\tilde{s}_i$, $i \in \mathbb{N}$, there exists a dual singular function $\tilde{s}_{-i}$ which is a kernel function of the adjoint operator L . Since L in (3.3) is a formal selfadjoint operator we get

$$t_{-i}(\varphi) = t_i(\varphi) \quad \forall i \in \mathbb{N} \ .$$

Now the description of the solutions of (3.4) is finished. Let $\tau(r) \in C^\infty((0,1))$ be a cut-off function, i.e.

$$\tau(r) = \begin{cases} 1 & \text{for } 0 < r < a \\ 0 & \text{for } b < r < 1 \end{cases} \ , \quad a < b \ ,$$

where a, b are arbitrary but fixed constants, and let $s_i^! = \tau\tilde{s}_i$, $i \in Z\{0\}$. Now the results of Maz'ja-Plamenevskij are summarized in the following

<u>Lemma</u>: Let $-2\alpha_2 - (\alpha-1)(p-2) < \beta < -2\alpha_1 - (\alpha-1)(p-2)$ $(\alpha_1 = \alpha\ !)$ and

$$\int_C r^{-(\alpha-1)(p-2)+\beta} |\tilde{f}|^2 \, dx < \infty \ , \quad \tilde{g} \in H^{1/2,2}(B) \ .$$

Then the weak solution v of (3.3) admits the singular expansion

$$v = k_v s + w_v \ , \quad k_v \in \mathbb{R} \ ,$$

where w_v satisfies the estimate

$$\int_C r^{(\alpha-1)(p-2)+\beta} |\nabla w_v|^2 \, dx < c \int_C r^{-(\alpha-1)(p-2)+\beta} |\tilde{f}|^2 \, dx + \\ + \|\tilde{g}\|^2_{1/2,2;B} \ .$$

k_v is given by

$$k_v = \int_C \{ \tilde{f}\nabla s'_{-1} - v\, L s'_{-1} \} \, dx \ .$$

The representation formula for k_v is also proved in Blum-Dobrowolski [4] for the case $L = -\Delta$.

Now we write

$$A(ks + w) = |k|^{p-2} Lw + R(w)$$

where the remainder term R satisfies the estimate

$$|(R(w),\psi)| \leq c \int_C \left\{ r^{(\alpha-1)(p-3)} |\nabla w|^2 + |\nabla w|^{p-1} \right\} |\nabla\psi| \, dx \qquad (3.5)$$

for all $\psi \in \overset{\circ}{C}{}^{\infty}(C)$. The constant c depends on k.

Now choose $R > 0$ sufficiently small such that

$$T_R u = ks + w_R$$

with $|w_R|$, $|\nabla w_R| \leq \varepsilon$ on B. We apply the Schauder fixed theorem to the problem

$$Lw_R = -|k|^{2-p} R(w_R) \quad \text{in } C \,, \quad ks + w_R = \text{given on } B \,.$$

Using the lemma stated above we obtain that

$$\int_C r^{(\alpha-1)(p-2)+\beta} |\nabla w|^2 \, dx < \infty$$

for $\beta > -2\alpha_2 - (\alpha-1)(p-2)$, which, by local regularity leads to the estimates

$$|w| \leq cr^{\alpha_2 - \varepsilon} \,, \quad |\nabla w| \leq cr^{\alpha_2 - 1 - \varepsilon} \quad \forall \varepsilon > 0 \,,$$

see Tolksdorf [13], [14]. This completes the proof of the theorem.

From the proof of Theorem 2 it follows that the regularity of w is given by the second eigenvalue of the linearized eigenvalue problem (3.4). This does not hold for the regular variational problem:

Example: Consider the problem

$$Au = -\operatorname{div} \left\{ (1 + |\nabla u|^2)^{(p-2)/2} \nabla u \right\} = 0 \quad \text{in } C \,,$$

and assume that $u = ks + w$ with $|w| \leq cr^{\alpha+\varepsilon}$, $|\nabla w| \leq cr^{\alpha-1+\varepsilon}$ etc. From

$$A(ks + w) = -k\,\Delta s + |k|^{p-2} Lw + \ldots = 0$$

it follows that $-\Delta s$ and Lw have the same power in r. Since

$$-\Delta s \sim r^{\alpha-2} \,, \quad Lw \sim r^{(\alpha-1)(p-2)+\alpha+\varepsilon-2}$$

we get $\varepsilon \leq -(\alpha-1)(p-2)$. For $\omega = 2\pi$ we have $\alpha = (p-1)/p$ and hence $\varepsilon \leq (p-2)/p$, which may be arbitrary small.

From the proof of Theorem 2 we conclude that there also exists a representation formulafor the stress intensity factor,

$$k|k|^{p-2} = \int_C \left\{ R(w)s'_{-1} - uLs'_{-1} \right\} dx \quad . \tag{3.6}$$

The estimate (3.5) shows that both integrals exist ($\alpha_{-1} = -\alpha - (\alpha-1)(p-2)$).

The inhomogeneous Pseudo-Laplace equation is treated in the following

<u>Theorem 3</u>: Let the assumptions of Theorem 2 be satisfied and let f be a function on C with $0 \leq f \leq cr^{(\alpha-1)(p-1)-1+\varepsilon}$. Then the assertion of Theorem 2 holds true for the inhomogeneous equation $Au = f$ in (3.1).

<u>Proof</u>: The steps 2 and 3 of the proof of the preceding theorem remain the same. Since f and g are positive functions we immediately obtain the estimate $k_1 s \leq u$ for $k_1 > 0$. For constructing a supersolution we set

$$v = k_\alpha s + k_\beta r^\beta t_\beta(\varphi) \quad , \quad \beta > \alpha \quad ,$$

and we get by Taylor's expansion

$$Av = |k_\alpha|^{p-2} L(k_\beta r^\beta t_\beta(\varphi)) + R(r^\beta t_\beta(\varphi)) \tag{3.7}$$

where the remainder satisfies

$$|R(r^\beta t_\beta(\varphi))| \leq c \left\{ |k_\alpha|^{p-3}|k_\beta|^2 + |k_\beta|^{p-1} \right\} \times r^{(\alpha-1)(p-3)+2(\beta-1)-1} \tag{3.8}$$

where c depends on α, β, p and t, t_β . Now

$$|k_\alpha|^{p-2} L(k_\beta r^\beta t_\beta(\varphi)) = |k_\alpha|^{p-2} k_\beta r^{(\alpha-1)(p-2)+\beta-2} \, D(\beta,\partial_\varphi)\, t_\beta \quad , \tag{3.9}$$

where $D(\beta,\partial_\varphi)$ is the linear one dimensional differential operator

$$\begin{aligned} D(\beta,\partial_\varphi)\, t &= -\beta(\beta+(p-2)(\alpha-1))\,(\alpha^2t^2+|\partial_\varphi t|^2)^{(p-2)/2}\, t_\beta \; + \\ &\quad - \partial_\varphi((\alpha^2t^2+|\partial_\varphi t|^2)^{(p-2)/2}\,\partial_\varphi t_\beta) - \alpha(p-2)((p-2)(\alpha-1)+\beta) \; \times \\ &\quad \times (\alpha^2t^2+|\partial_\varphi t|^2)^{(p-4)/2}\,(\alpha\beta t t_\beta + \partial_\varphi t \partial_\varphi t_\beta)\, t \quad . \end{aligned}$$

Thus $D(\beta,\partial_\varphi)$ is a parameter dependent operator with main part

$$(\alpha^2t^2+|\partial_\varphi t|^2)^{(p-2)/2} + (p-2)(\alpha^2t^2+|\partial_\varphi t|^2)^{(p-4)/2}\,|\partial_\varphi t|^2$$

strictly elliptic and independent of β . Therefore the problem

$$D(\beta,\partial_\varphi)\, t_\beta = 1 \quad \text{in} \quad (0,\omega) \quad , \quad t_\beta(0) = t_\beta(\omega) = 0 \tag{3.10}$$

is uniquely solvable except on isolated points $\beta \in \mathbb{C}$. Since $D(\beta,\partial_\varphi)$ coincides with the linearized nonlinear eigenvalue problem for $\beta = \alpha$ we can solve (3.10) for all $|\beta - \alpha| < \varepsilon$, $\beta \neq \alpha$, with ε sufficiently small. Depending on f we choose k_β **sufficiently** large and, by (3.7) -(3.9), we can determine k_α such that v is a supersolution of $Au=f$.

4. The ordinary Finite Element Method

Here we consider the problem

$$\begin{aligned} Au &= -\operatorname{div}\left\{ (K + |\nabla u|^2)^{(p-2)/2}\, \nabla u \right\} = f \quad \text{in } \Omega , \\ u &= 0 \quad \text{on } \partial\Omega \quad , \quad 2 \le p < \infty \quad , \quad K \ge 0 \end{aligned} \tag{4.1}$$

in a bounded domain $\Omega \subset \mathbb{R}^2$ which contains one corner with interior angle $\omega > \pi$. It is assumed that a singular expansion holds,

$$u = ks + w \quad , \quad k \in \mathbb{R}/\{0\} \tag{4.2}$$

where $s = r^\alpha t(\varphi)$ with $0 < \alpha < 1$, $|w| \le cr^{\alpha+\varepsilon}$, $|\nabla w| \le cr^{\alpha-1+\varepsilon}$ etc.

A simple calculation shows that for $u,v \in \mathring{H}^{1,p}(\Omega)$

$$(Au - Av, u - v) \ge c \begin{cases} \|u - v\|_{1,p}^p \\ \|u - v\|_{1,2}^2 \end{cases} , \tag{4.3}$$

$$(Au - Av, u - v) \le c \, \|1 + \nabla u + \nabla v\|_p^{p-2} \, \|u - v\|_{1,p}^2 \tag{4.4}$$

Since by (4.2) the regularity of (4.1) is weak we will only consider the FEM with linear elements, but all results remain valid for the higher order case. The space S^h is constructed by dividing Ω into triangles of size h such that the usual regularity condition is satisfied:

Each two triangles may only meet in common sides or in vertices. Each triangle contains a ball of radius h and is contained in a ball of radius ch .

Then the space of linear elements is defined by

$$S^h = \{ v_h \in C^0(\Omega) : \text{The restriction of } v_h \text{ to each triangle is linear and } v_h = 0 \text{ on } \partial\Omega \} .$$

In view of (4.2) we have the approximation properties

$$\|\nabla^k(u - I_h u)\|_q \leq ch^{\alpha-k+2/q} \quad , k = 0,1 \quad , \alpha + 2/q \leq 2 \tag{4.5}$$

where $I_h u \in S^h$ denotes the interpolant of u. The usual inverse estimates holds,

$$\|\nabla v_h\|_p \leq ch^{2/p - 2/q} \|\nabla v_h\|_q \quad \forall\, v_h \in S^h \;, 1 \leq q \leq p \leq \infty \;. \tag{4.6}$$

The FEM approximation $Pu \in S^h$ is defined by

$$(APu, v_h) = (f, v_h) \quad \forall\, v_h \in S^h \;. \tag{4.7}$$

By the monotonicity (4.3) this finite dimensional nonlinear equation possesses a unique solution. Denoting by $\gamma = \alpha - 1 + 2/p$ the convergence rate of the interpolation error (4.5) in $H^{1,p}$ we have

Theorem 4: For the solutions u of (4.1) and Pu of (4.7) the following error estimates hold

(i) $\|u - Pu\|_{1,p} \leq ch^{\gamma-\varepsilon}$,

(ii) $\|u - Pu\|_{1,2} \leq ch^{p\gamma/2 - \varepsilon}$,

(iii) $\|\nabla Pu\|_{2/(1-\alpha) - \varepsilon} \leq c$

for all $\varepsilon > 0$.

Remark: Since $\alpha = (p-1)/p$ for $\omega = 2\pi$ we obtain that $\|u - Pu\|_{1,2} \leq ch^{1/2 - \varepsilon}$ which is independent of p .

Proof: Assume that we have proved that

$$\|\nabla(u - Pu)\|_p \leq ch^{\beta} \quad , \beta < \gamma \;.$$

Then the inverse estimate (4.6) gives

$$\|\nabla Pu\|_s \leq ch^{2/s - 2/p + \beta} \quad , \tag{4.8}$$

which is bounded for $s = 2p/(2 - \beta p)$. Since

$$(Au - APu, u - Pu) = \int_\Omega a_{ij}^h \partial_i(u - Pu)\partial_j(u - Pu)\,dx$$

with $|a_{ij}^h| \leq c\left\{1 + |\nabla u|^{p-2} + |\nabla Pu|^{p-2}\right\}$ we can estimate by (4.3)

$$\begin{aligned}\|u - Pu\|_{1,p}^p &\leq c \int_\Omega a_{ij}^h \partial_i(u - Pu)\partial_j(u - Pu)\,dx \\ &\leq c \int_\Omega a_{ij}^h \partial_i(u - I_h u)\partial_j(u - I_h u)\,dx\end{aligned}$$

$$\leq \; c \, \| 1 + |\nabla u| + |\nabla Pu| \|^{p-2}_{r(p-2)/(r-2)} \, \|\nabla(u - I_h u)\|^2_r \quad .$$

From (4.8) we obtain that r can be chosen such that $r(p-2)/(r-2) = 2p/(2-\beta p)$ and hence

$$\|u - Pu\|_{1,p} \leq ch^{2\gamma/p + \beta(p-2)/p} \quad .$$

The sequence $\beta^+ = 2\gamma/p + \beta(p-2)/p$ is monotone increasing and converges to the fixed point $\beta = \gamma$. This finishes the proof of (i). The second assertion follows from (4.3) and the first from (4.8).

5. The Dual Singular Function Method

For introducing a general iteration idea we consider an arbitrary linear or nonlinear equation

$$Au = f \tag{5.1}$$

in a Banach space B , $A : B \to B^*$. It is assumed that the solution can be split into

$$u = ks + w \quad , \quad k \in \mathbb{R} \tag{5.2}$$

and that there exist formulas for computing the stress intensity factor $k \approx g_h^j(f,u)$, $j \in \mathbb{N}$, $h > 0$. Assuming that one has a sequence B_h of finite dimensional subspaces of B for $h \to 0$ one can try the following natural scheme

1) $k^o := 0$

2) Let $k^j \in \mathbb{R}$ be known. Then find $w_h^j \in B_h$ by

$$(A(k^j s + w_h^j, v_h) = (f, v_h) \qquad \forall v_h \in B_h \quad . \tag{5.3}$$

3) Set $u_h^j = k^j s + w_h^j$ and define $k^{j+1} := g_h^j(f, u^j)$.

4) Set $j := j+1$ and go to 2)

Depending on the functions g_h^j there is a great variety of possible methods (1.4). In order to become familar with the idea of the scheme we analyze the special case

> $k = g(f,u)$ is exact, independent of j and h and satisfies a Lipschitz-condition: $|g(f,v) - g(f,w)| \leq c \, \|v - w\|$. Furthermore, it is assumed that (Au,v) is scalar-product on the Hilbert-space $H := B$.

Denoting by $Pv \in B_h$ the usual FEM approximation to a $v \in B$ we obtain from the second step in (5.3)

$$w_h^j = Pu - k^j Ps$$

and, by noting that $u_h^j = k_h^j + w_h^j$

$$u_h^j = Pu + k^j(s - Ps) \ .$$

From (5.2) we get $Pu = k\,Ps + Pw$ and

$$u - u_h^j = (k - k^j)(s - Ps) + w - Pw \ .$$

Using the Lipschitz-condition for g we have $|k - k^j| \leq c \, \|u - u_h^j\|$ and, by assuming convergence $Ps \to s$ in B, $h \to 0$,

$$|k - k^\infty| + \|u - u^\infty\| \leq c \ \|w - Pw\| \ , \qquad (5.4)$$

for h sufficiently small.

In this simple case, the scheme (1.4) leads to approximates of the solution which converge as fast as the FEM approximation to the smoother part w. Moreover, an approximate stress intensity factor with the same accuracy is obtained. Obviously, the crucial point of the scheme is the construction of a function g which satisfies a Lipschitz-condition in the second variable, see [4], [5], [8] for further details in the linear case. Note that the idea of iteration may be neglected here. In fact, we have proved that the following method possesses a unique solution which satisfies the error estimate (5.4)

Find $u_h = k^h s + w_h$, $w_h \in B_h$, such that

$$(Au_h, v_h) = (f, v_h) \quad \forall v_h \in B_h \quad \text{and} \quad k^h = g(f, u_h) \ .$$

From the theoretical point of view it is more interesting to study functions g which depend on j and fail to have such a limit method. An example of such a scheme is given in [8].

Now let us consider the nonlinear equation

$$- \operatorname{div} \left\{ (K + |\nabla u|^2)^{(p-2)/2} \nabla u \right\} = f \quad \text{in} \ \Omega \ , \ K > 0 \ , \ 1 < p < \infty \ ,$$

$$u = 0 \quad \text{on} \quad \partial\Omega$$

and let us assume that, in the near of a corner, a singular expansion $u = ks + w$ holds with $s = \tau(r) r^\alpha t(\varphi)$, where $\tau(r)$ is a cut-off function such that $\tau(r) = 1$ in a neighbourhood of 1 and $s \in \overset{\circ}{H}{}^{1,p}(\Omega)$. All results of section 3 remain valid with the singular function s, especially we have the representation formula (3.6)

$$k|k|^{p-2} = \int_{\Omega} \{(f + R(w))s'_{-1} - uLs'_{-1}\} \, dx . \tag{5.5}$$

Here s'_{-1} denotes the first singular function of the dual operator L^*, i.e. $s'_{-1} = \tau(r)r^{\alpha_-}t(\varphi)$ with $\alpha_- = -\alpha - (\alpha-1)(p-2)$. This formula has the type $k^{p-1} = g(k,w,u)$ which satisfies a Lipschitz-condition with respect to u. The term $R(w)$ in (5.5) is, in the calculus of weighted Sobolev spaces, a compact perturbation with respect to the other terms. This motivates the use of (5.5) in the scheme (5.3) even in the case $1 < p < 2$. A convergence proof of this method would require very complicated estimates of the FEM in weighted Sobolev norms and is omitted here.

6. Numerical results

The numerical calculations refer to the regular variational problem for $p = 4$, i.e.

$$- \operatorname{div} \{ (1 + |\nabla u|^2) \nabla u \} = 1$$

in the slit domain $\Omega = (0,1)\times(0,1)/\{\frac{1}{2}\}\times(0,\frac{1}{2})$ which contains a corner with angle $\omega = 2\pi$. By the results of section 4 we expect a convergence rate $h^{1/2}$ in $H^{1,2}(\Omega)$ for the ordinary FEM. This can be improved to linear convergence in $L^2(\Omega)$ by using duality.

h^{-1}	FEM	DSFM	
	$\|u-P_h u\|$	$\|u-P\_u\|$	$\|k-k^h\|$
8	.0879	.1696	.3123
16	.0425	.0425	.0898
32	.0201	.0103	.0443
64	.0095	.0027	.0188

Here, the 8th iterate of (5.3), (5.5) is denoted by P_u. All numbers are relative errors, $\|\cdot\|$ denotes the $L^{\infty,loc}$-norm outside a neighbourhood of the corner.

REFERENCES

1. I.Babuska: Finite element methods for domains with corners, Computing 6, 264 - 273 (1970)
2. I.Babuska - W.C.Rheinboldt: Mathematical problems of computational decision in the finite element method, in: Proc.Conf.Rome 1975, Lecture notes in Mathematics 606, Springer 1975, 1 - 26
3. H.Blum: Zur Gitterverfeinerung für quasilineare Probleme auf Eckengebieten, to appear in ZAMM
4. H.Blum - M.Dobrowolski: On finite element methods for elliptic methods on domains with corners, Computing 28, 53 - 63 (1982)
5. M.Dobrowolski: Numerical approximation of elliptic interface problems, Habilitationsschrift, Bonn 1981
6. M.Dobrowolski: Nichtlineare Eckenprobleme und finite Elemente Methode, to appear in ZAMM
7. M.Dobrowolski: On the regularity of the Pseudo-Laplace equation in domains with corners, to appear in Comm.Diff.Equ.
8. M.Dobrowolski:On the numerical treatment of elliptic equations with singularities, to appear in Bonner Math.Sch.
9. G.Fix - S.Gulati - G.I.Warkof: On the use of singular functions with finite element approximation, J.Comp.Phys. 13, 209 - 238 (1973)
10. V.A.Kondrat'ev: Boundary value problems for elliptic equations in domains with conical or angular points, Trudy Mosc.Mat.Obsc. 16, 209 - 292 (1967) (=Tran.Mosc.Math.Soc. 16, 227 - 313 (1967))
11. V.G.Maz'ja - B.A.Plamenevskij: Coefficients in the asymptotics of the solutions of elliptic boundary value problems in a cone, Sem.LOMI 52, 110 - 127 (1975) (=J.Sov.Math. 9, 750 - 764 (1978))
12. A.H.Schatz - L.B.Wahlbin: Maximum norm estimates in the finite element method on plane polygonal domains I, II, I:Math.Comp. 32, 73 - 109 (1978), II: Math.Comp.33, 465 - 492 (1979)
13. P.Tolksdorf: On the Dirichletproblem for quasilinear equations in domains with conical boundary points, to appear in Com.Diff.Equ.
14. P.Tolksdorf: Regularity for a more general class of quasilinear elliptic equations, to appear in J.Diff.Equ.

SINGULARITIES IN ELLIPTIC PROBLEMS

M. DURAND
U.E.R. de Mathématiques
Université de Provence
13331 Marseille Cedex 3
France

We are concerned by the linear elasticity problem in the infinite space with a crack. To solve it, we suggest three different methods. The first one, due to A. Bamberger, transforms the elasticity problem into a variational one. The differential system is solved owing to the usual separation of waves in P-waves and S-waves. Its solution allows to obtain explicit expansionsof the bilinear form related to the variational problem. In our opinion, the framework of pseudo-differential operators makes the method clear.

By using the Wiener-Hopf method, as explained in the G.I. Eskin's book, C. Goudjo computes the singularities of the solution along the edge of the crack. Thus he obtains all the results about the regularity that are wanted.

At last, we use an approximation of the crack by regular closed curves, in order to define integral equationsthat solve approximated problems. Uniform a-priori estimates allow us to go to the limit. Then we obtain an integral operator as the limit of a sequence of pseudo-differential ones. These operators occur in the integral equations, and we are able to perform numerical computations by methods as the boundary element methods of W.L. Wendland and others.

In order to be quite clear, we recall some definitions :

$H_s(R^n)$ is the usual Sobolev space of order s ,

$H_s(\Gamma)$ is defined as usually for s positive integer; when s is positive, but no an integer, $H_s(\Gamma)$ is an interpolate of $H_{[s]}(\Gamma)$ and $H_{[s]+1}(\Gamma)$. Here Γ is a regular domain with boundary.

$\tilde{H}_s(\Gamma) = \{ u \in H_s(\Gamma) \ , \ \tilde{u} \in H_s(R^n) \}$, if $s > 0$. For every u in $\mathcal{D}'(\Gamma)$, we define $\tilde{u}$ as the distribution equal to u in Γ and zero elsewhere.

$H_s(\Gamma) = (H_{-s}(\Gamma))'$ if $s < 0$.

In $H_s(\Gamma)$, the norm $\|u\|_{H_s(\Gamma)}$ is equal to $\|\tilde{u}\|_s$.

Remark.- If $s > 0$ is not equal to $k+1/2$, k integer, $H_s(\Gamma)$ is the space $H^s(\Gamma)$ defined by Lions and Magenes. When $s = k=1/2$, k positive integer, $H_s(\Gamma)$ is their space $H^s_{oo}(\Gamma)$. For $s=-1/2$, $H_{-1/2}(\Gamma)$ contains (with a weaker norm) the Lions'space $H^{-1/2}(\Gamma)$.
Γ has two sides, positive and negative. if ϕ has some traces ϕ^+ and ϕ^- on Γ, we put $[\phi] = \phi^+ - \phi^- = Tr^+(\phi) - Tr^-(\phi)$.

I- A.BAMBERGER'S METHOD.

To simplify this paper, we suppose that the crack is a rectilinear curve Γ in the plane $\mathbb{R}^2$:

$$\Gamma = \{ (x_1,x_2) \in \mathbb{R}^2 \ ; \ 0 < x_1 < 1 \ , \ x_2 = 0 \} .$$

We seek for the solution of the following problem :

$$(I)_g \begin{cases} \text{Find } u_i \in H^1_S (R^2 \setminus \Gamma), \ i=1,2, \text{ such that} \\ \sum_j \partial_j \ \sigma_{ij}(u) + \rho\omega^2 u_i = 0 \ , \ u=(u_1,u_2) \ , \\ \sigma(u).n_{|\Gamma} = g \ , \ g \ \ H_{-1/2}(\Gamma). \end{cases}$$

Here $H^1_S(R^2 \setminus \Gamma) = \{ u \in \mathcal{S}' \ , \forall \phi \in \mathcal{D}(\mathbb{R}^2), \ \phi u \in H^1(\mathbb{R}^2 \setminus \Gamma)$ and u satis- -fies Sommerfeld's conditions $\}$.

σ_{ij} is the stress tensor, ε_{ij} will bethe strain tensor.

Because of the boundary condition on Γ, the solution u is a double layer potential with a density $\phi = [u] = u^+ - u^-$. Moreover we seek for solutions with finite energy, i.e. in H_1, then $\phi \in H_{1/2}(\Gamma)$ (if ϕ is continuous, we have $\phi(A) = \phi(B) = 0$, where A and B are the ends of Γ).

From the problem $(I)_g$, we deduce the following one :

$$(II)_\phi \begin{cases} \text{Find } u_i \in H^1_S(R^2 \setminus \Gamma), \ i =1,2, \text{ such that} \\ \sum_j \partial_j \ \sigma_{ij}(u) + \rho\omega^2 u_i = 0 \ , \\ [\sigma(u).n] = 0 \ , \\ [u] = \phi \in H_{-1/2}(\Gamma) \ . \end{cases}$$

If $u = \Phi(g)$ is a solution of $(I)_g$ and $\phi = [\Phi(g)]$, then u is a solution of $(II)_g$. Reciprocally if $u = G(\phi)$ is a solution of $(I)_\phi$ and $g = \sigma(u).n_{|\Gamma}$, then u is a solution of $(I)_g$.

The solution of $(II)_\phi$ has a simple expression :

$$(1) \quad u(x) = \int_\Gamma G_x(y) \ \phi(y) \ dy$$

where $G_x(y)$ is a tensor of order 2.

Via the Betti's formulae, a variational form is defined on the product $H_{1/2}(\Gamma) \times H_{1/2}(\Gamma)$:

$$b(\phi,\psi) = \sum_{i,j} \int_\Omega \sigma_{ij}(u)\, \varepsilon_{ij}(\bar{v})\, dx - \rho\omega^2 \sum_i \int_\Omega u_i \bar{v}_i \, dx$$

$$= \int_\Gamma (\sigma(u).n)\bar{\psi}\, d\gamma = \int_\Gamma (\sigma(\bar{v}).n)\phi \, d\gamma ,$$

where $\psi = [v]$.

If we define the linear form L_g, continuous on $H_{1/2}(\Gamma)$ and $\tilde{H}_{1/2}(\Gamma)$: $L_g(\psi) = \int_\Gamma g\bar{\psi}\, d\gamma$, we introduce the following variational problem :

$$(V) \; \left\{ \text{Find } \phi \in \tilde{H}_{1/2}(\Gamma) \text{ such that } b(\phi,\psi) = L(\psi) \text{ for every } \psi \in \tilde{H}_{1/2}(\Gamma). \right.$$

The aim of this work is to solve the problem (V) . Thus if g is given, ϕ is computed and the solution of the initial problem $(I)_g$ is given by the formula (1).

The Bamberger's idea is to extend the functions ϕ and ψ defined in $\tilde{H}_{1/2}(\Gamma)$ to functions $\tilde{\phi}$ and $\tilde{\psi}$ which belong to $H_{1/2}(\mathbb{R})$. Then a bilinear form b is defined on $H_{1/2}(\mathbb{R}) \times H_{1/2}(\mathbb{R})$:

$$b(\phi,\psi) = \int_{\mathbb{R}} (\sigma(u).n)\bar{\psi}\, dx .$$

We apply a Fourier transformation with respect to the first variable to the system of partial differential equations of the problem (II). We obtain the following system :

$$\begin{cases} -(\lambda+2\mu)\,\xi^2\,\hat{u}_1 + \dfrac{\partial^2 \hat{u}_1}{\partial x_2^2} - (\lambda+\mu)\, i\xi \, \dfrac{\partial \hat{u}_2}{\partial x_2} + \rho\omega^2 \hat{u}_1 = 0 , \\[2ex] -\mu\xi^2 \hat{u}_2 + (\lambda+2\mu)\dfrac{\partial^2 \hat{u}_2}{\partial x_2^2} - (\lambda+\mu)\, i\xi \, \dfrac{\partial \hat{u}_1}{\partial x_2} + \rho\omega^2 \hat{u}_2 = 0 \end{cases}$$

with $\hat{u}(\xi) = \int u(x_1,x_2)\, e^{-i\xi x_1} dx_1$.

Owing to the decomposition of u in longitudinal and transversal waves, this system is easily solved. Thus the stress on Γ can be computed and we obtain :

$$(\sigma(u).n)\hat{}\,(\xi) = T(\xi)\, \hat{\phi}(\xi) = \hat{g}_1(\xi) .$$

g_1 is the function defined on $\mathbb{R}$, equal to g on Γ and to the trace of $\sigma(u).n$ on $\mathbb{R} \setminus \Gamma$. The matrix T is diagonal, then the system is uncoupled.

The operator $T : \tilde{\phi} \longrightarrow g_1$ is elliptic pseudo-differential. Its expression is quite simple. For large ξ, we have

$$T_{11}(\xi) \sim \frac{\mu(\lambda+\mu)}{\lambda+2\mu} |\xi|$$

$$T_{22}(\xi) \sim \frac{\mu(\lambda+\mu)}{\lambda+2\mu} |\xi| ,$$

$$T_{12} = T_{21} = 0 .$$

The operator T is a derivation.

But the aim of the Bamberger's work is to solve the variational problem (V). After the extension to $\mathbb{R}$, the problem (V) is transformed in a new problem $(\tilde{V})$:

$$(\tilde{V}) \quad \begin{cases} \text{Find } l\phi \text{ in } H_{1/2}(\mathbb{R}) \text{ such that for every } \psi \in \tilde{H}_{1/2}(\Gamma) , \\ \tilde{b}(l\phi,\tilde{\psi}) = \int_{\mathbb{R}} (\sigma(u).n)\bar{\tilde{\psi}}\, dx = L(\tilde{\psi}) = \int_{\mathbb{R}} g_1\bar{\tilde{\psi}}\, dx . \end{cases}$$

As $\tilde{\psi}$ is zero out of Γ, the integration can be performed on Γ and we seek for a $\phi \in H_{1/2}(\Gamma)$ such that

$$b(\phi,\psi) = \int_{\Gamma} g\bar{\psi}\, dx , \text{ and}$$

$$b(\tilde{\phi},\tilde{\psi}) = b(\phi,\psi) = (1/2\pi) \int_{\mathbb{R}} T(\xi)\, \hat{\tilde{\phi}}(\xi)\, \bar{\hat{\tilde{\psi}}}(\xi)\, d\xi .$$

Such integral equations are very known. By a Fourier transformation, we obtain a classical integral equation and integrations by parts can be used to eliminate the singularities of the kernel.

This method seems very useful to perform explicit computations. But it needs a good framework of functional analysis which generates a lot of difficulties. For example the bilinear form b is not coercive on the space $\tilde{H}_{1/2}(\Gamma)$, that is the solution ϕ of the problem can be not in $\tilde{H}_{1/2}(\Gamma)$. Moreover explicit computations are not quite obvious, because in the problems encountered in mechanics, the operator T is known modulo regularizing operators. This is not a difficulty to compute the singularities, but to obtain explicit expressions of u these regularizing operators can "polluate" the computations.

To avoid the difficulties of the functional analysis of this problem, let us introduce two other methods which can be associated with that of A. Bamberger to solve the elasticity problem.

II- WIENER-HOPF METHOD.

C. Goudjo uses this method to solve elliptic problems in general spaces with a crack of codimension one. The framework of functional analysis is that of G.I. Eskin. An immediate applying is the study of the elasticity problem exposed in (I). C. Goudjo can compute the singularities explicitely. His method also allows to construct the operator $T : \phi \longrightarrow g$ from $\tilde{H}_{1/2}(\Gamma)$ into $H_{-1/2}(\Gamma)$, and to solve the corresponding problem.

To give an idea of his method and his results, we only consider a semi-infinite rectilinear crack : $\Gamma = \{ (x',x_n) \; ; \; x_n = 0 \; , \; x_{n-1} > 0 \}$.But the method also allows to consider bounded cracks defined by lipschitzian surfaces. To simplify once more, we consider the case of the Laplacian. Let be the problem ($\Omega = \mathbb{R}^2 \setminus \Gamma$) :

$$
(P) \begin{cases} \text{Find } u \text{ in the space } \mathcal{H} = \{ u \in \mathcal{S}'(\Omega) \; ; \; \forall \phi \in \mathcal{D}(\mathbb{R}^n), \; \phi u \in L^2(\Omega), \\ \qquad\qquad \Delta u \in L^2(\Omega) \} \text{, such that} \\ \Delta u = 0, \\ \left.\begin{array}{l} \dfrac{\partial u}{\partial x_n}(x',0_+) = g(x') \\ \dfrac{\partial u}{\partial x_n}(x',0_-) = h(x') \end{array}\right\} \quad x_{n-1} > 0 \; , \quad g,h \in \mathcal{S}(\mathbb{R}^{n-1}_+). \\ u = o(1) \quad \text{for } |x| \longrightarrow \infty . \end{cases}
$$

As G.I. Eskin we define the operator Π^+ by the relation :

$$\Pi^+ \hat{u}(\xi) = \mathcal{F}(Y(x_n).u)$$

where $\hat{u}(\xi) = \mathcal{F}(u(x)) = \int e^{ix\xi} u(x)\, dx$, Y is the Heaviside function. If $\hat{H}_s$ denotes the space of Fourier transforms of elements of H_s , the operator Π^+ is continuous from $\hat{H}_s$ into $\hat{H}_s$ when $|s| < 1/2$.
One key of the method is the following Paley-Wiener theorem :

Theorem.- Suppose $f \in \tilde{H}_s(\mathbb{R}^n)$. Then $\hat{f}(\xi',\xi_n+i\tau) = \mathcal{F}(f.e^{-x_n\tau})$ is a continuous function of τ for $\tau > 0$, with values in $\hat{H}_s$ that is analytic with respect to $\xi_n + i\tau$ in the half-plane $\tau > 0$ for almost all ξ' in $\mathbb{R}^{n-1}$ and satisfies :

$$\int_{-\infty}^{+\infty} |\hat{f}(\xi',\xi_n+i\tau)|^2 (1+|\xi'|+|\xi_n|+\tau)^{2s} \, d\xi' \, d\xi_n \leqslant C \quad , \tau \geqslant 0 ,$$

where C does not depend on τ.

Conversely, suppose for $\tau > 0$ we are given a function $\hat{f}(\xi',\xi_n+i\tau)$ that is locally integrable in $\mathbb{R}^n \times (0,\infty)$, is analytic with respect to

$\xi_n+i\tau$ for almost all $\xi' \in \mathbb{R}^{n-1}$, and satisfies the inequality above. Then there exists a function $f \in \tilde{H}_s$ such that $\hat{f}(\xi',\xi_n+i\tau) = \mathcal{F}(f e^{-x_n\tau})$.

Let u be a solution of the problem (P). By a partial Fourier transformation, the solution u can be written :

$$\hat{u}(\xi',x_n) = \begin{cases} A(\xi')\, e^{-x_n|\xi'|} & , x_n > 0 , \\ B(\xi')\, e^{x_n|\xi'|} & , x_n < 0 . \end{cases}$$

Now our problem is to find the functions A and B. Thanks to the continuity of u and $\partial_n u$ when $x_n = 0$ and $x_{n-1} < 0$, the supports of the following functions

$$\begin{cases} E_+ = u(x'',x_{n-1},0_+) - u(x'',x_{n-1},0_-) , \\ F_+ = \partial_n u(x'',x_{n-1},0_+) - \partial_n u(x'',x_{n-1},0_-) \end{cases}$$

are contained in $\mathbb{R}^{n-1}_+$.

Let ℓg and ℓh be the extensions of g and h in $\mathbb{R}^{n-1}$. The boundary conditions allow to define two functions G_- and H_- with support in $\mathbb{R}^{n-1}$:

$$\begin{cases} \hat{G}_-(\xi') = -|\xi'|\, A(\xi') - \widehat{\ell g}(\xi') , \\ \hat{H}_-(\xi') = |\xi'|\, B(\xi') - \widehat{\ell h}(\xi') . \end{cases}$$

By using the factorisation $|\xi'| = (\xi_{n-1}+i|\xi''|)^{1/2}(\xi_{n-1}-i|\xi''|)^{1/2}$

$$= q_+(\xi'',\xi_{n-1})q_-(\xi'',\xi_{n-1}) ,$$

the following relations are obtained after some elementary calculation :

$$\begin{cases} -q_+\hat{E}_+ = \dfrac{1}{q_-}(\widehat{\ell h} + \widehat{\ell g}) + \dfrac{1}{q_-}(\hat{H}_- + \hat{G}_-) , \\ \hat{F}_+ = \widehat{\ell g} - \widehat{\ell h} + \hat{G}_- - \hat{H}_- . \end{cases}$$

An application of the operator Π^+ give

$$\begin{cases} \hat{E}_+ = -\dfrac{1}{q_+}\, \Pi^+ \dfrac{\widehat{\ell h} + \widehat{\ell g}}{q_-} \\ \hat{F}_+ = \Pi^+ (\widehat{\ell g} - \widehat{\ell h}) . \end{cases}$$

At last, we obtain the expression of $\hat{u}$:

If $x_n > 0$, $\hat{u}_+(\xi',x_n) = \{-1/2\,\frac{1}{q_+}\, \Pi^+ \frac{\widehat{\ell h}+\widehat{\ell g}}{q_-} - 1/2\frac{1}{|\xi'|}\, \Pi^+ (\widehat{\ell g}-\widehat{\ell h})\}\, e^{-x_n|\xi'|}$.

The first term in the right part lies in $H_{3/2-\varepsilon}$, the second one is in $H_{2-\varepsilon}$.

If $x_n < 0$, $\hat{u}_-(\xi',x_n) = \{1/2\,\frac{1}{q_+}\,\Pi^+\,\frac{\widehat{\ell h} + \widehat{\ell g}}{q_-} - 1/2\,\frac{1}{|\xi'|}\,\Pi^+(\widehat{\ell g}-\widehat{\ell h})\}\, e^{-x_n|\xi'|}$.

The jump of u through the crack is then given by

$$\hat{\phi} = \hat{u}_+ - \hat{u}_- = -\frac{1}{q_+}\;\Pi^+\,\frac{\widehat{\ell h} + \widehat{\ell g}}{q_-} \;,$$

and if $g = h$, that is when the solution is a double layer potential as in the elasticity problem (I) ,

$$\hat{\phi} = -\frac{2}{q_+}\;\Pi^+\,\frac{\widehat{\ell g}}{q_-}\;.$$

We recall that by using the Helmholtz decomposition of a vector u ($u = u_p + u_s$, u_p potential vector, u_s solenoidal vector)the elasticity system is equivalent to the following one :

$$\begin{cases} \Delta u_p + k_1^2\, u_p = 0 \;, \; \mathrm{rot}\; u_p = 0 \;, \\ \Delta u_s + k_2^2\, u_s = 0 \;, \; \mathrm{div}\; u_s = 0 \;, \\ u = u_p + u_s \;, \; k_1^2 = \dfrac{\omega^2}{\lambda+2\mu} \;, \; k_2^2 = \dfrac{\omega^2}{\mu}\;. \end{cases}$$

Then the Goudjo's method is easily applyable to the problem of elasticity. Moreover he gives the regularity of the solution explicitely :
If $\vec{r}$ is the vector (x_{n-1},x_n) in the plane (x_n,x_{n-1}) , if α is the angle $(\vec{x}_{n-1},\vec{r})$

$$u = \frac{2}{\sqrt{\pi}}\sqrt{r}\,(\cos\frac{\alpha}{2} - \sin\frac{\alpha}{2})\gamma_o(\frac{\widehat{\ell h}+\widehat{\ell g}}{q_-})(x'') - \gamma_o(\frac{\widehat{\ell h}+\widehat{\ell g}}{q_-})(x'')v(x_{n-1},x_n)$$
$$+ w_o + w_1 \;,$$

with $v \in C^\infty(\overline{\mathbb{R}}^2_+)$, $w_1 \in H_{5/2-\varepsilon}(\Omega)$, $w_o \in H_{2-\varepsilon}(\Omega)$ (γ_o:first trace), and it is impossible to improve the regularity of the solution.

III- APPROXIMATION METHOD.

Here is a third method to solve the elasticity problem with a crack. The crack Γ is approached by a regular surface without boundary Γ_ε . To simplify the account, we suppose that we are concerned by the Helm-

-holtz equation in the complement of a lipschitzian surface in the space $\mathbb{R}^3$. But, as for the Goudjo's case, this limitation is dispensable,we can apply our results in $\mathbb{R}^n$ to general elliptic problems if an approximation of their Green's function is known.

Let $\Omega = \mathbb{R}^2 \setminus \Gamma$, and $V = \{ u \in \mathcal{S}'(\Omega) \ , \forall \phi \in \mathcal{D}(R^n),\ \phi u \in H_1(\Omega)\}$, we consider the following problem :

$$(D) \begin{cases} \text{Find } u \text{ in } V \text{ such that :} \\ (\Delta + k^2)\, u = 0 \quad \text{in } \Omega, \\ \partial_n^+ u|_\Gamma = \partial_n^- u|_\Gamma = \chi \ , \quad \chi \in H_{-1/2}(\Gamma) \ , \\ \text{Sommerfeld's conditions.} \end{cases}$$

An application of Green's and Plemelj's formulae transform this problem into the following one :

$$(K) \begin{cases} \text{For every } \chi \in H_{-1/2}(\Gamma) \ , \text{ find } \mu \in \tilde{H}_{1/2}(\Gamma) \text{ such that} \\ \lim\limits_{\substack{\tilde{x} \to x \\ \tilde{x} \notin \Gamma}} \partial_{n_x} \int_\Gamma \mu(y)\, \partial_{n_y} G(\tilde{x},y)\, d\Gamma(y) = -\chi(x) = K\mu(x). \end{cases}$$

To define the integral equation correctly, regular closed surfaces (i.e. without boundary) Γ_ε are introduced, that surround Γ and converge to it as ε goes to zero. If Ω'_ε denotes the interior domain corresponding to Γ_ε and Ω_ε the outside, two boundary value problems are introduced:

$$(D_\varepsilon) \begin{cases} (\Delta + k^2)\, u^\varepsilon = 0 \text{ in } \Omega_\varepsilon, \quad u^\varepsilon \in V_\varepsilon \text{ (space } V_\varepsilon \text{ corresponding to } \Omega_\varepsilon) \\ \partial_n u^\varepsilon|_{\Gamma_\varepsilon} = \chi^\varepsilon \in H_{-1/2}(\Gamma_\varepsilon) \ , \\ \text{Sommerfeld's conditions.} \end{cases}$$

$$(D'_\varepsilon) \begin{cases} (\Delta + k^2)\, v^\varepsilon = 0 \text{ in } \Omega'_\varepsilon \ , \quad v^\varepsilon \in H_1(\Omega_\varepsilon) \ , \\ \partial_n v^\varepsilon|_{\Gamma_\varepsilon} = \chi^\varepsilon \in H_{-1/2}(\Gamma_\varepsilon). \end{cases}$$

Thus a classical use of the Calderón's projectors allows to transform these two differential problems into the following integral one :

$$(I_\varepsilon) \ \{ \text{Find } \nu^\varepsilon \text{ in } H_{1/2}(\Gamma_\varepsilon) \text{ such that } K_\varepsilon \nu^\varepsilon = -\chi^\varepsilon \ .$$

K_ε is a pseudo-differential operator defined by the relation :

$$K_\varepsilon \nu^\varepsilon = \lim_{\substack{\tilde{x} \to x \\ \tilde{x} \notin \Gamma_\varepsilon}} \int_{\Gamma_\varepsilon} \partial_{n_{\varepsilon,x}} \partial_{n_{\varepsilon,}} G(\tilde{x},y)\, \nu^\varepsilon(y)\, d\Gamma_\varepsilon(y) ,$$

and $\nu^\varepsilon = Tr(u^\varepsilon) - Tr(v^\varepsilon) \in H_{1/2}(\Gamma_\varepsilon)$.

To study the limit of u^ε , an a priori uniform estimate is shown:

$$\|u^\varepsilon\|_{H^1_{\varepsilon,R}(\Delta)} \leqslant C \|\chi^\varepsilon\|_{H_{-1/2}(\Gamma_\varepsilon)} ,$$

where $\Omega_{R,\varepsilon} = \{x \in \Omega_\varepsilon , |x| < R\}$ and $\|u\|^2_{H^1_{\varepsilon,R(\Delta)}} = \|u\|^2 + \|\nabla u\|^2 + \|\Delta u\|^2$,

and $\|u\|$ is the norm in $L^2(\Omega_{R,\varepsilon})$.

Then the operator L defines the operator K . It is shown that $\mu = \nu^+ - \nu^-$ and the following result is obtained :

When ε goes to zero, the curves Γ_ε approach the closed curve γ which is a parametrization of the crack Γ (travelled twice), the operators K_ε tend to an operator L on γ , and this operator L defines the initial operator K on Γ . It is shown that the jump μ is equal to $\nu^+ - \nu^-$ and the following result is obtained :

Theorem.- The operator K is an isomorphism from $\widetilde{H}_{1/2}(\Gamma)$ onto $H_{-1/2}(\Gamma)$

Moreover in the prob em given by physicists, the function χ is as regular as we want, and μ is shown to lie in $H_{1-\varepsilon}(\Gamma)$. Then μ is continuous and is zero at the edges of Γ (that is, for example, an hypothesis in the Bamberger's work).

Numerical computations with the operator K are simple, for example by using boundary element methods.

REFERENCES.

Achenbach,J.D.,Wave propagation in elastic solids, North Holland, Amsterdam, 1973.

Bamberger,A., Approximation de la diffraction d'ondes élastiques. Une nouvelle approche. Rapports de l'Ecole Polytechnique n° 91, 96 and 98 , Palaiseau, 1983.

Durand,M., Layer potential and boundary value problems for the Helmholtz equation. Math. Meth. in the Appl. Sci. 5 (1983),389-421.

Eskin,G.I., Boundary value problems for elliptic pseudo-differential equations, Transl. of Math. monographs, A.M.S. ,1981.

Goudjo, C., To appear.

Lions,J.L.,Magenes,E.,Problèmes aux limites non homogènes et applications, vol I, Dunod, Paris, 1968.

Wendland,W.L., Boundary element methods and numerical techniques,ed. P. Filippi, in CISM courses and lectures, n°277, International centre for mechanical sciences, 135-216, Udine, 1983.

BOUNDARY INTEGRAL EQUATIONS FOR SOUND RADIATION BY A HARMONICALLY VIBRATING BAFFLED PLATE

Paul J.T.FILIPPI
Laboratoire de Mécanique et d'Acoustique
13277 Marseille cedex 9
- France -

SUMMARY :

It is shown that the displacement of the plate can be represented by a Green's formula, which involves the infinite fluid-loaded plate kernel. The boundary conditions lead to a system of two integral equations along the boundary of the plate : due to energy dissipation in the fluid, this system has always one and only one solution. A numerical approximation is obtained by a collocation method : the convergence is illustrated by a simple example.

1/ INTRODUCTION

The generation of sound by a vibrating structure like a diesel engine, a car, a washing machine or a refrigerator,etc... is a very complicated problem. In order to get a better understanding of the physical phenomena, and to derive simple practical methods of noise reduction, acousticians have looked at simple academic problems : the simplest vibrating structure is an infinite thin elastic plate, in a homogeneous perfect fluid occupying one or both of the two half-spaces separated by the plate plane.

A less simple structure is the baffled-plate. A thin elastic plate occupies a bounded domain Ω of the $z = 0$ plane, with a C^∞ boundary $\partial\Omega$. The plane complementary of $\overline{\Omega}$ is assumed to be perfectly rigid. The half-space $z > 0$ is occupied by a fluid, while the half-space $z < 0$ is a vacuum. It is assumed that the plate is excited by a harmonic ($e^{-i\omega t}$) force, with density F.

The sound pressure in the fluid can be explicitly expressed in terms of the plate displacement (solution of a Neumann problem). We are thus left with a unique unknown function, the plate displacement, which is solu-

tion of an integro-differential equation. Using the infinite fluid-loaded plate kernel, a Green's representation of the finite plate displacement can be used. So, the boundary conditions lead to a system of two integral equations along the boundary $\partial\Omega$ of the plate.
Section 2 is devoted to the statement of the problem, and the corresponding boundary integral equations are established. Then, the existence and uniqueness of the solution is shown. In section 3, a representation of the infinite fluid-loaded plate kernel is recalled : expressed in terms of Hankel functions and Laplace type integrals, it is very easy to compute numerically. Similar expressions of the first and second derivatives of this kernel are given. Section 4 deals with the numerical approximation adopted by the author, and the convergence of the algorithm is shown on a simple example.

2/ STATEMENT OF THE PROBLEM AND BOUNDARY INTEGRAL EQUATIONS

Let Ω be a bounded domain of the z = 0 plane, bounded by a C^∞ curve $\partial\Omega$ with exterior unit normal n . It is occupied by a thin elastic plate defined by E = Young modulus ; ν = Poisson ratio ; h = thickness ; m = surface mass density ; $D = Eh^3/12(1-\nu^2)$ = rigidity. The plane complementary $\complement\bar{\Omega}$ of $\bar{\Omega}$ is a perfectly rigid surface. The plate is clamped along $\partial\Omega$. The half-space $z < 0$ is a vacuum, while the half-space $z > 0$ is occupied by a homogeneous perfect fluid defined by the following parameters :
ρ_0 = density ; c_0 = sound speed.
Let p and u denote the sound pressure in the fluid, and the plate displacement, respectively. It is assumed that the plate is excited by a harmonic ($e^{-i\omega t}$) force, with density F. The unknown functions p and u are solution of the following equations :

$$(\Delta^2 - \lambda^4)u = \frac{F}{D} - \frac{Tr\, p}{D} \qquad \text{in } \Omega \qquad [1]$$

$$(\Delta + k^2)p = 0 \qquad \text{in } z > 0 \qquad [2]$$

$$Tr\, \partial_z p = \begin{cases} \omega^2 \rho_0 u & \text{in } \Omega \\ 0 & \text{in } \complement\bar{\Omega} \end{cases} \qquad [3]$$

$$Tr\, u = Tr\, \partial_n u = 0 \qquad \text{on } \partial\Omega \qquad [4]$$

Sommerfeld condition at infinity for p [5]
In these equations, $\lambda^4 = m\,\omega^2/D$, and $k = \omega/c_0$.

2-1 - The Green's representation of u, and the corresponding boundary integral equations :

Let G be the Neumann Green's function for the Helmholtz equation in $z>0$ that is :

$$G = -e^{ikr}/4\pi r$$

The sound pressure p is given by :

$$p = \omega^2 \rho_0 (u \otimes \delta_z) \underset{(3)}{*} G \qquad [6]$$

where $\otimes$ denotes the tensor product ; $\underset{(3)}{*}$ is the three-dimensional convolution product ; and u stands for the function which is equal to the plate displacement in Ω, and to zero in $\complement \bar{\Omega}$. Equation [1] thus becomes an integro-differential equation for u :

$$\left\{ (\Delta^2 - \lambda^4)\delta + \frac{\omega^2 \rho_0}{D} G \right\} \underset{(2)}{*} u = \frac{F}{D} \quad \text{in } \Omega \qquad [7]$$

where $\underset{(2)}{*}$ is the two-dimensional convolution product.
Let Γ be the infinite fluid-loaded plate Green's function, i.e. Γ is the solution of :

$$\left. \begin{array}{l} \left\{ (\Delta^2 - \lambda^4)\delta + \dfrac{\omega^2 \rho_0}{D} G \right\} \underset{(2)}{*} \Gamma = \dfrac{\delta}{D} \quad \text{in } R^2 \\ \text{Sommerfeld condition at infinity} \end{array} \right\} \qquad [8]$$

It can be established that the step formula of the in-vacuo plate operator $(\Delta^2 - \lambda^4)$ is valid for the fluid-loaded plate operator. Thus, the function u can be represented by a Green's formula :

$$u = u_0 + \Gamma \underset{(2)}{*} \left\{ \mu_1 \otimes \delta_{\partial\Omega} + \mu_2 \otimes \delta'_{\partial\Omega} \right\} \qquad [9]$$

with the following notations :

$$u_0 = \Gamma \underset{(2)}{*} F$$

$$\Gamma \underset{(2)}{*} (\mu_1 \otimes \delta_{\partial\Omega})(M) = \int_{\partial\Omega} \mu_1(P)\, \Gamma(M,P)\, dP$$

$$\Gamma \underset{(2)}{*} (\mu_2 \otimes \delta'_{\partial\Omega})(M) = -\int_{\partial\Omega} \mu_2(P)\, \partial_{n(P)} \Gamma(M,P)\, dP$$

The layer densities μ_1 and μ_2 are related to u by :

$$\mu_1 = -\mathrm{Tr}\, \partial_n \Delta u - (1-\nu)\, \partial_s \mathrm{Tr}\, \partial_n \partial_s u$$

$$\mu_2 = -\mathrm{Tr}\left[\Delta u - (1-\nu)\, \partial_s^2 u \right]$$

(s is the unit tangent vector). The boundary conditions [4] provide a system of two integral equations to determine μ_1 and μ_2 :

$$\left.\begin{aligned} &Tr\ \Gamma \underset{(2)}{*} \{\mu_1 \otimes \delta_{\partial\Omega} + \mu_2 \otimes \delta'_{\partial\Omega}\} = -Tr\ u_0 \\ &Tr\ \partial_n[\Gamma \underset{(2)}{*} \{\mu_1 \otimes \delta_{\partial\Omega} + \mu_2 \otimes \delta'_{\partial\Omega}\}] = -Tr\ \partial_n u_0 \end{aligned}\right\} \qquad [10]$$

2-2 - Uniqueness of the solution (p,u) of the system [1-5] :

Let $H(\Delta,\mathbb{R}^3_+)$ be the space of functions f which are integrable in $\mathbb{R}^3_+$ and such that Δf is square integrable too. The space $H^{loc}(\Delta, \bar{\mathbb{R}}^3_+)$ is the space of functions f such that :

$$f \in H^{loc}(\Delta,\bar{\mathbb{R}}^3_+) \text{ if } \varphi f \in H(\Delta,\mathbb{R}^3_+) \quad \text{for any } \varphi \in \mathcal{D}(\mathbb{R}^3) \qquad [11]$$

(see reference1)

Let $H(\Delta^2,\Omega)$ be the space of functions v such that $v \in L^2(\Omega)$ and $\Delta^2 v \in L^2(\Omega)$. We want to prove that [1-5] can have only one solution (p,u) in $H^{loc}(\Delta, \bar{\mathbb{R}}^3_+) x H(\Delta^2,\Omega)$. Assume that $F \equiv 0$, and that there exist a solution (p,u) in $H^{loc}(\Delta, \bar{\mathbb{R}}^3_+) x H(\Delta^2,\Omega)$ satisfying the system [1-5].Let V be a half-ball centered at the origin, and with radius R. It is bounded by a half-sphere W_1, and a disk W_2 in the $z = 0$ plane. It is assumed that Ω is included in W_2. The following equality is valid :

$$\int_V [\bar{p}(\Delta p + k^2 p) - p(\Delta \bar{p} + k^2 \bar{p})] = \int_{W_1 \cup W_2} (Tr\,p\ Tr\,\partial_n \bar{p} - Tr\,\bar{p}\ Tr\,\partial_n p) \qquad [12]$$

(n is the outgoing normal).

Because of equation [2] , the left hand side of [12] is zero. Using the boundary conditions [3] , and the Sommerfeld condition

$$\partial_R p - ikp = o(R^{-1}) \qquad \text{for } R \gg 1,$$

one gets :

$$2ik \int_{W_1} |Tr\,p|^2 + o(1) + \omega^2 \rho_0 \int_\Omega (Tr\,p\ \bar{u} - Tr\,\bar{p}\ u) \qquad [13]$$

Then, use is made of the plate equation [1] and the boundary conditions [4] , to get :

$$\int_\Omega (Tr\,p\ \bar{u} - Tr\,\bar{p}\ u) = -D \int_\Omega [\bar{u}(\Delta^2 - \lambda^4)u - u(\Delta^2 - \lambda^4)\bar{u}] = 0$$

Consequently, [13] becomes :

$$\int_{W_1} |Tr\,p|^2 = o(1) \qquad [14]$$

Because $(p,u) \in H^{loc}(\Delta, \overline{\mathbb{R}}^3_+) x H(\Delta^2, \Omega)$, the representation [6] of p is well-defined. Furthermore, this expression is defined in the whole space, and is symetrical in z. So, if W'_1 is the sphere with radius R, one has :

$$\int_{W'_1} |\mathrm{Tr}\, p|^2 = o(1) \qquad [15]$$

Because of the Rellich's theorem, we can conclude that $p \equiv 0$ outside the ball of radius R. The analyticity properties of [6] implies that $p \equiv 0$ in the whole space, and, consequently, $\mathrm{Tr}\ \partial_z p \equiv 0$ in Ω . Then, $u \equiv 0$ in Ω . This prove the uniqueness of the solution (p,u) of the system [1-5].

2-3 - <u>Existence of the solution of [1-5]</u> :

Let us consider the boundary value problem [7,4] satisfied by the plate displacement u, and the classical plate problem :

$$\left.\begin{array}{ll} (\Delta^2 - \lambda^4)\, u = \dfrac{F}{D} & \text{in } \Omega \\ \mathrm{Tr}\, u = \mathrm{Tr}\, \partial_n u = 0 & \text{on } \partial\Omega \end{array}\right\} \qquad [16]$$

The integral operator $u \to \frac{\omega^2 \rho_o}{D} G \underset{(2)}{*} u$ is compact. So, these two problems have the same index, which is zero.

Furthermore, the solution u of [7,4] is unique (if it exists). Indeed, let u' be another solution, to which corresponds p' given by :

$$p' = \omega^2 \rho_o (u \otimes \delta_z) \underset{(3)}{*} G.$$

The two pairs of functions (p,u) and (p',u') satisfy the same system of equations. The uniqueness of (p,u) proved in sub-section 2-2 implies

u = u'

Then, the uniqueness of the solution u of the boundary value problem [7,4] implies the existence of u, and consequently that of p.

<u>3/ THE INFINITE FLUID-LOADED GREEN'S FUNCTION</u>

The solution Γ of equation [8] has obviously a cylindrical symetry. Thus, the method classically used (see, for example {2}) is to use the Fourier transform. Then, the problem is to invert the Fourier transform of Γ in such a way that the representation obtained can be computed very fast. Using a method first developped by Weyl {3} , it has been shown [4] that Γ can be expressed as follows :

$$\Gamma = -\sum_{j=1}^{5} \frac{\alpha_j^2 - k^2}{4L'(\alpha_j^2)} \left\{ iD(\alpha_j^4 - \lambda^4) + \frac{\omega^2 \rho_o}{\beta_j} Y(\alpha_j) \right\} H_o(\alpha_j \rho)$$

$$+ \frac{\omega^2 \rho_o}{2\pi} e^{ik\rho} \sum_{j=1}^{5} \frac{\alpha_j^2 - k^2}{L'(\alpha_j^2)} \int_0^\infty \frac{e^{-k\rho t}}{\beta_j \, \varepsilon_o(\alpha_j, t) \, W_o(\beta_j, t)} dt \qquad [17]$$

In this expression, ρ is the radial co-ordinate and $H_o(z)$ stands for the zeroth order cylindrical Hankel function of the first kind. The other quantities are defined as follows :

- α_j^2 are the five roots of the polynomial
  $L(u) = \omega^4 \rho_o^2 - D^2(u-k^2)(u^2-\lambda^4)^2$;
- $W_o(\beta_j, t) = \sqrt{\frac{\beta_j^2}{k^2} + 2it - t^2}$;
- α_j is positive if α_j^2 is real ; Im $\alpha_j > 0$ if α_j^2 is complex ;
- $\beta_j^2 = k^2 - \alpha_j^2$; Im $\beta_j > 0$;
- $L'(u) = \partial L/\partial u$;
- $Y(\alpha_j)$ and $\varepsilon_o(\alpha_j, t)$ are given by :
  a/ $Y(\alpha_j) = \varepsilon_o(\alpha_j, t) = -1$ if α_j is real and $\alpha_j^2 > k^2$;
  b/ $Y(\alpha_j) = \varepsilon_o(\alpha_j, t) = -1$ if Im $\alpha_j \neq 0$ and Re$\alpha_j < 0$;
  c/ If Im $\alpha_j \neq 0$ and Re$\alpha_j > 0$, let $t_o = (1/2)\text{Im } \alpha_j^2/k^2$, and
  $U = \text{Re}(1 - \alpha_j^2/k^2) - \frac{1}{4}\text{Im } \alpha_j^2/k^2$; then :
  for $U > 0$, $Y(\alpha_j) = \varepsilon_o(\alpha_j, t) = 1$
  for $U < 0$, $Y(\alpha_j) = -1$, $\varepsilon_o(\alpha_j, t) = 1 \; \forall \, t \in [0, t_o[$, $\varepsilon_o(\alpha_j, t) = -1 \; \forall \, t \in [t_o, \infty[$.

Expression [17] of Γ seems to be the most suitable form for numerical purposes. Indeed, the Hankel functions $H_o(z)$ are easily computed by an ascending series for $|z| < 2$, and a Padé approximant for $|z| > 2$. For the Laplace type integrals, a Gauss method on automatically adjusted width intervals, as described in reference {5} , is used.
It is easily seen that Γ has the same kind of singularities as the in-vacuo plate kernel : Γ is regular at the origin ; its first derivative is zero ; and its second derivative has a logarithmic behaviour.
In the boundary integral equations system [10] , the first and second derivatives of Γ are needed. By derivating expression [17] , one gets :

$$\frac{\partial \Gamma}{\partial \rho} = \sum_{j=1}^{5} \frac{\alpha_j^2 - k^2}{4L'(\alpha_j^2)} \left\{ iD(\alpha_j^4 - \lambda^4) + \frac{\omega^2 \rho_0}{\beta_j} Y(\alpha_j) \right\} \alpha_j H_1(\alpha_j \rho) + \frac{\omega^2 \rho_0}{2\pi} k e^{ik\rho} \sum_{j=1}^{5} \frac{\alpha_j^2 - k^2}{L'(\alpha_j^2)} \int_0^\infty \frac{(i-t)\, e^{-k\rho t}}{\beta_j \, \mathcal{E}_0(\alpha_j, t) W_0(\beta_j, t)} \, dt \qquad [18]$$

$$\frac{\partial^2 \Gamma}{\partial \rho^2} = \sum_{j=1}^{5} \frac{\alpha_j^2 - k^2}{4L'(\alpha_j^2)} \left\{ iD(\alpha_j^4 - \lambda^4) + \frac{\omega^2 \rho_0}{\beta_j} Y(\alpha_j) \right\} \alpha_j^2 \left\{ H_0(\alpha_j \rho) - \frac{1}{\alpha_j \rho} H_1(\alpha_j \rho) \right\} + \frac{\omega^2 \rho_0}{2\pi} k^2 e^{ik\rho} \sum_{j=1}^{5} \frac{\alpha_j^2 - k^2}{L'(\alpha_j^2)} \int_0^\infty \frac{(i-t)^2 \, e^{-k\rho t}}{\beta_j \, \mathcal{E}_0(\alpha_j, t) W_0(\beta_j, t)} \, dt \qquad [19]$$

In these expressions, $H_1(z)$ is first order cylindrical Hankel function of the first kind. The functions $\partial_\rho \Gamma$ and $\partial^2_{\rho^2}\Gamma$ are computed by the same method as Γ.

4/ NUMERICAL APPROXIMATION

The system of boundary integral equations [10] is solved by a collocation method. Let M_i ($i = 1,2,\ldots N$) be the collocation points, the curvilinear abscissae of which are s_i. Let $F_i(s)$ be the set of functions defined by:

$$\begin{aligned} F_i(s) &= \left\{ \left| \frac{s-s_i}{s_i - s_{i-1}} \right| - 1 \right\}^2 \left\{ 2 \left| \frac{s-s_i}{s_i - s_{i-1}} \right| + 1 \right\} && \text{for } s \in [s_{i-1}, s_i[\\ &= \left\{ \left| \frac{s-s_i}{s_{i+1} - s_i} \right| - 1 \right\}^2 \left\{ 2 \left| \frac{s-s_i}{s_{i+1} - s_i} \right| + 1 \right\} && \text{for } s \in [s_i, s_{i+1}[\\ &= 0 && \text{for } s \notin [s_{i-1}, s_{i+1}] \end{aligned} \qquad [20]$$

These functions are of class C^1 (more or less regular approximation functions could be choosen). The unknown functions μ_1 and μ_2 are approximated by :

$$\begin{aligned} \mu_1(s) &= \sum_{i=1}^{N} \mu_{1i} \, F_i(s) \\ \mu_2(s) &= \sum_{i=1}^{N} \mu_{2i} \, F_i(s) \end{aligned} \qquad [21]$$

Thus, system [10] is replaced by the 2Nx2N-system of algebraïc linear equations :

$$\begin{pmatrix} K_1 & K_2 \\ K_3 & K_4 \end{pmatrix}\begin{pmatrix} \mu_1 i \\ \mu_2 i \end{pmatrix} = \begin{pmatrix} -\pi\mathcal{I}\, u_o(M_j) \\ -\pi\mathcal{I}\,\partial_n u_o(M_j) \end{pmatrix} \qquad [22]$$

in which $K_\ell (\ell = 1,\ldots 4)$ are the four NxN-matrices defined as follows :

$$K_{1\,j}^{\;i} \simeq \int_{S_{i-1}}^{S_{i+1}} F_i(S)\, \Gamma[M_j, P(s)]\, ds$$

$$K_{2\,j}^{\;i} \simeq \int_{S_{i-1}}^{S_{i+1}} F_i(s)\, \partial_{n(P)} \Gamma[M_j, P(s)]\, ds$$

$$K_{3\,j}^{\;i} \simeq \int_{S_{i-1}}^{S_{i+1}} F_i(s)\, \partial_{n(M_j)} \Gamma[M_j, P(s)]\, ds \qquad [23]$$

$$K_{4\,j}^{\;i} \simeq \int_{S_{i-1}}^{S_{i+1}} F_i(s)\, \partial_{n(M_j)} \partial_{n(P)} \Gamma[M_j, P(s)]\, ds$$

In [23] , the integrals are approximated by a 16-points Gauss rule. The theoretical convergence of this procedure has not been proved. Nevertheless, it seems that the theories developped by W. Wendland {6} and the Darmstadt school can provide the desired result.
An experimental proof of the convergence is shown on the following example. The plate is a steel one, 5 cm thick. It occupies a circular domain with radius 100 cm. It is excited by a point unit force δ_S located at S(x = 50., y = 0.), and with frequency 62.5 Hz. The fluid is water. The mechanical characteristics of this system are : $E = 2.16\ 10^{12}$ cgs ; $\nu = 0.276$; $m/h = 7.84\ g/cm^3$; $\rho_o = 1.g/cm^3$; $c_o = 1.49\ 10^5$ m/s. Figures 1 and 2 show the approximations of μ_1 and μ_2 corresponding to N = 4,6,9 and 12. It must be remarked that, for N < 12, the linear system [22] can be solved with a simple precision subroutine. But, for larger values of N, a double precision subroutine has been necessary.

5/ CONCLUDING REMARKS

In acoustics, one is mainly interested in the far-field directivity pattern, that is the leading term of the large distance asymptotic series of the pressure :

$$p \simeq -\omega^2 \rho_0 \frac{e^{ikR}}{2\pi R} \hat{u}(k\sin\theta \sin\phi, k\sin\theta\cos\phi)$$

in which (R,Θ,Φ) are the spherical co-ordinates of the observation point, and $\hat{u}(\xi,\eta)$ is the Fourier transform of u. The fonction $\hat{u}$ implies to compute the Fourier transform of the layer densities $\mu_1 \otimes \delta_{\partial\Omega}$ and $\mu_2 \otimes \delta'_{\partial\Omega}$: a numerical approximation is readily obtained by integrating the approximations $\Sigma\, \mu_1^i F_i$ and $\Sigma\, \mu_2^i F_i$ with a Simpson's rule, for example. So, for engineering applications, the results proposed here are quite satisfactory.

But, from a mathematical and numerical point of view, it seems useful to answer some other questions. First, the convergence of the numerical technique has to be proved. It would be interesting to check if other approximation functions F_i could not increase the accuracy of the numerical results. Instead of the integral equations system [10] which is of the first kind, it is possible to use equations of the second kind: it is useful to see if such a system is more stable and can avoid the use of a double precision subroutine to invert the approximating system of algebraïc equations. Hence, the hypothesis of a C^∞ boundary $\partial\Omega$ must be relaxed, the contours of practical interest being, in general, piecewise-C^∞ ones. So, the functions μ_1 and μ_2 can be singular at the corners (see P. Grisvard {7}), and the corresponding singularities must be introduced in the numerical approximations.

This work has been supported by the "Direction des Recherches et Etudes Techniques (Délégation Générale de l'Armement)", convention n° 82/302.

BIBLIOGRAPHY

{1} C.H. WILCOX, 1975, *Scattering theory for the d'Alembert equation in exterior domains*, Lecture Notes in Mathematics n° 442, Springer-Verlag, Berlin-Heidelberg-New York.

{2} L.M. BREKHOVSKIKH, 1960, *Waves in layered media*, New-York, Academic Press.

{3} H. WEYL, 1919, *Annales der Physic, Leipzig*, 60, 481-500. Ausbreitung elektromagnetischer Weller über einen ebene Leiter.

{4} P.J.T. FILIPPI, 1983, *Journal of Sound and Vibration*, (to appear). Sound radiation by baffled plates and related boundary integral equations.

{5} N. BAKHVALOV, 1976, *Méthodes numériques*, Mir, Moscow.

{6} W.L. WENDLAND, 1983, in *Theoretical acoustics and numerical techniques*, CISM courses and Lectures n° 277, Springer-Verlag, Wien-New

York.

{7} P. GRISVARD, 1980 , *Boundary value problems in non-smooth domains*, Lecture Note 19, University of Maryland, MD 20742.

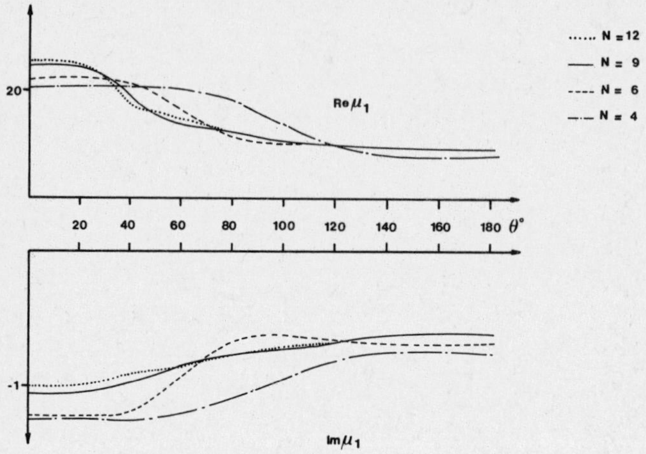

- Figure 1 -

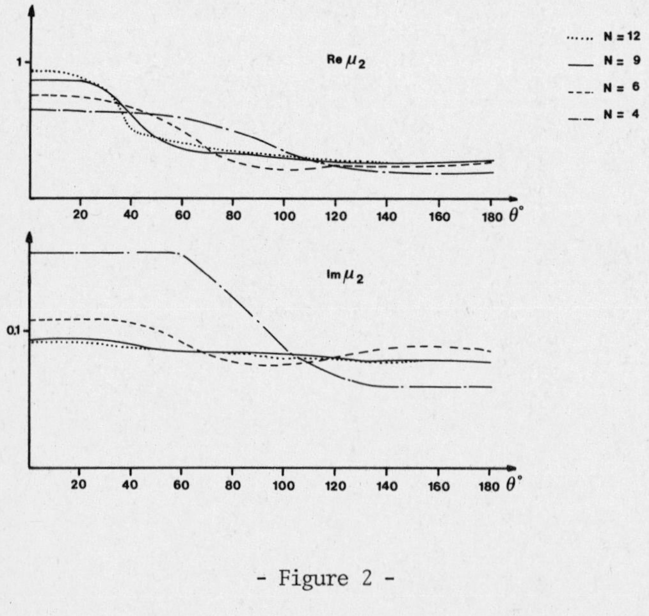

- Figure 2 -

EIGENFUNCTION EXPANSIONS FOR NON SELF ADJOINT OPERATORS AND SEPARATION OF VARIABLES

G.Geymonat
Politecnico
Corso Duca degli Abruzzi 24
10129 Torino
Italia

P.Grisvard
I.M.S.P.
Parc Valrose
06034 NICE Cedex
France

Abstract : We consider an unbounded operator A in a Hilbert space H. We do not assume self adjointness but we assume that the inverse of A belongs to a Carleman class $\mathscr{L}_p(H)$ with a small enough p. In addition we assume that the resolvent of A has minimal growth along the negative real axis. Then we show that the solution u of the abstract elliptic problem

$$u''(t) - A\,u(t) = 0, \quad t > 0$$

with the boundary condition u(0)=x may be expanded in a series of generalized eigenfunctions of A for large values of t.

This is applied to operators arising from two point boundary value problems and leads to explicit separation of the variables for the solution of elliptic boundary value problems in polygons near the vertices.
Examples are the biharmonic (hence elasticity system) in a polygon with the usual boundary conditions even in the case of fractures. Other examples are the Stokes equations and the Stokes-Beltrami equation in special geometries.
This research has been mainly motivated by various papers of **D.Joseph** (especially ref. [7],[8] below). The main results are presented in the short note [5].

Résumé : Soit A un opérateur non borné dans l'espace de Hilbert H. On ne le suppose pas nécessairement auto-adjoint mais on suppose que son inverse appartient à une classe de Carleman $\mathscr{L}_p(H)$ convenable. De plus on suppose que la résolvante de A a la décroissance maximale le long de l'axe réel négatif. Sous ces hypothèses on montre que la solution u de l'équation opérationnelle elliptique

$$u''(t) - A\,u(t) = 0, \quad t > 0$$

avec la condition au bord u(0)=x, admet un développement en série de fonctions propres généralisées de A pour t assez grand.

On applique ce résultat abstrait à des opérateurs provenant de problèmes aux limites pour des équations différentielles ordinaires et ceci conduit à la séparation des variables pour les solutions de certains problèmes elliptiques dans des polygones, au voisinage des coins.
Les exemples possibles sont le bilaplacien (donc l'élasticité plane) dans un polygone avec les conditions aux limites habituelles y compris dans le cas de fissures. On considère aussi les équations de Stokes et de Stokes-Beltrami dans des géométries particulières.
Cette recherche a été motivée pour l'essentiel par les travaux de **D.Joseph** (en particulier [7], [8]). L'essentiel a été résumé dans la note [5].

Sunto : Consideriamo un operatore lineare A non limitato nello spazio di Hilbert H. Non supponiamo che A sia autoaggiunto ma soltanto che abbia inverso in una opportuna classe di Carleman $\mathscr{L}_p(H)$. Inoltre supponiamo che l'operatore risolvente di A abbia decrescenza massimale sull'asse reale negativo. Con queste ipotesi si prova che la soluzione u dell'equazione ellittica astratta

$$u''(t) - A\,u(t) = 0, \quad t > 0$$

colla condizione al bordo u(0)=x, può essere sviluppata in una serie di autofunzioni generalizzate di A per t abbastanza grande.

Questi risultati astratti vengono applicati a operatori definiti mediante problemi al contorno per equazioni differenziali e questo implica la possibilità di separare le variabili nelle soluzioni di problemi ellitici in poligoni vicino agli angoli. Gli esempi sono il problema biarmonico (quindi l'elasticità) colle condizioni al contorno usuali in un poligono anche nel caso della fessura. Vengono anche considerate le equazioni di Stokes e di Stokes-Beltrami in domini particolari.
Questo lavoro é stato fortemente motivato dai risultati di **D.Joseph** ([7] and[8] fra l'altro). L'articolo presente corrisponde alla breve nota [5].

Organisation of this paper.

1. How to separate variables

Let us start from a very well known and simple minded example. Let u be a harmonic function in the sector D defined in polar coordinates as follows

$$D = \{(r\cos\theta,\ r\sin\theta);\quad 0 < r < \rho\ ,\ 0 < \theta < \omega\}$$

where ρ is strictly positive and ω belongs to $]0,2\pi]$. Indeed the case when $\omega = 2\pi$ corresponds to a slit domain. Assume that u vanishes on the arms of D namely

$$u(r,0) = u(r,\omega) = 0,\qquad 0 < r < \rho .$$

To be precise we also assume that

$$u \in H^1(D).$$

Then u may be expanded as follows

$$(1.1)\qquad u(r,\theta) = \sum_{k\geq 1} c_k\, r^{k\pi/\omega}\ \sin k\pi\theta/\omega$$

In other words u is a series of particular solutions of the form $\varphi(r)\psi(\theta)$ i.e. with separated variables (in polar coordinates).

We aim at deriving similar expansions for biharmonic functions and for solutions of the Lamé's equations.

Let us go back to the proof of (1.1). One rewrites the Laplace equation in polar coordinates as follows

$$(1.2)\qquad \left(r\frac{\partial}{\partial r}\right)^2 u + \left(\frac{\partial}{\partial\theta}\right)^2 u = 0.$$

Then the basic fact is that the operator in θ together with the Dirichlet boundary conditions generates a self-adjoint operator that may be diagonalized in $L^2(I)$ where $I =]0,\omega[$. To be more precise we set

$$A = -\partial^2/\partial\theta^2$$

$$H = L^2(I)$$

$$D_A = H^2(I) \cap H^1_0(I).$$

Then A is self-adjoint and has a compact resolvent. Its normalized eigenfunctions $\sqrt{\frac{2}{\omega}}\sin k\pi\frac{\theta}{\omega}$, $k \geqslant 1$, define a basis of H and this implies (1.1).

Now let us attempt to follow the same road for expanding biharmonic functions. Therefore we consider a biharmonic function

$$u \in H^2(D)$$

such that u and its normal derivative vanish on both arms of D namely

$$u(r,0) = u(r,\omega) = \frac{\partial u}{\partial\theta}(r,0) = \frac{\partial u}{\partial\theta}(r,\omega) = 0,\qquad 0 < r < \rho.$$

We rewrite the biharmonic equation as a second order system in polar coordinates as follows :

$$(1.3)\qquad \left(r\frac{\partial}{\partial r}\right)^2\psi - A\psi = 0\ ,\qquad 0 < r < \rho ,$$

where

$$(1.4) \qquad A = \begin{bmatrix} 0 & 1 \\ -\left(\frac{\partial^2}{\partial\theta^2}+1\right)^2 & -2\left(\frac{\partial^2}{\partial\theta^2}-1\right) \end{bmatrix}$$

and

$$\psi = [u/r \;;\; (r\frac{\partial}{\partial r})^2 (u/r)]^T .$$

Taking into account the boundary conditions this operator, is not formally self-adjoint since one easily checks the existence of non real eigenvalues. It is not even formally normal since it possesses non semi-simple eigenvalues for some values of ω. We shall conveniently work out the properties of A in the following framework

$$H = H_o^2(I) \times L^2(I)$$
$$D_A = \{H^4(I) \cap H_o^2(I)\} \times L^2(I).$$

The only obvious property of A is the compactness of its resolvent.

Nevertheless we shall achieve the expansion of ψ solution of (1.3) in generalized eigenfunctions of A for small values of r at least. For some technical reasons it is convenient to perform the following change of variable

$$(1.5) \qquad \frac{r}{\rho} = e^{-t},$$

which reduces equation (1.3) to the form

$$(1.6) \qquad D_t^2\,\psi - A\,\psi = 0\,, \quad 0 < t.$$

The above concrete examples make it sensible to assume that ψ is given for $r=\rho$ or equivalently for $t=0$. Therefore we add the boundary condition

$$(1.7) \qquad \psi(0) = x$$

with a given $x \in H$.

2. Existence and uniqueness for the abstract problem

It is easily derived with the help of the theory of analytic semi-groups. From now on we assume the following :

(H1) A is closed and is densely defined.

(H2) For every $\lambda \leqslant 0$, $(A-\lambda I)$ is invertible and there exists a constant M s.t.

$$\|(A-\lambda I)^{-1}\| \leqslant M/(|\lambda| + 1).$$

Under these assumptions one can define a square-root $A^{\frac{1}{2}}$ s.t. $-A^{\frac{1}{2}}$ generates an analytic semi-group $e^{-tA^{\frac{1}{2}}}$. In addition this semi-group is uniformly bounded. Therefore assuming $x \in D_{A^{1/2}}$ problem (1.6)(1.7) has the unique solution

$$(2.1) \qquad \psi(t) = e^{-tA^{\frac{1}{2}}}\,x$$

in a suitable class of solutions e.g. :

$$(2.2)\qquad \begin{cases} \psi \in L^2(]0,\infty[;D_A 1/2), \ \psi' \in L^2(]0,\infty[;H), \\ \psi' \in L^2(]\varepsilon,\infty[;D_A) \text{ for every } \varepsilon > 0, \\ \psi(t) \in D_A \quad \text{a.e.} \end{cases}$$

This provides us with an explicit representation of ψ in terms of x:

$$(2.3)\qquad \psi(t) = \frac{1}{2i\pi}\int_\gamma e^{-t\sqrt{\lambda}}(A - \lambda I)^{-1}\ x\ d\lambda$$

where γ is a suitable curve in the complex plane linking $\infty e^{-i(\pi-\delta)}$ to $\infty e^{i(\pi-\delta)}$ for some $\delta > 0$ within the resolvent set of A and having index +1 with respect to any point of the negative real axis $]-\infty,0]$ the cut for defining the function $\sqrt{\lambda}$. Indeed we observe that due to **(H2)** the resolvent set of A includes a neighborhood of the origin together with a sector around the negative real axis.

The actual derivation of a series expansion for $\psi(t)$ using (2.3) will involve very refined assumptions on the compactness of the resolvent of A. These assumptions are phrased in terms of Carleman classes and Fredholm determinants.

3. Carleman classes and Fredholm determinant

Consider a pair of Hilbert spaces $\mathcal{H}_1$ and $\mathcal{H}_2$. Let T be a compact linear operator from $\mathcal{H}_1$ to $\mathcal{H}_2$. Clearly $\sqrt{T^*T}$ is a compact symmetric operator in $\mathcal{H}_1$. Denote by μ_k, k=1,2,... the sequence of its eigenvalues in decreasing order and repeated according multiplicities. T is said to belong to $\mathcal{L}_p(\mathcal{H}_1,\mathcal{H}_2)$ if

$$\sum_{k\geq 1} \mu_k^p < +\infty .$$

Here p need not be larger than or equal to one. From now on we assume

$$\textbf{(H3)}\qquad A^{-1} \in \mathcal{L}_p(\mathcal{H},\mathcal{H}) \ \underline{\text{for every}}\ \ p > \tfrac{1}{2}.$$

This assumption is easily checked on the examples in section 1. Indeed one knows from **El Kolli** [4] and **Triebel** [13] that

$$I \in \mathcal{L}_p(H^{k+2}(I)\ ,\ H^K(I))$$

for every $p > \frac{1}{2}$. Unfortunately a similar inclusion <u>does not hold</u> for $p=\frac{1}{2}$.

Now A^{-1} being compact A has only point spectrum. Let us denote by λ_k, k=1,2,... the eigenvalues of A. Clearly $|\lambda_k| \to \infty$ as $k \to \infty$. Here one must make clear the definition of <u>multiplicity</u> ν_k for the eigenvalue λ_k. For every k we set

(3.1) $$\mathcal{H}_k = \bigcup_{m\geq 1} \ker(A-\lambda_k I)^m$$

This is a finite dimensional subspace of $\mathcal{H}$ whose dimension we denote by ν_k. Vectors in $\mathcal{H}_k$ are called generalized eigenvectors of A. Accordingly y belongs to $\mathcal{H}_k$ iff there exists $m \geq 1$ s.t.

(3.2) $$(A - \lambda_k I)^m y = 0.$$

Such an operator A has a Fredholm determinant

(3.3) $$d(\lambda) = \prod_{k\geq 1} (1-\lambda/\lambda_k)^{\nu_k}$$

This is an entire function of exponential type p for every $p > \frac{1}{2}$ which vanishes exactly to the order ν_k at λ_k for every k. This Fredholm determinant is the key tool for the expansion of $\psi(t)$.

4. The central result

Theorem : **Let us assume that** (H1)(H2)(H3) **are fulfilled and that there exists a constant** K **s.t.**

(4.1) $$d(\lambda) = O(e^{K\sqrt{|\lambda|}}).$$

Then there exists an increasing sequence $r_j \nearrow +\infty$ **and** $t \geqslant 0$ **such that for every** $x \in D_{A^{1/4}}$ **there exists a sequence of vectors**

$$x_k \in \mathcal{H}_k$$

s.t.

(4.2) $$\psi(t) = \sum_{j=1}^{\infty} \left(\sum_{r_j < |\lambda_k| < r_{j+1}} e^{-tA^{\frac{1}{2}}} x_k \right)$$

At first sight this does not look very much like separation of variables. In order to make things more obvious let us look at one of the terms in such a series. Denote by e_j, $1 \leqslant j \leqslant \nu_k$ a basis of $\mathcal{H}_k$ that reduces the restriction of A to $\mathcal{H}_k$ to normal form. Assuming

(4.3) $$x_k = \sum_{j=1}^{\nu_k} \alpha_j \ e_j$$

there exist polynomials p_k of degree $\nu_k - 1$ s.t.

(4.4) $$e^{-tA^{\frac{1}{2}}} x_k = e^{-t\sqrt{\lambda_k}} \sum_{j=1}^{\nu_k} p_j(t) \ e_j.$$

In practise as in the examples of section 1, e_j is a function of $\theta \in I$ and performing the inverse change of variable of **(1.5)** one sees that ψ is a series of terms like

$$r^{\sqrt{\lambda_k}} p_j(-\mathrm{Log}\ \frac{r}{\rho}) \ e_j(\theta).$$

This shows separation of variables in polar coordinates.

The conditional convergence i.e. the necessity of a particular grouping in the series **(4.2)** in order to achieve convergence is due to the technique applied (see next section). However a similar phenomena has been shown by **Dauge** [2] when dealing with high order elliptic problems in polygons thus using the fact that A is a differential operator.

For being able to apply the abstract result in practice it is useful to provide a formula for x_k. The spectral projector on $\mathcal{H}_k$ may be defined as

(4.5) $$P_k = \frac{1}{2i\pi} \int_{C_k} (\lambda I-A)^{-1} d\lambda$$

where C_k is a circle with center at λ_k positively oriented and such that no other eigenvalue lies inside or on the circle. With this notation one has

(4.6) $$x_k = P_k\ x$$

Let us focus on the particular case when λ_k is semi-simple i.e. when

$$\mathcal{H}_k = \ker(A-\lambda_k I)$$

This occurs very often in practice. Then A reduces to multiplication by λ_k on $\mathcal{H}_k$ and therefore

$$e^{-tA^{\frac{1}{2}}} x_k = e^{-t\sqrt{\lambda_k}}\, x_k.$$

Finally when λ_k is simple i.e. $\nu_k=1$ it is easy to make P_k explicit. Indeed one has

$$P_k x = \frac{(x\ ,\psi_k)}{(\Phi_k\ ,\psi_k)}\ \Phi_k$$

where $A\,\Phi_k = \lambda_k \Phi_k$, $A^* \psi_k = \overline{\lambda}_k\ \psi_k$. Accordingly when all eigenvalues are simple the expansion **(4.2)** turns out to be <u>biorthogonal</u> expansion

(4.7) $$\psi(t) = \sum_{j=1}^{\infty} \left(\sum_{r_j<|\lambda_k|<r_{j+1}} e^{-t\sqrt{\lambda_k}} \frac{(x\ ;\psi_k)}{(\Phi_k\ ,\ \psi_k)}\ \Phi_k \right)$$

5. <u>Outline of a proof</u>

We approximate ψ by a sequence ψ_j defined by a Dunford integral similar to **(2.3)** where integration is performed on a bounded curve γ_j that we shall define below.

First of all we need to make precise the sequence r_j involved in the above statement. We apply the minimum modulus theorem to the entire function d which fulfills **(4.1)**. Therefore for a given $\varepsilon > 0$ there exists a sequence $r_j \nearrow +\infty$ such that

$$|d(\lambda)| \geqslant e^{-(K+\varepsilon)\sqrt{|\lambda|}} \qquad (5.1)$$

for $|\lambda| = r_j$, 1,2,... . This allows us to define γ_j as coinciding with γ for $|\lambda| < r_j$. The rest of γ_j is an arc of circle with center at the origin, radius r_j and lying in the connected component S of $\mathbb{C}$ whose boundary is γ and which contains the spectrum of A. This is certainly clearer on the figure.

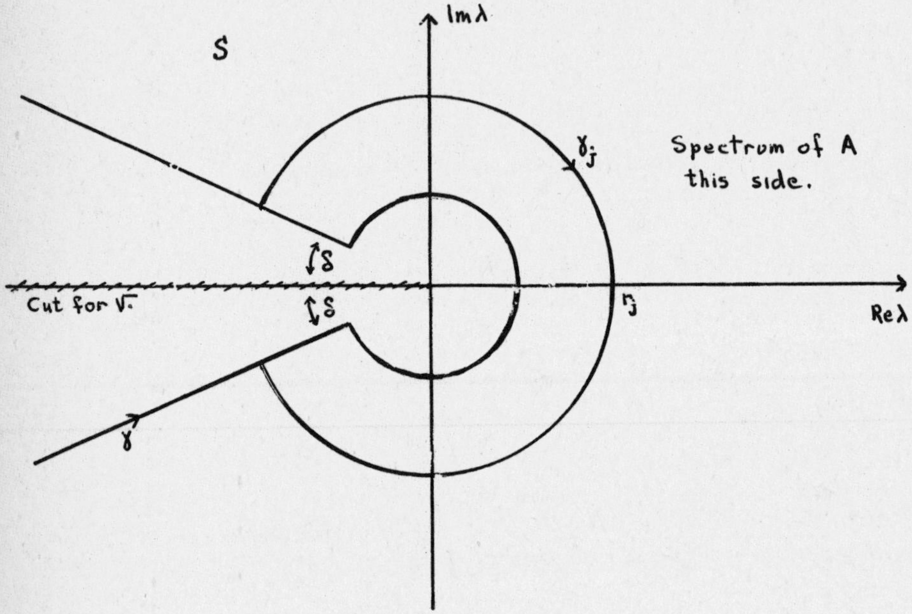

Due to inequality **(5.1)** no eigenvalue λ_k has radius r_j and therefore $(A-\lambda I)^{-1}$ is analytic on γ_j. We define ψ_j as

$$\psi_j(t) = \frac{1}{2i\pi} \int_{\gamma_j} e^{-t\sqrt{\lambda}} \, (A-\lambda I)^{-1} \, x \, d\lambda . \qquad (5.2)$$

Using Cauchy theorem and **(4.5)** one easily checks that

$$\psi_j(t) = \sum_{\ell=1}^{j} \sum_{r_\ell < |\lambda_k| < r_{\ell+1}} e^{-tA^{\frac{1}{2}}} P_k \, x \qquad (5.3)$$

and by **(4.6)** this shows that ψ_j is a partial sum of the series **(4.2)**.

We are left with showing that $\psi_j(t)$ converges to $\psi(t)$ as $j \longrightarrow \infty$ and this amounts to derive suitable estimates of

$$e^{-t\sqrt{\lambda}} \, (A-\lambda I)^{-1} \, x$$

on $\gamma_j \setminus \gamma$.

Due to assumption **(H3)** the Carleman-Hadamard inequality holds. In other words for every $p > \frac{1}{2}$ there exists a constant Γ_p s.t.

$$\|d(\lambda)(I-\lambda A^{-1})^{-1}\| \leqslant e^{\Gamma_p |\lambda|^p} \tag{5.4}$$

This is derived in **Dunford-Schwartz** [3] and **Gohberg-Krein** [6]. This estimate holds in particular in S. However due to **(H1)** one has

$$\|(I-\lambda A^{-1})^{-1}\| = O(1)$$

on ∂S. Together with **(4.1)** this shows that

$$\|d(\lambda)(I-\lambda A^{-1})^{-1}\| = O(e^{K\sqrt{|\lambda|}}) \tag{5.5}$$

on ∂S. Now the function

$$d(\lambda)(I-\lambda A^{-1})^{-1}$$

is entire and consequently we can apply the Phragmen-Lindelöf principle since the angle of the sector corresponding to S is strictly less than 2π. This shows that the estimate **(5.5)** holds in the whole of S and in particular on $\gamma_j \setminus \gamma$.

From the lousy identity

$$(I-\lambda A^{-1})^{-1} = A(A-\lambda I)^{-1} = I + \lambda (A-\lambda I)^{-1}$$

one derives at once the following estimate on $\gamma_j \setminus \gamma$:

$$\|(A-\lambda I)^{-1}\| \leqslant \frac{1}{r_j}\left(1 + \frac{O(e^{K\sqrt{r_j}})}{|d(r_j)|}\right) .$$

Then we finally obtain the estimate

$$\|e^{-t\sqrt{\lambda}}(A-\lambda I)^{-1} x\| \leqslant \frac{1}{r_j} e^{-t \operatorname{Re}\sqrt{\lambda}}\left[1 + O(e^{[2K+\varepsilon]\sqrt{r_j}})\right] \|x\|$$

that shows the convergence of $\psi_j(t)$ to $\psi(t)$ for large enough t independently of x.

6. How to calculate a Fredholm determinant in practise

In order to make technical facts a little bit easier we gonna consider here a rather simple minded example related to the biharmonic. Namely we consider the bending of semi-infinite strip shaped plate. Let us consider a biharmonic function belonging to $H^2(\Omega)$ where Ω denotes

$$\Omega = \{(t,y) \ ; \ |y| < 1 \ , \ t > 0\}$$

with the boundary conditions

$$u(t,\pm 1) = \frac{\partial u}{\partial y}(t,\pm 1) = 0, \quad t > 0.$$

With the help of our theorem we shall show separation of variables (in cartesian coordinates) for large values of t.

First of all we reduce our problem to the form **(1.6)** by introducing the following

(6.1)
$$A = \begin{bmatrix} 0 & 1 \\ -\frac{\partial^4}{\partial y^4} & -2\frac{\partial^2}{\partial y^2} \end{bmatrix}$$

and

$$\psi = \left[u, \frac{\partial^2 u}{\partial t^2}\right]^T .$$

We realize A as unbounded operator in the same space $\mathscr{H}$ and with the same domain D_A as in section 1 when dealing with the biharmonic in a sector. As expectable the operator A here and the one in section 1 have same principal part. This is why most of the calculations below are also relevant for **(1.4)**.

The assumptions **(H1)(H2)** and **(H3)** are easily checked while the inequality **(4.1)** deserves a proof. Recalling **(3.3)** we must calculate the eigenvalues λ_k and their multiplicities ν_k. Accordingly let us consider the equation

(6.2)
$$A\,v - \lambda\,v = f$$

that governs the resolvent. Eliminating the second component v_2 of v we are looking for

$$v_1 \in H^4(I) \cap H_0^2(I)$$

a solution of the equation

(6.3)
$$-v_1^{(IV)} - 2v_1'' - \lambda^2 v_1 = f_2 + 2f_1'' + \lambda f_1 .$$

Elementary explicit calculations show that the Green operator is the quotient of an integral operator depending analytically on λ by the entire function

(6.4)
$$W(\lambda) = \frac{1}{\lambda}\left\{1 - \left[\frac{\sin 2\sqrt{\lambda}}{2\sqrt{\lambda}}\right]^2\right\} .$$

Thus the eigenvalues λ_k are the roots of W. In addition all those roots are simple. This implies that the poles of the resolvent operator

$$(A - \lambda I)^{-1}$$

are simple. This implies that all the eigenvalues are semi-simple (cf. **Taylor** [12]). In other words $\mathscr{H}_k$ is just the kernel of $A - \lambda_k I$. Direct explicit calculations show that this kernel is one-dimensional. Summing up we have checked that d and W have the same roots λ_k, k=1,2,... with the same multiplicities.

On the other hand it follows from **(H3)** that d is an entire function of exponential type p for every $p > \frac{1}{2}$. The same is obviously true for W. Applying the uniqueness theorem in **Boas** [1] we conclude that d and W are proportional. Therefore any growth condition on W is fulfilled by d and this is why we have

$$d(\lambda) = \left(0 \; e^{4\sqrt{|\lambda|}}\right)$$

This is inequality **(4.1)** with K=4.

Straightforward application of the above theorem provides a (conditionnaly convergent) series representation of u, for large enough t, whose terms are

$$\alpha_k \, e^{-t\sqrt{\lambda_k}} \left\{ \left(\frac{\sin 2\sqrt{\lambda_k}}{\sqrt{\lambda_k}} - 2\cos 2\sqrt{\lambda_k} \right) (y+1) \sin\left[\sqrt{\lambda_k}\,(y+1)\right] \right.$$
$$\left. -2\sin 2\sqrt{\lambda_k} \left(\frac{\sin \sqrt{\lambda_k}\,(y+1)}{\sqrt{\lambda_k}} - (y+1)\cos\left[\sqrt{\lambda_k}\,(y+1)\right] \right) \right\}$$

where λ_k is solution of

$$\begin{cases} \sin 2\sqrt{\lambda_k} = \pm 2\sqrt{\lambda_k} \\ \mathrm{Re}\,\sqrt{\lambda_k} > 0. \end{cases}$$

Furthermore the identity **(4.7)** also provides an explicit value for α_k depending on u(0,y) and $\frac{\partial^2 u}{\partial t^2}(0,y)$. Actually one can check on these explicit formulas the unconditional convergence of the series expansion for every $t \geqslant 0$. This is achieved in **Joseph** [7][8] .

To better situate this with respect to Joseph's works let us emphasize that our theorem is a mere procedure to produce explicit representation formulas. Of course convergence is easily checked directly although not built in the theorem. The applicability of our theorem depends on our ability to identify the Fredholm determinant d with the Wronskian W that appears as denominator in the Green operator.

7. Back to the biharmonic in a sector

Let us now apply our theorem to the operator **(1.4)**. From the assumption that u belongs to $H^2(D)$ and from the boundary conditions one deduces that

(7.1) $$r^{-2+|\alpha|} D^\alpha u \in L^2(D), \quad |\alpha| \leqslant 2.$$

This is just applying the Hardy inequality.
Next using **Kondratiev** [9] we also derive that

(7.2) $$r^{-2+|\alpha|} D^\alpha u \in L^2(D), \quad |\alpha| \leqslant 4.$$

Taking into account the definition of ψ one easily checks that such ψ belongs to the class of solutions defined by **(2.2)**.

Again the assumptions **(H1)(H2)(H3)** are easily checked. This is true especially because of the similarity with the operator in section 6.

In order to check the inequality **(4.1)** we follow the same steps as in section 6. The equation

(7.3) $$A\,v - \lambda v = f$$

implies that we look for

$$v_1 \in H^4(I) \cap H_0^2(I)$$

a solution of

(7.4) $$-v_1^{(IV)} - 2(\lambda+1)v_1'' + (\lambda-1)^2 v_1 = f_2 + 2f_1'' + (\lambda-2)f_1$$

Here the Green operator is an integral operator whose kernel is the quotient of an entire function of λ by the entire function

(7.5) $$W(\lambda) = \frac{\sin^2\sqrt{\lambda}\,\omega - \lambda\sin^2\omega}{\lambda-1}$$

The roots of W are well known. In particular they have been thoroughly investigated by **Lozi** [10] **Seif** [11] . It turns out that only a finite number of roots are double and this is the maximal multiplicity that occurs for such roots. In addition there is only a sequence of exceptional values ω_ℓ , $\ell = 1,2,\ldots$ for which there exist double roots. For the sake of brevity let us assume that we stay away from these values. Unfortunately this rules out the case of a crack when $\omega = 2\pi$. This very important special case will be taken care of in a separate section (n°8).

For simple roots everything is easy. The resolvent has only simple poles and from **Taylor** [12] we conclude that

$$\mathcal{H}_k = \ker(A - \lambda_k I).$$

Explicit calculations show that this kernel has always dimension one. Thus $\nu_k = 1$ and the functions d and W have the same roots which are all simple. They are both entire functions of order p , for every $p > \frac{1}{2}$ and the uniqueness result in **Boas** [1] implies that they are multiple of one another. Accordingly we have the estimate

(7.6) $$d(\lambda) = O\left(e^{2\omega\sqrt{|\lambda|}}\right)$$

which allows us to apply our abstract theorem.
This leads to a (conditionnally convergent) series expansion of u for small enough $r \geqslant 0$ whose terms are of the following form

$$\alpha_k\, r^{1+\sqrt{\lambda_k}} \Big\{ (\sqrt{\lambda_k}\,\sin\omega\cos\sqrt{\lambda_k}\,\omega - \sin\sqrt{\lambda_k}\,\omega\cos\omega)\,\sin\theta\,\sin\sqrt{\lambda_k}\,\theta$$
$$-\sin\omega\sin\sqrt{\lambda_k}\,\omega\,(\sqrt{\lambda_k}\,\sin\theta\cos\sqrt{\lambda_k}\,\theta - \sin\sqrt{\lambda_k}\,\theta\cos\theta)\Big\}$$

where λ_k is solution of

$$\begin{cases} \sin\omega\sqrt{\lambda_k} = \pm(\sin\omega)\sqrt{\lambda_k} \\ \mathrm{Re}\sqrt{\lambda_k} > 0 \ ; \ \lambda_k \neq 1. \end{cases}$$

Again here identity **(4.7)** provides explicit values for α_k and unconditional convergence may be checked directly for every $r \leqslant \rho$ as in Joseph's works.

8. Biharmonic and craks

Let us now focus on the particular case $\omega = 2\pi$. Here a lot of simplifications occur. Starting again from **(7.4)** one shows that the denominator of the Green function is now

(8.1)
$$W(\lambda) = \sin 2\pi\sqrt{\lambda}\ .$$

Thus the resolvent has only simple poles and

$$\mathcal{H}_k = \ker(A - \lambda_k I).$$

Direct calculation shows that

$$\begin{cases} \nu_k = \dim \mathcal{H}_k = 1 & \text{if } \lambda_k = 1 \\ \nu_k = \dim \mathcal{H}_k = 2 & \text{if } \lambda_k = (\frac{\ell}{2})^2,\ \ell \text{ integer} \neq 2. \end{cases}$$

Therefore applying Boas theorem shows that d and W^2 have same growth. Accordingly one has

(8.2)
$$d(\lambda) = O(e^{4\pi\sqrt{|\lambda|}}).$$

The abstract theorem leads to the (conditionally convergent) series expansion of u for small enough r, whose terms are

$$\begin{cases} \alpha_\ell \ r^{\frac{\ell}{2}+1} \sin\theta \sin\frac{\ell}{2}\theta \ , & \ell \text{ integer } \geqslant 1 \\ \beta_\ell \ r^{\frac{\ell}{2}+1} \left\{\frac{\ell}{2}\sin\theta\cos\frac{\ell}{2}\theta - \sin\frac{\ell}{2}\theta\cos\theta\right\}, & \ell \text{ integer} \geqslant 1 \text{ but } \neq 2. \end{cases}$$

9. Other examples

In short let us mention other examples where our theorem may be applied. They are all taken from Joseph's work.

Proceeding as in section 6 one easily separates variables for an axisymmetric Stokes flow between coaxial cylinders.

Introducing weighted Sobolev spaces for $\mathcal{H}$ and D_A makes the theorem applicable to separation of variables for an axisymmetric Stokes flow in one cylinder. Here the inequality **(4.1)** is obtained by comparing the Fredholm determinant to a Wronskian W whose values are

$$W(\lambda)= b\sqrt{\lambda}\,[J_0^2(b\sqrt{\lambda})+ J_1(b\sqrt{\lambda})] - 2J_0(b\sqrt{\lambda})\, J_1(b\sqrt{\lambda})$$

where b is the radius of the cylinder and J_0 & J_1 are the usual Bessel functions.

Also Stokes flows in cones with axisymmetry lead to a Stokes-Beltrami equation for which separation of variable turns to be possible. Here two weighted Sobolev spaces are used.

References.

[1] **Boas,R.P.**, Entire functions, Academic Press, New York, 1954

[2] **Dauge,M.**, Second membre analytique pour un problème elliptique d'ordre 2m sur un polygone, Séminaire d'Analyse, Univ. de Nantes, 1981/2

[3] **Dunford,N.**, **J.Schwartz**, Linear operators, PartII, Interscience Publishers, New York, 1963

[4] **El Kolli,A.**, n^{ième} épaisseur dans les espaces de Sobolev, J. for Approximation theory, 1974

[5] **Geymonat,G.**, **P.Grisvard**, Diagonalisation d'opérateurs non autoadjoints et séparation des variables, CRAS, Paris, 1983, t.296, Série I, p.809-812

[6] **Gohberg,I.C.**, **M.G.Krein**, Introduction to the theory of linear non self-adjoint operators, A.M.S., Providence, 1969

[7] **Joseph,D.D.**, A new separation of variables theory for problems of Stokes flow and elasticity, Proc. of "Trends in applications of pure mathematics to mechanics", Pitman, London, 1979

[8] **Joseph,D.D.**, The convergence of biorthogonal series for biharmonic and Stokes flow edge problems, Part I, SIAM J. of Applied Maths, vol.33, n°2 1977, p.337-347

[9] **Kondratiev,V.A.**, Boundary value problems for elliptic equations in domains with conical or angular points, Transactions of the Moscow Math. Soc. 1967, p.227-313

[10] **Lozi,R.**, Résultats numériques de régularité du problème de Stokes et du Laplacien itéré dans un polygone, RAIRO, Analyse Numérique, 12, n°3, 1978, p.267-282

[11] **Seif,J.B.**, On the Green function for the biharmonic equation in an infinite wedge, Transactions A.M.S., 182, 1973, p.241-260

[12] **Taylor,A.E.**, Introduction to Functional analysis, J.Wiley

[13] **Triebel,H.**, Interpolation theory, function spaces, differential operators, North Holland, 1978.

CASTIGLIANO AND SOBOLEV

F. Hartmann
Structural Engineering Department
University of Dortmund

If the complementary strain energy of a mechanical system is expressed in terms of the n displacements, δ_i, corresponding to a system of n prescribed forces, P_i, the first partial derivative of the complementary strain energy with respect to any of these forces, P_i, is equal to the displacement, δ_i, at point i in the direction of P_i.

This Theorem is known as Castigliano's First Theorem. Substituting the displacement term for the force term and replacing the complementary strain energy by the strain energy (which is the same in linear mechanics); i.e. calculating the derivative of the strain energy with respect to δ_i) one obtains Castigliano's Second Theorem.

In mechanics it is assumed that these two theorems are theorems of universal standing. But this is not true as we want to show in the following. We must distinguish between problems with finite energy and problems with infinite energy.

To start let us repeat the derivation of Castigliano's Theorem for a simple model problem. (We put no emphasis on the distinction between Castigliano's 1st and 2nd Theorem because these are, essentially, equivalent statements. Hence, we speak simply of Castigliano's Theorem).

The displacement u(y) of the truss element in Fig.1 is the solution of the bvp

$$- EA\, u''(y) = \delta_o(y-x)\, P \qquad u(0) = u(l) = 0$$

and as such the sum

$$u(y) = \{g_o(y,x) + u_R(y)\}\, P$$

of a fundamental solution, $g_o(y,x)$ and a regular homogeneous solution $u_R(y)$.

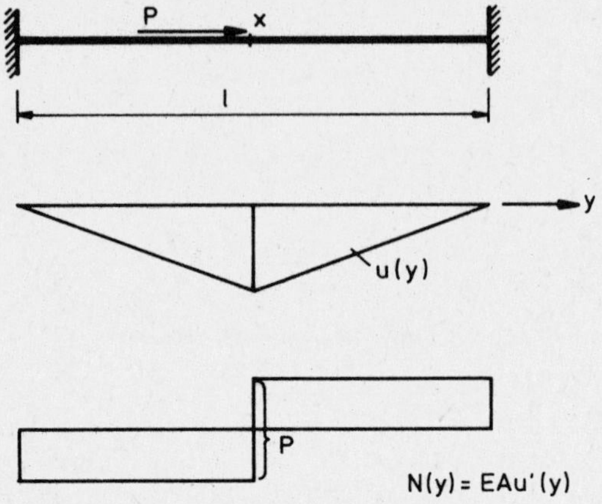

Fig. 1

The key to Castigliano's Theorem is the first identity of the governing operator in question. In the case of the operator $-EA\, d^2/dx^2$ this identity reads

$$\frac{1}{2}\, G(u,u) = \frac{1}{2}\, \{\int_0^l -EA\, u''\, u\, dy + [Nu]_o^l - \int_0^l \frac{N^2}{EA}\, dx\} = 0 \quad \forall\, u \in C^2[0,l]$$

It expresses the fact that the exterior eigenwork, the sum of the first two terms, is equal to the internal energy, the term preceded by a minus sign.

As the displacement u of the truss element does not belong to $C^2[0,l]$ (the first derivative jumps at x) we first formulate this identity, this energy balance, in the domain

$$\Omega_\varepsilon = [0, x - \varepsilon] \cup [x + \varepsilon, l]$$

the interval $[0,l]$ minus a small neighborhood $(x-\varepsilon, x+\varepsilon)$ of the source point x,

$$\frac{1}{2}\, G_\varepsilon(u,u) = \frac{1}{2}\, \{\int_{\Omega_\varepsilon} -EA\, u''\, u\, dy + [Nu]_o^{x-\varepsilon} - [Nu]_{x+\varepsilon}^{l} + \int_{\Omega_\varepsilon} \frac{N^2}{EA}\, dy\}$$

$$= \frac{1}{2} \{ N(x-\varepsilon)u(x-\varepsilon) - N(x+\varepsilon)u(x+\varepsilon) - \int_{\Omega_\varepsilon} \frac{N^2}{EA} dy\} = 0$$

and we, then, let ε shrink to zero. This renders

$$\lim_{\varepsilon \to 0} \frac{1}{2} G_\varepsilon(u,u) = \frac{1}{2} P\, u(x) - \frac{1}{2} \int_0^1 \frac{N^2}{EA} dy = 0$$

and, therewith, because of

$$u(x) = \{g_0(x,x) + u_R(x)\} P$$

also Castigliano's First Theorem.

$$u(x) = \frac{\partial}{\partial P} \frac{1}{2} \int_0^1 \frac{N^2}{EA} dy$$

The operators which govern the displacement of the single structural elements in classical mechanics are linear elliptic operators of degree 2 or 4, resp. (By degree we mean the maximum order of the derivatives in an equation).

Operators of second degree are

Truss element $\quad - EA\, u'' = p$

Membrane $\quad - N\, \Delta w = p$

and the following systems (summation over repeated indices is implied)

Elastic body, elastic plate

$$- \mu\, \Delta u_i - \frac{\mu}{1-2\nu}\, u_{j,ji} = p_i \qquad i,j = 1,2,3 \text{ (3-D) or } i,j = 1,2 \text{ (2-D)}$$

Reissner plate

$$- K(\frac{1-\nu}{2})[\{\varphi_{\alpha,\beta} + \varphi_{\beta,\alpha} + \frac{2\nu}{1-\nu}\varphi_{\gamma,\gamma}\,\delta_{\alpha\beta}\}_{,\beta} - \bar{\lambda}^2(\varphi_\alpha + w_{,\alpha})] = p_\alpha$$

$$\alpha, \beta = 1,2$$

$$- K(\frac{1-\nu}{2})^2\, \bar{\lambda}^2(\varphi_\alpha + w_{,\alpha})_{,\alpha} = p_3$$

Equations of fourth degree are

Beam $\quad EI\, w^{IV} = p$

Kirchhoff plate $K\,\Delta\Delta w = p$

Let us, hence, replace the simple bvp of the truss element by the more general equations

$$D\,u = \delta_i(y-x)\,F_i \qquad i \text{ fixed} \tag{1}$$

$$\partial^j u\,\big|_\Gamma = 0 \qquad 0 \leqslant j \leqslant m-1 \tag{2}$$

where D denotes a linear elliptic operator of degree 2m with constant coefficients and the small ∂^j denote operators of degree j.

The lower terms are 'displacements'

$$\partial^j u \qquad 0 \leqslant j \leqslant m-1$$

and the upper terms 'forces'

$$\partial^j u \qquad m \leqslant j \leqslant 2m-1$$

In case the operator has degree 2 ($m = 1$) the Dirac functions of degree $i = 0,1,\ldots 2m-1$ represent

δ_o = concentrated force, $\qquad \delta_1$ = jump in displacement

and in case the operator has degree 4 ($m = 2$)

δ_o = concentrated force $\qquad \delta_1$ = couple

δ_2 = jump in rotation $\qquad \delta_3$ = jump in displacement

F_i is the intensity of the singularity.

We call a function $u(y)$ a solution of Equ.(1) if it is a homogeneous solution at all points $y \neq x$,

$$D\,u = 0 \qquad y \neq x\,, \qquad y \in \Omega$$

and satisfies the 2m equations

$$\lim_{\varepsilon\to 0}\int_{\Gamma_{N_\varepsilon}(x)} \partial^{2m-1-j}u\, ds = \begin{cases} F_i & \text{if } i = j \\ 0 & \text{if } i \neq j \end{cases} \qquad 0 \leqslant j \leqslant 2m-1 \qquad (3)$$

where $\Gamma_{N_\varepsilon}(x)$ is the boundary of an ε-neighborhood of the source point, see Fig.2.

In the case of a Kirchhoff plate loaded with a concentrated force as in Fig.2 the conditions in (3) would , e.g., read

$$\lim_{\varepsilon\to 0}\int_{\Gamma_{N_\varepsilon}(\underline{x})} \partial^3 w\, ds = P, \qquad \lim_{\varepsilon\to 0}\int_{\Gamma_{N_\varepsilon}(\underline{x})} \partial^j w\, ds = 0 \qquad j = 0,1,2$$

where $\partial^3 w = V_n$ is the Kirchhoff shear and $\partial^0 w = w$, $\partial^1 w = \partial w/\partial n$ and $\partial^2 w = M_n$ are the deflection, the normal derivative and the bending moment, resp.

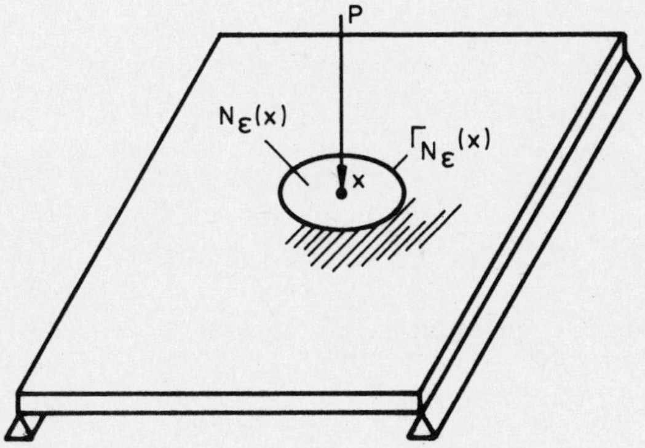

Fig. 2

The first identity for an operator D as introduced above has the form

$$\frac{1}{2}\, G(u,u) = \frac{1}{2}\, \{ \int_\Omega Du \cdot u\, d\Omega - \sum_{i=1}^{m} (-1)^i \int_\Gamma \partial^{2m-i}u \cdot \partial^{i-1}u\, ds - E(u,u) \} = 0$$

$$\forall\, u \in H^{2m}(\Omega)$$

We may assume that the internal energy, 1/2 E(u,u), of the operator D in question is a continuous

$$|E(u,\hat{u})| < c_1 \|u\|_m \cdot \|\hat{u}\|_m \qquad \forall\, u,\hat{u} \in H_0^m(\Omega)$$

as well as coercive

$$\frac{1}{2}\, E(u,u) > c_2\ ||u||^2_m \qquad \forall\ u\ \in\ \overset{\circ}{H}{}^m(\Omega)$$

bilinear form on $\overset{\circ}{H}{}^m(\Omega)$.

In the case of elastic bodies, $\underline{u} = \{u_1, u_2, u_3\}$ and Reisssner plates, $\underline{u} = \{\varphi_1, \varphi_2, w\}$ the pertinent energy space for which these inequalities hold true is the triple product

$$(\overset{\circ}{H}{}^1(\Omega))^3 = \overset{\circ}{H}{}^1(\Omega)\ x\ \overset{\circ}{H}{}^1(\Omega)\ x\ \overset{\circ}{H}{}^1(\Omega)$$

This is demonstrated for elastic bodies in [1]. In the case of Reissner plates the proof of the coerciveness

$$\frac{1}{2}\, E(u,u) = \frac{1}{2}\, K\ (1-\nu)\int_\Omega [\ \varphi^2_{1',1} + \frac{1}{2}(\varphi_{1',2} + \varphi_{2',1})^2 + \varphi^2_{2',2})$$
$$- \frac{\nu}{1-\nu}(\varphi_{1',1} + \varphi_{2',2})^2 + \frac{\bar{\lambda}^2}{2}\{(\varphi_1 + w,_1)^2 + (\varphi_2 + w,_2)^2\}]d\,\Omega$$
$$> c_3\ (||\varphi_1||^2_1 + ||\varphi_2||^2_1 + ||w||^2_1)$$

is done with the inequalities of Poincaré and Young.

We may assume that the solution of the bvp (1),(2) has the form

$$u(y) = \{g_i(y,x) + u_{R_i}(y)\}\ F_i$$

where $g_i(y,x)$ is a fundamental solution corresponding to the Dirac function δ_i and its companion u_{R_i} a regular, homogeneous solution.

The formulation of the energy balance for such a function proceeds, because $g_i(y,x)$ does not belong to $H^{2m}(\Omega)$, in two steps.

We first formulate the identity in a domain $\Omega_\varepsilon = \Omega - N_\varepsilon(x)$ where u is regular

$$\frac{1}{2}\, G_\varepsilon(u,u) = \frac{1}{2}\,\{ \int_{\Omega_\varepsilon} Du\cdot u\ d\Omega - \sum_{i=1}^{m}\ (-1)^i \int_\Gamma \partial^{2m-i}u\cdot\partial^{i-1}u\ ds$$
$$- \sum_{i\,=\,1}^{m} (-1)^i \int_{\Gamma_{N_\varepsilon}(x)} \partial^{2m-i}u\cdot\partial^{i-1}u\ ds\ -\ E(u,u)_{\Omega_\varepsilon}\,\}$$

$$= \frac{1}{2} \{ \sum_{i=1}^{m} (-1)^i \int_{\Gamma_{N\varepsilon}(x)} \partial^{2m-i} u \cdot \partial^{i-1} u \, ds - E(u,u)_{\Omega_\varepsilon} \} = 0$$

(note that $D\,u = 0$ in Ω_ε and $\partial^i u|_\Gamma = 0, \ 0 \leqslant i \leqslant m-1,$)

and we then let the radius ε of the hole $N_\varepsilon(x)$

$$N_\varepsilon(x) = \{y \in \Omega | \ |y-x| \leqslant \varepsilon\}$$

shrink to zero. If this second step yields in the limit an expression as

$$\lim_{\varepsilon \to o} \frac{1}{2} G_\varepsilon(u,u) = \frac{1}{2} \ \partial^i u(x) \, F_i - \frac{1}{2} E(u,u) = 0$$

then, because of

$$u(x) = \{g_i(x,x) + u_{R_i}(x)\} \, F_i$$

Castigliano's First Theorem is obtained.

$$\partial^i u(x) = \frac{\partial}{\partial F_i} \frac{1}{2} E(u,u)$$

Obviously, for Castigliano's Theorem to hold the conjugated quantity $\partial^i u(x)$ must be bounded and the internal energy $1/2\ E(u,u)$ of the solution must be finite. Both conditions are equivalent. (A quantity $\partial^i u(x)$ is said to be conjugated to $\delta_j(y-x)$ if $i = j$)

As the functions with finite energy are the functions in $H_o^m(\Omega)$

$$u \in H_o^m(\Omega) \ \rightarrow \ \lim_{\varepsilon \to o} \frac{1}{2} E(u,u)_{\Omega_\varepsilon} = \frac{1}{2} E(u,u) < \infty$$

and also, as we shall see, the functions with pointwise bounded derivatives $\partial^i u$, the cardinal question is: does the bvp (1), (2) have a solution in $H_o^m(\Omega)$?

To this end let us consider the weak formulation of our bvp, that is the problem:

Find $u \in H_o^m(\Omega)$ such that

$$E(u,\hat{u}) = \int_\Omega F_i \delta_i(y-x) \, \hat{u}(y) \, d\Omega_y = \partial^i \hat{u}(x) \, F_i \qquad \forall \ \hat{u} \in H_o^m(\Omega)$$

The weak problem, according to the Lax-Milgram Theorem, has a solution in $H^m_o(\Omega)$ if the Dirac function $\delta_i(y-x)$ belongs to $H^{-m}(\Omega)$, the dual space of $H^m_o(\Omega)$. This in turn depends on

Sobolev's Embedding Theorem, [1]

If $\Omega \subset R^n$ is a bounded domain which satisfies the cone hypothesis and if the index m of the Sobolev space $H^m(\Omega)$ exceeds n/2

$$m > n/2$$

then $H^m(\Omega) \subset C(\bar{\Omega})$ and, furthermore, there exists a constant c_4, depending only on the domain Ω and the index m of the Sobolev space such that

$$\max_{x \in \bar{\Omega}} |u(x)| < c_4 ||u||_m \qquad \forall\ u \in H^m(\Omega)$$

□

Now, let u an arbitrary function in $H^m_o(\Omega)$ then $\partial^i u$ belongs to $H^{m-i}_o(\Omega)$ and if $m - i > n/2$ then

$$\max_{x \in \Omega} |\partial^i u(x)| < c_4 ||\partial^i u||_{m-i}$$

But as

$$||\partial^i u||_{m-i} < c_5 ||u||_m$$

we may continue and state

$$\max_{x \in \Omega} |\partial^i u(x)| < c_4 ||\partial^i u||_{m-i} < c_4 c_5 ||u||_m$$

That is, Dirac's function δ_i is a bounded continuous functional on $H^m_o(\Omega)$.

This is true as long as the index m of the Sobolev space minus the index i of the Delta function exceeds n/2. That is, Castigliano's Theorem depends on the inequality

$$m - i > n/2$$

In the following table this inequality is evaluated for operators of degree 2 (m=1) and of degree 4 (m=2).

			n = 1 Bar Beam	n = 2 Plate	n = 3 Body
m = 1					
	$\delta_0 \in H^{-1}$	$1-0 > n/2$	yes	no	no
	$\delta_1 \in H^{-1}$	$1-1 > n/2$	no	no	no
m = 2					
	$\delta_0 \in H^{-2}$	$2-0 > n/2$	yes	yes	yes
	$\delta_1 \in H^{-2}$	$2-1 > n/2$	yes	no	no
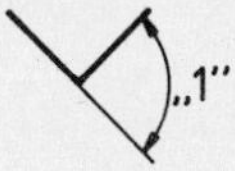	$\delta_2 \in H^{-2}$	$2-2 > n/2$	no	no	no
	$\delta_3 \in H^{-2}$	$2-3 > n/2$	no	no	no

If the answer is 'yes' then the energy if finite and the conjugated quantity is bounded, Castigliano's Theorem applies. If 'no' then the energy and the conjugated quantity are both infinite, Castigliano's Theorem does not apply.

Consider, e.g. the results in row (1). According to this row the energy of a bar, n = 1, loaded with a concentrated force is finite but the energy is infinite if the same load acts on an elastic plate, a Reissner plate, a membrane or on an elastic body.

With operators of degree 4 ($m = 2$) the situation improves, somewhat. The energy corresponding to a concentrated force is finite, in all dimensions. Now it is the couple ($= \delta_2$) which causes a state of stress with infinite energy in two dimensions (Kirchhoff plate) but finite energy in one dimension (beam).

If the singularity is a jump in displacement or a rotation ($= \delta_1$ if $m = 1$ and δ_2 or δ_3 if $m = 2$) then the internal energy is always infinite. This is what we must expect if we 'mistreat' an elastic medium. The forces needed to contort, e.g., a beam are infinite.

The simple problem of a prestressed membrane perhaps best illustrates the situation when the energy is infinite.

The first identity of the Laplace operator Δ which governs the deflection of the membrane is

$$\frac{1}{2} G(w,w) = \frac{1}{2} \{ \int_\Omega -\Delta w \, w \, d\Omega + \int_\Gamma \frac{\partial w}{\partial n} \, w \, ds - \int_\Omega \text{grad}w \cdot \text{grad}w \, d\Omega \} = 0$$

According to this equation the force term on the boundary of a membrane is the normal derivative $\partial w / \partial n$. This we easily understand if we study Fig.3. The greater the load the greater the tensile forces (the normal derivative) needed to keep the membrane in place (needed to follow the deflection).

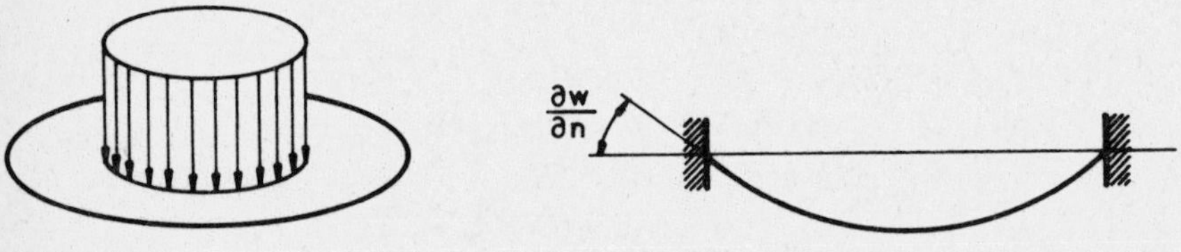

Fig. 3

Suppose now the prestressed ($N = 1$) unit circle is loaded with a concentrated force $P = 2\pi$ at its center $\underline{x} = \underline{0}$, see Fig. 4a. The corresponding bvp

$$N \, \Delta w = \delta_0(\underline{y} - \underline{0}) \, 2\pi \qquad w = 0 \quad \text{on } \Gamma, \qquad N = 1$$

has the solution w = - ln r.

$$\text{energy} = \frac{1}{2} \int_{\Omega} (w_{,1}^2 + w_{,2}^2) \, d\Omega = \frac{1}{2} \int_{\Omega} \frac{1}{r^2} \, d\Omega =$$

$$= \frac{1}{2} \int_0^R \frac{1}{r} \, dr = \infty$$

-lnr

displacement

$\frac{1}{r^2}$

energy - surface

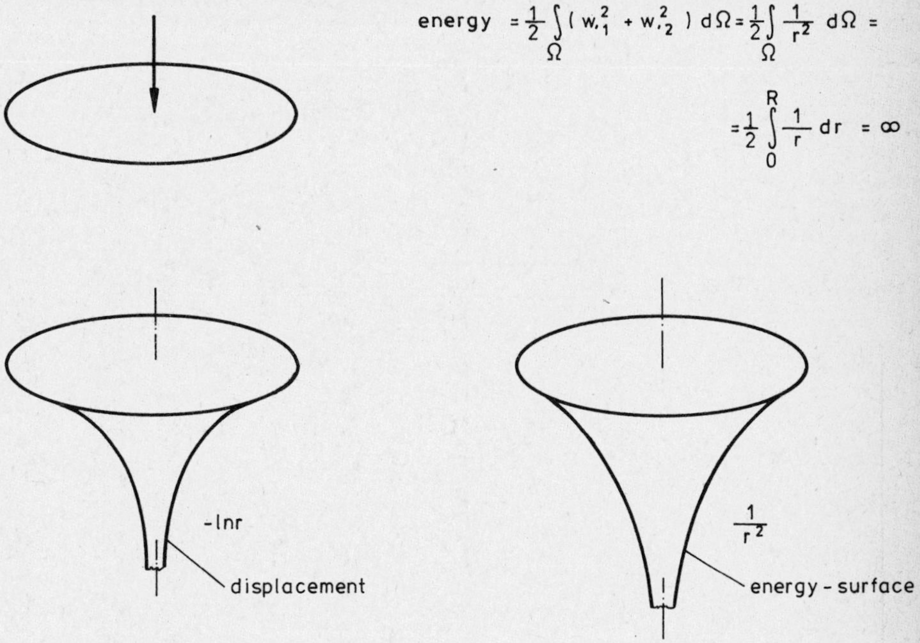

Fig. 4

The internal energy is defined as

$$\frac{1}{2} E(w,w) = \frac{1}{2} \int_{\Omega} \text{gradw} \cdot \text{gradw} \, d\Omega = \frac{1}{2} \int_{\Omega} (w_{,1}^2 + w_{,2}^2) \, d\Omega$$

Hence, its value in the ring $\epsilon < r < 1$ is

$$\frac{1}{2} E(w,w)_{\Omega_\epsilon} = \frac{1}{2} \int_\epsilon^1 \int_0^{2\pi} \frac{1}{r^2} (\cos^2\varphi + \sin^2\varphi) r \, dr \, d\varphi = \pi \int_\epsilon^1 \frac{1}{r} \, dr = \pi \ln \frac{1}{\epsilon}$$

and consequently it tends to infinity if we close the ring

$$\lim_{\epsilon \to 0} \frac{1}{2} E(w,w)_{\Omega_\epsilon} = \lim_{\epsilon \to 0} \pi \ln \frac{1}{\epsilon} = + \infty$$

The volume under the energy-surface has no finite measure. In agreement with this the displacement at the center is unbounded, $w(\underline{0}) = - \ln 0$. The energy balance is $1/2 \, G(w,w) = + \infty - \infty = 0$.

Illustrative examples are also the four singularities of a beam, see

Fig.5.

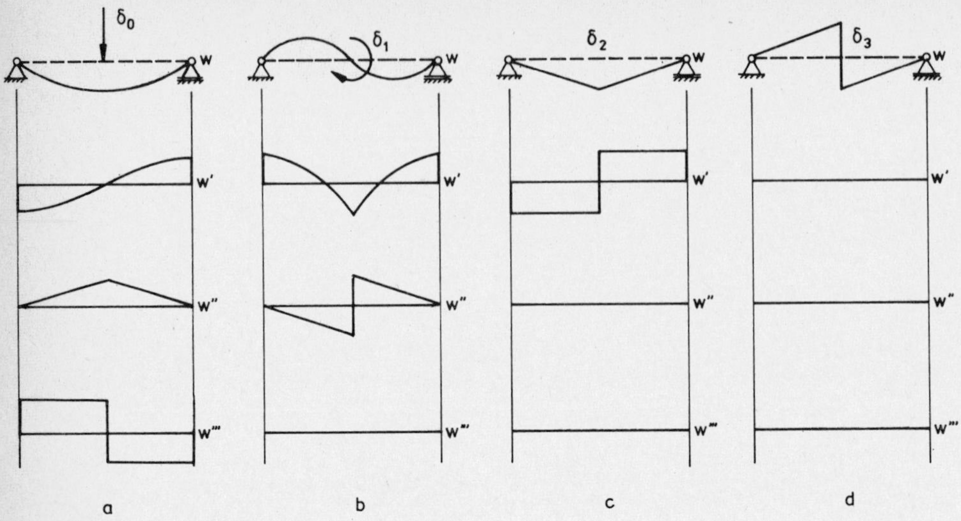

Fig. 5

The energy of a beam (EI = 1) is the integral of $(w'')^2$. Hence, the energy is finite if the area of the w"-diagram is finite as in the case of δ_o and δ_1. But what about δ_2 or δ_3? No w"-diagram exists.

In this case, e.g. in the case of δ_2, we expand the deflection, the function with the bend, see Fig.5c, into a Fourier series

$$w(x) = \frac{\pi}{2} - \frac{4}{\pi}\left(\cos 1x + \frac{1}{3^2}\cos 3x + \frac{1}{5^2}\cos 5x + \dots\right)$$

and we remember that the integral of the square of a Fourier series is just the sum of the coefficients squared. Hence, the energy, the square of the second derivative, becomes

$$\frac{1}{2}\int (w'')^2\,dx = \frac{8}{\pi^2}\,(1 + 1 + 1 + \dots)$$

and this, evidently, is infinite.

The best insight into the physics behind Sobolev's Embedding Theorem provides, we think, the following comparative study of a membrane, a Kirchhoff plate and an elastic body, each loaded with a concentrated unit force P = 1.

The normal derivative, $\partial w / \partial n$ (= force per unit length) of a membrane loaded with a concentrated force P = 1 must satisfy the equation

$$\lim_{\varepsilon \to 0} \int_{\Gamma_{N_\varepsilon}(\underline{x})} \frac{\partial w}{\partial \nu} \, ds = \lim_{\varepsilon \to 0} \int_0^{2\pi} \frac{\partial w}{\partial \nu} \, \varepsilon \, d\varphi = 1 \qquad (4)$$

the Kirchhoff-shear of a plate the equation

$$\lim_{\varepsilon \to 0} \int_{\Gamma_{N_\varepsilon}(\underline{x})} V_\nu \, ds = \lim_{\varepsilon \to 0} \int_0^{2\pi} V_\nu \, \varepsilon \, d\varphi = 1 \qquad (5)$$

and the traction vector $\underline{\tau}(\underline{u})$ of an elastic body loaded with a concentrated force $\underline{P} = \underline{e}_1 = \{1,0,0\}$ at $\underline{x}$ the equation

$$\lim_{\varepsilon \to 0} \int_{\Gamma_{N_\varepsilon}(\underline{x})} \underline{\tau}(\underline{u}) \, ds = \lim_{\varepsilon \to 0} \int_0^{\pi} \int_0^{2\pi} \underline{\tau}(\underline{u}) \, \varepsilon^2 \sin\vartheta \, d\varphi \, d\vartheta = \underline{e}_1 \qquad (6)$$

In two dimensions $\Gamma_{N_\varepsilon}(\underline{x})$ is a circle whose measure is

$$\text{mes } \Gamma_{N_\varepsilon}(\underline{x}) = \int_{\Gamma_{N_\varepsilon}} ds = 2\pi\varepsilon$$

and in three dimensions a sphere

$$\text{mes } \Gamma_{N_\varepsilon}(\underline{x}) = \int_{\Gamma_{N_\varepsilon}} ds = 4\pi\varepsilon^2$$

Consequently for Equs.(4) and (5) to hold the membrane forces and the Kirchhoff-shear must tend to infinity as ε^{-1} to balance the shrinking size, $2\pi\varepsilon$, of $\Gamma_{N_\varepsilon}(\underline{x})$ when ε tends to zero, see Fig.6.

$$\frac{\partial w}{\partial \nu} = O(\varepsilon^{-1}), \qquad V_\nu = O(\varepsilon^{-1})$$

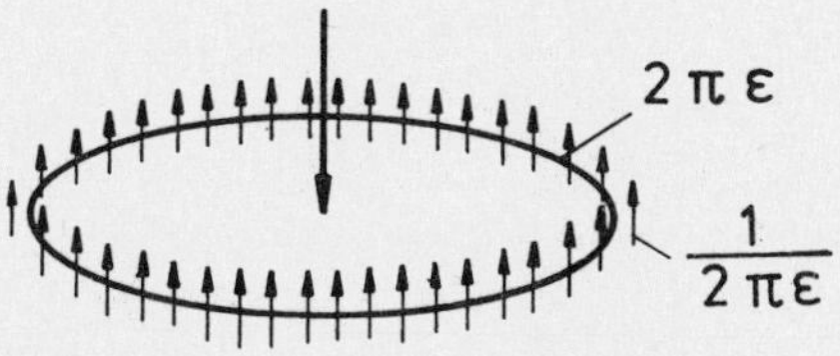

Fig. 6

Similarily for Equ.(6) to hold the traction vector must behave as

$$\underline{\tau}(\underline{u}) = O(\varepsilon^{-2})$$

We, thus, come to understand why the number n appears in Sobolev's inequality. It is the measure of $\Gamma_{N\varepsilon}(x)$ and, therefore, the dimension, n, of the continuum which determines how fast the stresses must tend to infinity.

The importance of the second number, m, in Sobolev's inequality is understood if we consider the following:

The forces $\partial w/\partial n$ of a membrane are essentially the first derivatives of the deflection w and, therefore, the deflection w the integral of the forces. If we place a concentrated force on a membrane then the normal derivative must behave as

$$\frac{\partial w}{\partial n} = O(r^{-1})$$

and, hence, the deflection as

$$w = O(\ln r)$$

Or consider an elastic body loaded with a concentrated force. The stresses are the first derivatives of the displacement field

$$(\underline{\tau}(\underline{u}))_i = 2\mu u_{i},_{j} n_j + \frac{\mu\nu}{1-2} n_i u_{j},_{j} + \mu e_{ijk} n_j e_{klm} u_{m},_{l}$$

and, hence, $\underline{u}$ the integral of the stresses. But because the stresses in the vicinity of a concentrated force behave as r^{-2} the displacements must behave as r^{-1},

$$\underline{u} = O(r^{-1})$$

The force and displacement terms of equations of second degree (m = 1) are only one differentiation apart while they lie three differentiations apart if the operator is of fourth degree (m = 2).

Integrating r^{-1} once we still have a singular function, namely ln r, but integrating it thrice we obtain a well behaved function

$$\iiint \frac{1}{r}\,dr\,dr\,dr = \frac{1}{2} r^2(\ln r - \frac{3}{2})$$

Note that

$$\lim_{r \to 0} r^2 \ln r = 0$$

The following table illustrates these facts.

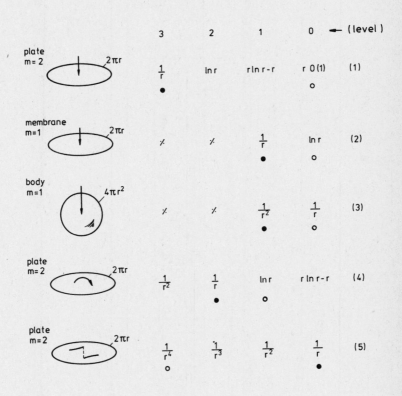

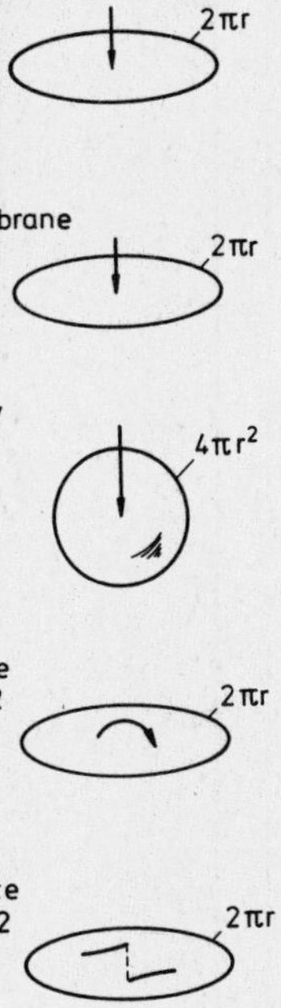

	3	2	1	0 ← (level)	
plate m = 2	$\frac{1}{r}$ ●	$\ln r$	$r \ln r - r$	$r\ O(1)$ ○	(1)
membrane m = 1	./.	./.	$\frac{1}{r}$ ●	$\ln r$ ○	(2)
body m = 1	./.	./.	$\frac{1}{r^2}$ ●	$\frac{1}{r}$ ○	(3)
plate m = 2	$\frac{1}{r^2}$	$\frac{1}{r}$ ●	$\ln r$ ○	$r \ln r - r$	(4)
plate m = 2	$\frac{1}{r^4}$ ○	$\frac{1}{r^3}$	$\frac{1}{r^2}$	$\frac{1}{r}$ ●	(5)

● = original singularity, ○ = conjugated quantity

The third number which appears in Sobolev's inequality is i, the degree or the level of the singularity. If i is low, say i = 0, then the singularity is a force, i.e. a function on level 3 (in case m = 2). The conjugated quantity is the displacement, i.e. a function on level 0 and consequently the distance between the two levels is at its maximum. There are enough integrations in between to dampen the singularity r^{-1} (see the first line).

If i increases then the place of the singularity r^{-1} moves to the right, e.g. from level 3 to level 2 (we load the plate with a couple, see line 4) while the place of the conjugated quantity moves, at the same time, to the left, from level 0 to level 1. The distance becomes smaller, only one integration separates the two conjugated quantities and this is no longer sufficient to transform the singularity r^{-1} into a well behaved function. We only obtain ln r for the normal derivative, the conjugated quantity.

If the level of the singularity is higher than the level of the conjugated quantity - this happens, e.g., if we force upon the plate a jump in displacement, see line 5, - then we must perform differentiations instead of integrations to calculate the conjugated quantity, the Kirchhoff-shear. This is, with respect to Castigliano's Theorem, ill luck because then the behaviour of the conjugated quantity is even worse than the behaviour of the original singularity which causes the disturbance.

References

[1] G.Fichera, Existence Theorems in Elasticity, in: Encyclopedia of Physics, Chief Editor S. Flügge, Volume VIa/2, Mechanics of Solids II, Editor C. Truesdell, Springer-Verlag Berlin, Heidelberg, New York, 1972

Acknowledgments

The author wants to thank Dr. Blum from Bonn for his help in getting some mathematical details straight.

Final Remark

Castigliano's Theorem should not be confused with Castigliano's Principle. The latter states that the potential energy attains its minimum at the equilibrium configuration.

AN INTEGRAL EQUATION FORMULATION FOR A BOUNDARY VALUE PROBLEM OF ELASTICITY IN THE DOMAIN EXTERIOR TO AN ARC*

George C. Hsiao, Ernst P. Stephan and Wolfgang L. Wendland

Department of Mathematical Sciences, University of Delaware,
Newark, Delaware 19711, U.S.A.
Department of Mathematics, Georgia Institute of Technology ,
Atlanta, Georgia 30332, U.S.A., and
Fachbereich Mathematik , Technische Hochschule Darmstadt ,
D-6100 Darmstadt, FRG .

Abstract: We consider here a Dirichlet problem for the two-dimensional linear elasticity equation in the domain exterior to an open arc in the plane. It is shown that the problem can be reduced to a system of boundary integral equations with the unknown density function being the jump of stresses across the arc. Existence, uniqueness as well as regularity results for the solution to the boundary integral equations are established in appropriate Sobolev spaces. In particular, asymptotic expansions concerning the singular behavior for the solution near the tips of the arc are obtained. By adding special singular elements to the regular splines as test and trial functions, an augmented Galerkin procedure is used for the corresponding boundary integral equations to obtain a quasi-optimal rate of convergence for the approximate solutions.

1. Introduction

This paper extends the results of the recent works [17][23][25][26] by the authors on the crack and screen problems for the Laplace and Helmholtz equation as well as on the exterior elasticity problems with a regular smooth boundary. Throughout the paper, let Γ be an open arc in the plane $\mathbb{R}^2$. We consider here the boundary value problem consisting of the linear elasticity equation for the displacement field $\underset{\sim}{u}$:

$$\mu\Delta\underset{\sim}{u} + (\lambda+\mu) \text{ grad div } \underset{\sim}{u} = \underset{\sim}{0} \quad \text{in} \quad \Omega_\Gamma = \mathbb{R}^2 \setminus \overline{\Gamma} \qquad \text{(E)}$$

together with the boundary condition

$$\underset{\sim}{u}|_\Gamma = \underset{\sim}{g} \, , \qquad \text{(B)}$$

* This work was supported by the "Alexander von Humboldt-Stiftung", FRG .

where $\mu > 0$ and $\lambda > -\mu$ are given Lâme constants [10], and $\underset{\sim}{g}$ is a prescribed smooth function. In addition we assume that $\underset{\sim}{u}-\underset{\sim}{r}$ is *regular* at infinity. Following [17] [19], by this we mean

$$D^{\alpha}(\underset{\sim}{u}-\underset{\sim}{r}) = O(|\underset{\sim}{x}|^{-\alpha-1}) \qquad \alpha = 0,1, \quad \text{as} \quad |x| \to \infty \tag{C}$$

with $D = \frac{\partial}{\partial x_i}$. More precisely, let us represent the rigid motion $\underset{\sim}{r}(x)$ in the form:

$$\underset{\sim}{r}(x) = \omega_1\hat{e}_1 + \omega_2\hat{e}_2 + \omega_3(x_2\hat{e}_1 - x_1\hat{e}_2)$$

where $\hat{e}_i$ denotes the unit vectors in $\mathbb{R}^2$ and ω_i's are unknown constants. As indicated in [14] [17], condition (C) implies that

$$\int_{\Gamma} [T(\underset{\sim}{u})]ds_y = \underset{\sim}{0} \tag{C_1}$$

and in order to ensure the uniqueness, we further impose the equilibrium condition of vanishing total momentum

$$\int_{\Gamma} (y_2\hat{e}_1 - y_1\hat{e}_2) \cdot [T(\underset{\sim}{u})]ds_y = 0 \tag{C_2}$$

which will become transparent later (see (2.12)). We note that condition (C_2) will not be needed in the case when ω_3 is given [17]. Here in the formulation, as will be seen, $[T(\underset{\sim}{u})]$ stands for the jump of traction $T(\underset{\sim}{u})$ across Γ ,

$$T(\underset{\sim}{u}) := 2\mu \frac{\partial}{\partial\hat{n}} \underset{\sim}{u} + \lambda\hat{n} \operatorname{div} \underset{\sim}{u} + \mu\hat{n} \times \operatorname{curl} \underset{\sim}{u} \tag{1.1}$$

with $\hat{n}$ being the unit normal to Γ , and $\operatorname{curl} \underset{\sim}{u} := \operatorname{curl} (u_1,u_2,0)$.

In the following we shall refer to the problem defined by (E),(B), (C_1) and (C_2) as the crack problem (P). Such crack problems arise if e.g. an inlet of rigid material is immersed at Γ into the elastic material occupying Ω_Γ .

Our aim is to develop a solution procedure for (P) by making use of an integral equation method which allows us to obtain the explicit singular behavior of the "stress" near the tips of Γ . Following [17] [26], we reduce the problem (P) to a system of boundary integral equations of the first kind [11][13][28] with the jump of traction across Γ as the unknown. These boundary integral equations are derived by the "direct approach" based on the Betti formula. By using the method of local Mellin transform as in [3]-[8], and the calculus of pseudodifferential operators [9][22], we establish existence, uniqueness and regularity results for the solution of our boundary integral equations. In particular, we are able to obtain appropriate asymptotic expansions for the jump of tractions near the tips of Γ . The latter provides us useful informations concerning numerical treatment such as the Galerkin scheme for our boundary integral equations. In fact, in our augmented boundary element method, we use, as in [20] [26] [27] [29] in addition to the regular finite elements, appropriate singular elements concentrated near the tips and improve significantly the asymptotic convergence rates of our approximate solutions [15].

It should be emphasized that since our boundary integral equations are derived directly from the Betti formula, physically, the boundary charges are precisely the jumps of tractions across Γ . From our boundary element method using augmented test and trial function spaces with the appropriate singular elements, we are able to compute both approximate boundary charges and the stress intensity factors simultaneously. Hence, our asymptotic error estimates in [15] include explicit estimates for the stress intensity factors, as well.

In this paper, we shall present only the main idea and some of the results and leave the details to [15].

2. Integral Representation

We begin with the variational formulation for the problem (P). We then derive the integral representation for the variational solution by the direct method based on the Betti formula. In order to characterize the variational solution of (P), we introduce the function space $H_c^1(\Omega_\Gamma)$, the completion of all C^∞-functions $\underset{\sim}{f}(x)$ of the form

$$\underset{\sim}{f}(x) = \underset{\sim}{f}_o(x) + \underset{\sim}{r}(x) \tag{2.1}$$

with respect to the norm $\|\cdot\|_{1,c}$ defined by

$$\| \underset{\sim}{f} \|_{1,c} := \{\int_{\Omega_\Gamma} E(\underset{\sim}{f},\underset{\sim}{f})dx + \int_\Gamma |\underset{\sim}{f}|^2 ds\}^{1/2} \quad , \tag{2.2}$$

where $\underset{\sim}{f}_o$ is regular at infinity and $\underset{\sim}{r}$ denotes a rigid motion of the form:

$$\underset{\sim}{r}(x) = \omega_1\hat{e}_1 + \omega_2\hat{e}_2 + \omega_3(x_2\hat{e}_1 - x_1\hat{e}_2) =: M(x)\underset{\sim}{\omega} \tag{2.3}$$

with $\underset{\sim}{\omega} = (\omega_1,\omega_2,\omega_3)^T$ and $M(x)$ the corresponding 2×3 matrix. Here

$$E(\underset{\sim}{f},\underset{\sim}{g}) := (\lambda+\mu)\,\mathrm{div}\,\underset{\sim}{f}\,\mathrm{div}\,\underset{\sim}{g} + \frac{1}{2}\mu \sum_{j\neq k}^{2} \left(\frac{\partial f_j}{\partial x_k} + \frac{\partial f_k}{\partial x_j}\right)\left(\frac{\partial g_j}{\partial x_k} + \frac{\partial g_k}{\partial x_j}\right)$$
$$+ \frac{1}{2}\mu \sum_{j,k=1}^{2} \left(\frac{\partial f_j}{\partial x_k} - \frac{\partial f_k}{\partial x_j}\right)\left(\frac{\partial g_j}{\partial x_k} - \frac{\partial g_k}{\partial x_j}\right) \qquad (2.4)$$

is a bilinear form for the derivatives of $\underset{\sim}{f}$ and $\underset{\sim}{g}$ (see [19]). In addition, we denote by $\mathring{H}_c^1(\Omega_\Gamma)$ the closed suspace of $H_c^1(\Omega_\Gamma)$ such that

$$\mathring{H}_c^1(\Omega_\Gamma) = \{\underset{\sim}{f} \in H_c^1(\Omega_\Gamma) \mid \underset{\sim}{f}|_\Gamma = \underset{\sim}{0}\} .$$

For the equivalent variational formulation of the crack problem (P), let $\underset{\sim}{h} \in H_c^1(\Omega_\Gamma)$ be an extension of the boundary values of $\underset{\sim}{g} \in H^{1/2}(\Gamma)$ with $\underset{\sim}{h}|_\Gamma = \underset{\sim}{g}$, and be regular at infinity. Then the problem (P) is equivalent to the *variational problem: for given* $\underset{\sim}{h} \in H_c^1(\Omega_\Gamma)$ *, find a function* $\underset{\sim}{u} \in H_c^1(\Omega_\Gamma)$ *such that* $\underset{\sim}{u}-\underset{\sim}{h} \in \mathring{H}_c^1(\Omega_\Gamma)$ *and*

$$B(\underset{\sim}{u},\underset{\sim}{\phi}) := \int_{\Omega_\Gamma} E(\underset{\sim}{u},\underset{\sim}{\phi})\,dx = 0 \qquad (2.5)$$

for all $\underset{\sim}{\phi} \in \mathring{H}_c^1(\Omega_\Gamma)$. In view of Korn's inequality [10] and the Riesz-Fréchet representation theorem it is easy to see that there exists exactly one solution $\underset{\sim}{u} \in H_c^1(\Omega_\Gamma)$ of the problem (P) given by the variational problem.

In order to derive an integral representation of the variational solution of the problem (P), as in [26], we extend Γ to an arbitrary smooth simple closed curve $\dot{G}_1$, and denote by G_1 the bounded domain inside $\dot{G}_1$. We use the notation [v] to denote the jump $v_- - v_+$ of a function v across $\dot{G}_1$. Here the subscripts $-, +$ denote the limits taken from G_1 and $\mathbb{R}^2 \setminus \overline{G}_1$ respectively. For later use, let B_R be a circle with radius R sufficiently large enough to enclose $\overline{G}_1$. The domain bounded by $\dot{G}_1$ and the boundary $\dot{B}_R$ of B_R will be denoted by G_2 . The boundary $\dot{G}_2$ of G_2 consists of Γ together with $\dot{G}_1 \setminus \Gamma$ and $\dot{B}_R$ (see Figure 1).

Figure 1:

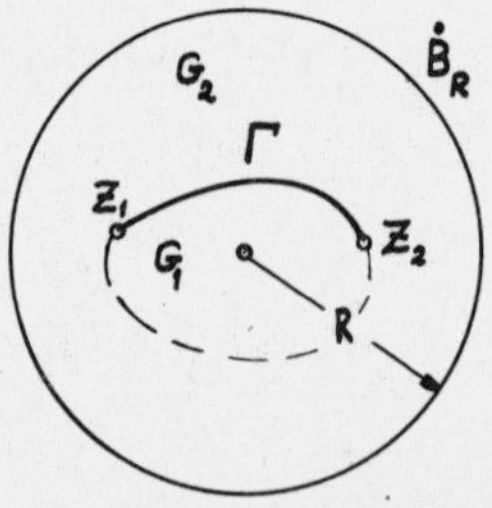

In what follows, let $H^s(\dot{G}_1)$ be defined as the trace of $H^{s+1/2}(\mathbb{R}^2)$ for $s > 0$, as $L^2(\dot{G}_1)$ for $s = 0$, and as the dual space of $H^{-s}(\dot{G}_1)$ for $s < 0$. For $s \geq 0$, $H^s(\Gamma)$ denotes the usual trace space of $H^s(\dot{G}_1)$ on Γ and $\tilde{H}^s(\Gamma)$ is defined by

$$\tilde{H}^s(\Gamma) := \{\underset{\sim}{f} = \underset{\sim}{f}'|_\Gamma : \underset{\sim}{f}' \in H^s(\dot{G}_1)\ ,\ \underset{\sim}{f}'|_{\dot{G}_1\setminus\Gamma} = \underset{\sim}{0}\}$$

equipped with the topology of $H^s(\dot{G}_1)$. For $s < 0$, we define

$$H^s(\Gamma) := (\tilde{H}^{-s}(\Gamma))' \text{ and } \tilde{H}^s(\Gamma) := (H^{-s}(\Gamma))'$$

by duality with respect to the $L^2(\Gamma)$ scalar product. It is clear that from the definition, for $s > 0$

$$\tilde{H}^{-s}(\Gamma) = \{\underset{\sim}{f} \in H^{-s}(\dot{G}_1) \mid \mathrm{supp}(\underset{\sim}{f}) \subset \overline{\Gamma}\}$$

which is also the completion of $C_0^\infty(\Gamma)$ with respect to the norm of $H^{-s}(\dot{G}_1)$ (see [1], [12, Theorem 2.5.1, p. 51 ff]).

We now state some properties concerning the solution $\underset{\sim}{u} \in H_c^2(\Omega_\Gamma)$ of (2.5):

Lemma 2.1. *Let* $\underset{\sim}{u} \in H_c^1(\Omega_\Gamma)$ *be the solution of* (2.5). *Then* $\underset{\sim}{u}|_{\dot{G}_1} \in H^{1/2}(\dot{G}_1)$ *and* $\underset{\sim}{u}|_\Gamma \in H^{1/2}(\Gamma)$. *For the traction* $T(\underset{\sim}{u})$, *we have*

$$T(\underset{\sim}{u})|_{\dot{G}} \in H^{-1/2}(\dot{G}_1) \quad \textit{and} \quad T(\underset{\sim}{u})|_\Gamma \in (\tilde{H}^{1/2}(\Gamma))' ;$$

moreover, if we denote by $[T(\underset{\sim}{u})] = T(\underset{\sim}{u})_- - T(\underset{\sim}{u})_+$ *the jump of the traction acoross* Γ, *then we have*

$$[T(\underset{\sim}{u})]|_\Gamma \in \tilde{H}^{-1/2}(\Gamma) .$$

Here in the definition of $T(\underset{\sim}{u})$ *(see* (1.1)*) the exterior normal derivative to* $\dot{G}_1$ *is used.*

The proof of Lemma 2.1 is similar to the two-dimensional screen problem of the Laplacian. It is based on the trace theorem [21] and Weyl's lemma; and we omit the details here [15].

For the integral representation of the solution $\underset{\sim}{u}$, we need the fundamental solution of (E), the Kelvin matrix (see e.g. [2])

$$\underset{\approx}{\gamma}(y,x) = \frac{\lambda+3\mu}{4\pi\mu(\lambda+2\mu)} \{\log\frac{1}{|x-y|}\underset{\approx}{I} + \frac{\lambda+\mu}{\lambda+3\mu}\frac{(x-y)(x-y)^T}{|x-y|^2}\} , \tag{2.6}$$

and the corresponding stress matrix on $\dot{G}_j$ [17]

$$\underset{\approx}{\gamma}_1(y,x) := (T_y \underset{\approx}{\gamma}(y,x))^T$$
$$= \frac{\mu}{8\pi(\lambda+2\mu)} \{ I + \frac{2(\lambda+\mu)}{\mu|x-y|^2}(x-y)(x-y)^T) \frac{\partial}{\partial n_y} + \begin{pmatrix} 0 & -1 \\ 1 & 0 \end{pmatrix} \frac{\partial}{\partial s_y} \} \log \frac{1}{|x-y|} , \tag{2.7}$$

where ^{T} stands for the transpose. By applying the Betti formula to the variational solution $\underset{\sim}{u}$ in G_i , $i = 1$ and 2 , we obtain

$$\alpha_j \underset{\sim}{u}(x) = - \int_{\dot{G}_j} \underset{\approx}{\gamma}_1(y,x)\underset{\sim}{u}(y)ds_y + \int_{\dot{G}_j} \underset{\approx}{\gamma}(y,x)\ T(\underset{\sim}{u})(y)ds_y \tag{2.8}$$

for fixed $x \in G_1$ with $\alpha_1 = 1$ and $\alpha_2 = 0$. These representations hold for $\underset{\sim}{u} \in H_c^1(\Omega_\Gamma)$, since from Lemma 2.1 , we have $\underset{\sim}{u}_{|\dot{G}_j} \in H^{1/2}(\dot{G}_j)$ and $T(\underset{\sim}{u})_{|\dot{G}_j} \in H^{-1/2}(\dot{G}_j)$. Now from (2.8) it follows easily that

$$\underset{\sim}{u}(x) = - \int_{\dot{B}_R} \underset{\approx}{\gamma}_1(y,x)\underset{\sim}{u}(y)ds_y + \int_{\dot{B}_R} \underset{\approx}{\gamma}(y,x)T(\underset{\sim}{u})(y)ds_y + \int_\Gamma \underset{\approx}{\gamma}(y,x)[T(\underset{\sim}{u})](y)ds_y \tag{2.9}$$

for fixed $x \in G_1$. We note that $[T(\underset{\sim}{u})]_{|\dot{G}_1 \setminus \Gamma} = 0$.

If $\underset{\sim}{u} - M(x)\underset{\sim}{\omega}$ is regular at infinity, one can show that the first two terms tend to $M(x)\underset{\sim}{\omega}$ as $R \to \infty$ [17]. Thus, we arrive at the representation:

$$\underset{\sim}{u}(x) = \int_\Gamma \underset{\approx}{\gamma}(y,x)[T(\underset{\sim}{u})](y)ds_y + M(x)\underset{\sim}{\omega} , \qquad x \in G_1 . \tag{2.10}$$

Clearly in a similar manner, one can show that the same representation holds for $x \in G_2$ with arbitrary R .

We summarize the foregoing results in the following theorem.

Theorem 2.2. *Suppose* $\underset{\sim}{u} \in H_c^1(\Omega_\Gamma)$ *is a variational solution of* (P) . *Then* $\underset{\sim}{u}$ *admits the integral representation:*

$$\underset{\sim}{u}(x) = \int_\Gamma \underset{\approx}{\gamma}(y,x)[T(\underset{\sim}{u})]ds_y + M(x)\underset{\sim}{\omega} , \qquad x \in \mathbb{R}^2 \setminus \bar{\Gamma} \tag{2.11}$$

where $[T(\underset{\sim}{u})]_{|\Gamma} \in \tilde{H}^{-1/2}(\Gamma)$ *is the jump of traction across* Γ , *satis-*

fying the condition (C_1) *, and* $M(x)\underset{\sim}{\omega}$ *corresponds to the rigid motion of* $\underset{\sim}{u}$ *at infinity.*

We remark that from the representation (2.11), one may derive the work formula as in [17]:

$$\int_{\Omega_\Gamma} E(\underset{\sim}{w},\underset{\sim}{u})dx = -\int_\Gamma \underset{\sim}{w}\cdot[T(\underset{\sim}{u})]ds + \int_\Gamma ((M(y)\underset{\sim}{\Omega}(\underset{\sim}{w}))\cdot[T(\underset{\sim}{u})]ds_y \tag{2.12}$$

for $\underset{\sim}{w} \in H_c^1(\Omega_\Gamma)$, where $\underset{\sim}{u} \in H_c^1(\Omega_\Gamma)$ is a variational solution of (P) and $M(y)\underset{\sim}{\Omega}(\underset{\sim}{w})$ corresponds to the rigid motion of $\underset{\sim}{w}$ at infinity. This work formula indicates that one indeed needs the additional condition such as (C_2) in order to ensure the uniqueness of the solution of the problem (P) .

3. Boundary Integral Equations.

We now reduce the variational boundary value problem of section 2 to equivalent boundary integral equations for the jump of traction, $[T(\underset{\sim}{u})]$ across Γ . This can be achieved from the integral representation (2.11) by letting x tend to Γ . In fact, the following result can be established.

Theorem 3.1. *Let* $\underset{\sim}{g} \in H^{1/2}(\Gamma)$ *be given. Then* $\underset{\sim}{u} \in H_c^1(\Omega_\Gamma)$ *is the variational solution of* (P) *if and only if* $[T(\underset{\sim}{u})]_{|\Gamma} \in \tilde{H}^{-1/2}(\Gamma)$, $\underset{\sim}{\omega} \in \mathbb{R}^3$ *solve the integral equations*

$$\begin{aligned} &\int_\Gamma \underset{\approx}{\gamma}(y,x)[T(\underset{\sim}{u})](y)ds_y + M(x)\underset{\sim}{\omega} = \underset{\sim}{g} \\ &\int_\Gamma [T(\underset{\sim}{u})]ds = \underset{\sim}{0} \quad \textit{and} \quad \int_\Gamma \underset{\sim}{m}_3(y)\cdot[T(\underset{\sim}{u})]ds = \underset{\sim}{0}\ . \end{aligned} \tag{3.1}$$

for $x \in \Gamma$ *, where* $\underset{\sim}{m}_3(y) := y_2\hat{e}_1 - y_1\hat{e}_2$.

The necessity follows clearly from the derivation of the integral representation (2.11). It remains only to show the last condition in (3.1). This follows immediately from (2.12) together with the condition $\int_\Gamma [T(\underset{\sim}{u})]ds = \underset{\sim}{0}$, if $\underset{\sim}{w}$ in (2.12) is replaced by $\underset{\sim}{\phi} \in \mathring{H}_c^1(\Omega_\Gamma)$. For the sufficiency, we refer to [15] for the details.

In order to guarantee that the system (1.3) is always solvable, we now need some properties concerning the integral operators in (3.1). For convenience, let us denote by V_Γ the boundary integral operator defined by

$$V_\Gamma \underset{\sim}{\phi}(x) := \int_\Gamma \underset{\approx}{\gamma}(y,x)\underset{\sim}{\phi}(y)ds_y , \qquad x \in \Gamma \tag{3.2}$$

and by Λ_Γ , the functional defined by

$$\Lambda_\Gamma \underset{\sim}{\phi} := \int_\Gamma M^T(y)\underset{\sim}{\phi}(y)ds_y \tag{3.3}$$

where $M^T(y)$ is the transpose of the matrix $M(y)$ in (2.3) , that is,

$$M^T(y) = \begin{pmatrix} 1 & 0 \\ 0 & 1 \\ y_2 & -y_1 \end{pmatrix} . \tag{3.4}$$

We also introduce the matrix operator A_Γ :

$$A_\Gamma(\underset{\sim}{\phi},\underset{\sim}{\omega}) := \begin{pmatrix} V_\Gamma & M \\ \Lambda_\Gamma & 0 \end{pmatrix} \begin{pmatrix} \underset{\sim}{\phi} \\ \underset{\sim}{\omega} \end{pmatrix} . \tag{3.5}$$

Then, as one expected from [26] and [29], the operator A_Γ here also satisfies a Gårding inequality in the energy space $\tilde{H}^{-1/2}(\Gamma)\times\mathbb{R}^3$. This means that A_Γ is a Fredholm operator of index zero [16] [18] [24] and hence together with the uniqueness of the solution of (3.1), it implies that (3.1) is always *uniquely solvable*. In the theorem below, the relevant mapping properties of the boundary integral operator A_Γ are included. These properties will be established in [15].

<u>Theorem 3.2.</u> *The matrix operator* A_Γ *and its adjoint* A_Γ^* *with respect to the duality*

$$\langle(\underset{\sim}{\psi},\underset{\sim}{\kappa}),(\underset{\sim}{\phi},\underset{\sim}{\omega})\rangle_{L^2(\Gamma)\times\mathbb{R}^3} := (\underset{\sim}{\psi},\underset{\sim}{\phi})_{L^2(\Gamma)} + \underset{\sim}{\kappa}\cdot\underset{\approx}{\omega}$$

both are continuous and bijective mappings:

$$\tilde{H}^s(\Gamma) \times \mathbb{R}^3 \rightarrow H^{s+1}(\Gamma) \times \mathbb{R}^3 \quad \text{for} \quad -1 < s < 0 .$$

Moreover, the operator A_Γ *satisfies a Gårding inequality in* $\tilde{H}^{-1/2}(\Gamma)\times\mathbb{R}^3$, i.e. *there exists a constant* $\gamma > 0$ *and a continuous mapping*

$$C_\Gamma : \tilde{H}^{-1/2}(\Gamma)\times\mathbb{R}^3 \rightarrow H^{1/2+\varepsilon}(\Gamma)\times\mathbb{R}^3 \qquad \textit{for some} \quad \varepsilon > 0$$

such that the inequality

$$<(A_\Gamma + C_\Gamma)(\underset{\sim}{\phi},\underset{\sim}{\omega}),(\underset{\sim}{\phi},\underset{\sim}{\omega})>_{L^2(\Gamma)\times\mathbb{R}^3} \geq \gamma\{\|\underset{\sim}{\phi}\|^2_{\tilde{H}^{-1/2}(\Gamma)} + |\underset{\sim}{\omega}|^2\}$$

holds for all $(\underset{\sim}{\phi},\underset{\sim}{\omega}) \in \tilde{H}^{-1/2}(\Gamma) \times \mathbb{R}^3$.

We now come to the point of our main concern - the singularity of $[T(\underset{\sim}{u})]$ near the crack tips Z_i . As in the case of potential theory [23] [25] [26], the variational solution $\underset{\sim}{u}$ of (2.5) has in general umbounded traction $[T(\underset{\sim}{u})]$ even for C^∞-data. Hence, one will not be able to improve the approximation of $[T(\underset{\sim}{u})]$ by the conventional constructive methods such as finite element or boundary element methods without any modifications, since for better approximations, one generally requires higher regularity of the exact solution $[T(\underset{\sim}{u})]$. For this purpose, it is best to decompose $[T(\underset{\sim}{u})]$ into special singular terms concentrated near the tips Z_i and a regular remainder. In this way, as in [4] [20] [23] [25] [26] and [29], one may then augment the finite element spaces of test and trial functions with appropriate global singular elements, according to the special forms of singular terms in $[T(\underset{\sim}{u})]$, to improve the order of convergence of the approximations.

The following result concerns the asymptotic behavior of the exact solution $[T(\underset{\sim}{u})]$ of (3.1) near the crack tips. The analysis here follows [3]-[8] by using the Mellin transform technique, which gives us the the appropriate function spaces together with the exact form of the singularities at the tips.

Theorem 3.3. *For* $|\sigma| < \frac{1}{2}$, *let* $g \in H^{s+\sigma}(\Gamma)$, $s = \frac{3}{2}$ *or* $\frac{5}{2}$ *be given. Then the solution* $[T(\underset{\sim}{u})]_{|\Gamma} \in \tilde{H}^{-1/2}(\Gamma)$ *(and* $\underset{\sim}{\omega} \in \mathbb{R}^3$*) of the integral equations* (3.1) *admits the asymptotic representation form near the endpoints* $Z_i \in \bar{\Gamma}$ *:*

(a) $$[T(\underset{\sim}{u})]_{|\Gamma} = \sum_{i=1}^{2} \underset{\sim}{\alpha}_i\, \rho_i^{-1/2}\chi_i + \underset{\sim}{\psi}_o \qquad \text{for } s = \frac{3}{2}$$

with $\psi_o \in \tilde{H}^{1/2+\sigma}(\Gamma)$, $\underset{\sim}{\alpha}_i \in \mathbb{R}^2$,

(b) $$[T(\underset{\sim}{u})]_{|\Gamma} = \sum_{i=1}^{2} (\underset{\sim}{\alpha}_i\rho_i^{-1/2} + \underset{\sim}{\beta}_i\rho_i^{1/2})\chi_i + \underset{\sim}{\psi}_i \qquad \text{for } s = \frac{5}{2}$$

with $\underset{\sim}{\psi}_1 = \tilde{H}^{3/2+\sigma}(\Gamma)$, $\underset{\sim}{\alpha}_i, \underset{\sim}{\beta}_i \in \mathbb{R}^2$.

Here ρ_i *denotes the distance between* $x \in \Gamma$ *and* Z_i *, while* χ_i *is a* C^∞ *cut-off function such that* $0 \leq \chi_1 \leq 1$, $\chi_1 = 1$ *near* Z_i

and $\chi_i = 0$ *elsewhere*, $i = 1,2$.

We remark that the coefficients $\underset{\sim}{\alpha}_i$ and $\underset{\sim}{\beta}_i$ in the above theorem are indeed the *stress intensity factors* as in the crack problem, since $T(\underset{\sim}{u})$ is the traction. We further comment that the regularity of the remainder term in $[T(\underset{\sim}{u})]_{|\Gamma}$ may be improved as the given data $\underset{\sim}{g}$ becomes smoother, however, in general the solution $[T(\underset{\sim}{u})]_{|\Gamma}$ possesses singularities and is always *unbounded* at the tips.

To incorporate the expansions of $[T(\underset{\sim}{u})]_{|\Gamma}$ into the augmented Sobolev spaces for the purpose of boundary element approximation, we need mapping properties of $[T(\underset{\sim}{u})]$ such as in Theorem 3.2. Let us begin with the following defintion:

<u>Definition 3.4.</u>

a) *For* $s < 1$, *we define*

$$Z^s(\Gamma) := \{\underset{\sim}{\psi} = \sum_{i=1}^{2} \underset{\sim}{\alpha}_i \rho_i^{-1/2} \chi_i + \underset{\sim}{\psi}_o \mid \underset{\sim}{\alpha}_i \in \mathbb{R}^2,\ \underset{\sim}{\psi}_o \in \tilde{H}^s(\Gamma)\}$$

equipped with

$$\|\underset{\sim}{\psi}\|_{Z^s(\Gamma)} := \begin{cases} \sum_{i=1}^{2} |\underset{\sim}{\alpha}_i| + \|\underset{\sim}{\psi}_o\|_{\tilde{H}^s(\Gamma)} & \text{for } 0 \le s < 1 \\ \|\underset{\sim}{\psi}\|_{\tilde{H}^s(\Gamma)} & \text{for } s < 0 \end{cases}$$

b) *For* $1 \le s < 2$, *we define*

$$Z^s(\Gamma) := \{\underset{\sim}{\psi} = \sum_{i=1}^{2} (\underset{\sim}{\alpha}_i \rho_i^{-1/2} + \underset{\sim}{\beta}_i \rho_i^{1/2}) \chi_i + \underset{\sim}{\psi}_1 \mid \underset{\sim}{\alpha}_i, \underset{\sim}{\beta}_i \in \mathbb{R}^2,\ \underset{\sim}{\psi}_1 \in \tilde{H}^s(\Gamma)\}$$

equipped with

$$\|\underset{\sim}{\psi}\|_{Z^s(\Gamma)} := \sum_{i=1}^{2} |\underset{\sim}{\alpha}_i| + |\underset{\sim}{\beta}_i| + \|\underset{\sim}{\psi}_1\|_{\tilde{H}^s(\Gamma)}\ .$$

These augmented spaces allow us to extend the mapping properties in Theorem 3.2 to higher order spaces. The following results are similar to those in [8],[26] and [29].

<u>Theorem 3.4.</u> *For fixed* $\sigma, |\sigma| < \frac{1}{2}$, *the operator* A_Γ *defined by* (3.5) *possesses the following mapping properties:*

$$A_\Gamma : Z^{1/2+\sigma}(\Gamma) \times \mathbb{R}^3 \to H^{3/2+\sigma}(\Gamma) \times \mathbb{R}^3 \quad \textit{and}$$

$$A_\Gamma : Z^{3/2+\sigma}(\Gamma)\times\mathbb{R}^3 \rightarrow H^{5/2+\sigma}(\Gamma)\times\mathbb{R}^3$$

with

$$\{\alpha_1,\alpha_2,\psi_o,\omega\} \mapsto \begin{pmatrix} V_\Gamma(\sum_{i=1}^{2}\alpha_i\rho_i^{-1/2}\chi_i+\psi_o) + M\omega \\ \Lambda_\Gamma(\sum_{i=1}^{2}\alpha_i\rho_i^{-1/2}\chi_i+\psi_o) \end{pmatrix} = \begin{pmatrix} g \\ b \end{pmatrix}$$

and (3.6)

$$\{\alpha_1,\alpha_2,\beta_1,\beta_2,\psi_1,\omega\} \mapsto \begin{pmatrix} V_\Gamma(\sum_{i=1}^{2}(\alpha_i\rho_i^{-1/2}+\beta_i\rho_i^{1/2})\chi_i+\psi_1) + M\omega \\ \Lambda_\Gamma(\sum_{i=1}^{2}(\alpha_i\rho_i^{-1/2}+\beta_i\rho_i^{1/2})\chi_i+\psi_1) \end{pmatrix} = \begin{pmatrix} g \\ b \end{pmatrix},$$

respectively, are continuous and bijective. Furthermore, there hold the corresponding à priori estimates:

$$\|\psi\|_{Z^{1/2+\sigma}(\Gamma)} + |\omega| \le c\{\|g\|_{H^{3/2+\sigma}(\Gamma)} + |b|\},$$

and

$$\|\psi\|_{Z^{3/2+\sigma}(\Gamma)} + |\omega| \le c\{\|g\|_{H^{5/2+\sigma}(\Gamma)} + |b|\} .$$

We remark that in particular, one may take $b = 0$ in the above theorem. Then (3.6) coincides with the integral equation (3.1) with

$$[T(u)]_{|\Gamma} = \sum_{i=1}^{2}\alpha_i\rho_i^{-1/2}\chi_i+\psi_o \quad \text{or} \quad [T(u)]_{|\Gamma} = \sum_{i=1}^{2}(\alpha_i\rho_i^{-1/2}+\beta_i\rho^{1/2})\chi_i+\psi_1 ,$$

depending on the given data $g \in H^{3/2+\sigma}(\Gamma)$ or $g \in H^{5/2+\sigma}(\Gamma)$. It is this system (3.6) that we have solved in [15] for α_1,ψ_o,ω and $\alpha_i,\beta_i,\psi_1,\omega$ by an augmented boundary element method originally employed for a closed smooth curve [20] [27] [29]. In our augmented method we use besides regular splines for ψ_o,ψ_1, the special singular elements $\rho_i^{-1/2}\chi_i$, $\rho_i^{1/2}\chi_i$ as in Theorem 3.3 and 3.4. Our procedure has the advantage that we are able not only to obtain higher rates of convergence but also to compute the stress intensity factors simultaneously together with the approximate desired boundary charges $[T(u)]_{|\Gamma}$. The details of our procedure and error estimates are available in [15] .

REFERENCES.

[1] R.A. Adams, Sobolev Spaces, Academic Press, New York 1975.

[2] J.F. Ahner and G.C. Hsiao, On the two-dimensional exterior boundary-value problems of elasticity, SIAM J. Appl. Math. 31 (1976) pp. 677-685.

[3] M. Costabel, Boundary integral operators on curved polygons, Ann. Mat. Pura Appl. 33 (1983) pp. 305-326.

[4] M. Costabel and E. Stephan, Boundary integral equations for mixed boundary value problems in polygonal domains and Galerkin approximation, Banach Center Publications, Warsaw, to appear (Preprint 593, FB Mathematik, TH Darmstadt 1981).

[5] M. Costabel and E. Stephan, Curvature terms in the asymptotic expansion for solutions of boundary integral equations on curved polygons, J. Integral Equations 5 (1983) pp. 353-371.

[6] M. Costabel and E. Stephan, A direct boundary integral equation method for transmission problems, to appear in J. Math. Anal. Appl. (Preprint 753, FB Mathematik, TH Darmstadt 1982).

[7] M. Costabel and E. Stephan, The method of Mellin transformation for boundary integral equations on curves with corners, in: Numerical Solution of Singular Integral Equations (ed. A. Gerasoulis,R.Vichnevetsky) IMACS, Rutgers Univ. (1984) pp. 95-102.

[8] M. Costabel, E. Stephan and W.L. Wendland, On the boundary integral equations of the first kind for the bi-Laplacian in a polygonal plane domain, Ann. Scuola Norm. Sup. Pisa, Ser. IV 10 (1983) pp 197-242.

[9] G.I. Eskin, Boundary Problems for Elliptic Pseudo-Differential Operators, Transl. of Math. Mon., American Math. Soc. 52, Providence, Rhode Island (1981).

[10] G. Fichera, Existence theorems in elasticity. Unilateral constraints in elasticity. - Handbuch der Physik (S. Flügge ed.) Berlin - Heidelberg - New York, 1972, VI a/2, pp. 347-424.

[11] G. Fichera, Linear elliptic equations of higher order in two independent variables and singular integral equations,in: Proc. Conf. Partial Differential Equations and Conf. Mechanics. University Wisconsin Press (1961) pp. 55-80.

[12] L. Hörmander, Linear Partial Differential Operators, Springer-Verlag, Berlin (1969).

[13] G.C. Hsiao and R.C. MacCamy, Solution of boundary value problems by integral equations of the first kind, SIAM Rev. 15 (1973) pp. 687-705.

[14] G.C. Hsiao, P. Kopp and W.L. Wendland, Some applications of a Galerkin-collocation method for boundary integral equations of the first kind, Math. Meth. Appl. Sci., 6 (1984), to appear.

[15] G.C. Hsiao, E. Stephan and W.L. Wendland, A boundary element approach to fracture mechanics, in preparation.

[16] G.C. Hsiao and W.L. Wendland, A finite element method for some integral equations of the first kind, J. Math. Anal. Appl. 58 (1977) pp. 449-481.

[17] G.C. Hsiao, W.L. Wendland, On a boundary integral method for some exterior problems in elasticity, to appear in Akad. Dokl. Nauk SSSR (Preprint 769, FB Mathematik, TH Darmstadt 1983).

[18] G.C. Hsiao, W.L. Wendland, The Aubin-Nitsche lemma for integral equations, J. Integral Equations 3 (1981) pp. 299-315.

[19] V.D. Kupradze, Potential methods in the Theory of Elasticity. Jerusalem, Israel Program Scientific Transl., 1965.

[20] U. Lamp, K,-T. Schleicher, E. Stephan and W.L. Wendland, Galerkin collocation for an improved boundary element method for a plane mixed boundary value problem, to appear in Computing.

[21] I.L. Lions, E. Magenes, Non-homogeneous boundary value problems and applications I, Berlin - Heidelberg - New York, Springer (1972).

[22] R. Seeley, Topics in pseudo-differential operators, in: Pseudo-Differential Operators (ed. L. Nirenberg) C.I.M.E., Cremonese, Roma (1969) pp. 169-305.

[23] E. Stephan, Boundary integral equations for mixed boundary value problems, screen and transmission problems in $\mathbb{R}^3$, Habilitationsschrift (Darmstadt) (1984).

[24] E. Stephan, W.L. Wendland, Remarks to Galerkin and least squares methods with finite elements for general elliptic problems, Lecture Notes Math. 564 Springer, Berlin (1976) pp. 461-471, Manuscripta Geodaesica 1 (1976) pp. 93-123.

[25] E. Stephan and W.L. Wendland, Boundary element method for membrane and torsion crack problems, Computer Meth. in Appl. Eng., 36 (1983) pp. 331-358.

[26] E. Stephan and W.L. Wendland, An augmented Galerkin procedure for the boundary integral method applied to two-dimensional screen and crack problems, Applicable Analysis, to appear (Preprint 802, FB Mathematik, TH Darmstadt 1984).

[27] W.L. Wendland, I. Asymptotic convergence of boundary element methods; II. Integral equation methods for mixed boundary value problems, in: Lectures on the Numerical Solution of Partial Differential Equations (ed. I. Babuška, I.-P. Liu, J. Osborn) Lecture Notes, University of Maryland, Dept. Mathematics, College Park, Md. USA #20 (1981) pp. 435-528.

[28] W.L. Wendland, On applications and the convergence of boundary integral methods, in: Treatment of Integral Equations by Numerical Methods (ed. C.T. Baker and G.F. Miller) Academic Press, London (1982) pp. 465-476.

[29] W.L. Wendland, E. Stephan and G.C. Hsiao, On the integral equation method for the plane mixed boundary value problem of the Laplacian, Math. Meth. in Appl. Sci. 1 (1979) pp. 265-321.

SCALAR DECOMPOSITION OF ELASTIC WAVES SCATTERED BY PLANAR CRACKS

Steen Krenk
Risø National Laboratory
Roskilde, Denmark

1 INTRODUCTION

It is well known that the scattered field from a crack in an infinite homogeneous elastic medium can be represented in terms of the displacement discontinuities across the crack, Δu_j [1]. The resulting stress field can also be evaluated, and in principle the limit values at the crack faces provide three integral equations for Δu_j [2]. However, these integral equations contain non-integrable singularities and they will in general be coupled.

In the case of coplanar cracks, i.e. one or more cracks located in the same plane, it follows from symmetry that the the normal displacement discontinuity Δu_3 is determined solely by the normal stress of the incident wave. In the shear problem the arbitrariness of the orientation of the coordinate system can be avoided by replacing the unknown tangential discontinuity vector Δu_α by its divergence $\Delta u_{\alpha,\alpha}$ and its rotation $e_{\alpha\beta}\Delta u_{\alpha,\beta}$. This choice of unknown functions leads to a scalar form of the integral equations.

Also the scattered far field is described conveniently in terms of the scalar discontinuities Δu_3, $\Delta u_{\alpha,\alpha}$ and $e_{\alpha\beta}\Delta u_{\alpha,\beta}$. The rotation discontinuity $e_{\alpha\beta}\Delta u_{\alpha,\beta}$ only produces SH waves, while the discontinuities Δu_3 and $\Delta u_{\alpha,\alpha}$ combine in a symmetric way to produce P and SV waves.

Finally the scattering cross section of plane elastic waves is derived by using the asymptotic far field representation to transform the power integral over the crack faces to a point property of the scattered far field in the forward direction.

2 THE REPRESENTATION INTEGRAL

Consider an infinite homogeneous elastic medium with displacements u_j and elasticity tensor C_{ijkl}, where the subscripts refer to cartesian coordinates x_j, $j = 1,2,3$. Complex notation is used, and the time factor $e^{-i\omega t}$ is generally omitted. The equation of motion then is

$$C_{ijkl}u_{k,lj} + \rho\omega^2 u_i + f_i = 0 \tag{2.1}$$

where a comma denotes differentiation, and summation over repeated subscripts is implied. f_i is the volume force.

It is well known that the displacement field inside a closed surface S can be expressed in the form [1]

$$u_m(x_r) = \int_V U_{mi}(x_r-y_r)f_i(y_r)\,dV + C_{ijkl}\int_S \{U_{im}(x_r-y_r)\frac{\partial}{\partial y_l}u_k(y_r) - \frac{\partial}{\partial y_l}U_{km}(x_r-y_r)u_i(y_r)\}n_j dS \tag{2.2}$$

where $u_i = U_{im}(x_r)$ is the displacement produced at x_r by a unit force in the m direction at the origin. Thus $U_{im}(x_r)$ is the solution of (2.1) for $f_i = \delta_{im}\delta(x_r)$.

In scattering problems it is convenient to write the total displacement field as the sum of an incident field and a scattered field,

$$u_m(x_r) = u_m^i(x_r) + u_m^s(x_r) \tag{2.3}$$

The incident field in turn is the sum of contributions from sources at infinity - plane waves - and sources f_m at finite distance. The plane waves correspond to the surface integral in (2.2) evaluated on a sphere of infinite radius, and (2.2) and (2.3) can therefore be combined to the following expression for the scattered field.

$$u_m^s(x_r) = C_{ijkl}\int_{S_o} \{ U_{im}(x_r-y_r)\frac{\partial}{\partial y_l}u_k(y_r) - \frac{\partial}{\partial y_l}U_{km}(x_r-y_r)u_i(y_r) \} n_j\, dS \tag{2.4}$$

where the integration surface S_o is the finite part of the original surface.

In the particular case of cracks of vanishing thickness the surface S_o consists of matching sides S_+ and S_- with different limiting values of $u_j(x_r)$ and $\partial u_j(x_r)/\partial x_1$. (2.4) can then be expressed as an integral over S_+ containing the discontinuities

$$\Delta u_i(y_r) = u_i^+(y_r) - u_i^-(y_r) \tag{2.5}$$

$$n_j C_{ijkl} \Delta \frac{\partial}{\partial y_1} u_k(y_r) = n_j C_{ijkl} \{ (\frac{\partial}{\partial y_1} u_k(y_r))^+ - (\frac{\partial}{\partial y_1} u_k(y_r))^- \} \tag{2.6}$$

In the following the crack faces are considered to be stress free, and consequently the discontinuity (2.6) vanishes. Upon changing the argument of differentiation (2.4) then becomes

$$u_m^s(x_r) = C_{ijkl} \frac{\partial}{\partial x_1} \int_{S_+} U_{km}(x_r - y_r) \Delta u_i(y_r) n_j \, dS \tag{2.7}$$

In (2.7) the normal n_j is directed from the elastic medium at S_+ into the crack.

For isotropic materials the elasticity tensor is

$$C_{ijkl} = \lambda \delta_{ij} \delta_{kl} + \mu (\delta_{ik} \delta_{jl} + \delta_{il} \delta_{jk}) \tag{2.8}$$

and as shown e.g. in [3]

$$U_{jm}(x_r - y_r) = \frac{i}{4\pi\rho\omega^2} \{ \delta_{jm} k^3 h_o(kr) - \frac{\partial}{\partial x_j} \frac{\partial}{\partial x_m} (h h_o(hr) - k h_o(kr)) \} \tag{2.9}$$

Here

$$r = |x_r - y_r| = ((x_r - y_r)(x_r - y_r))^{\frac{1}{2}} \tag{2.10}$$

h and k are the wave numbers for longitudinal and transverse waves, respectively,

$$h^2 = \frac{\rho\omega^2}{\lambda + 2\mu} = \left(\frac{\omega}{c_L}\right)^2 \quad , \quad k^2 = \frac{\rho\omega^2}{\mu} = \left(\frac{\omega}{c_T}\right)^2 \tag{2.11}$$

c_L and c_T are the longitudinal and transverse wave speed.

Finally

$$h_o(kr) = \frac{e^{ikr}}{ikr} \tag{2.12}$$

is the spherical Hankel function of the first kind.

3 REPRESENTATION BY SCALARS

In the special case of planar cracks it is convenient to choose the coordinate system such that the crack is located in the y_1y_2 plane, and Greek subscripts with range 1,2 are introduced to describe vector properties in the plane of the crack.

When the contributions from $\Delta u_3(y_\gamma)$ and $\Delta u_\alpha(y_\gamma)$ are separated and use is made of (2.8) the representation integral (2.7) becomes

$$\begin{aligned} u_m^s(x_r) = & - \int_S \{(\lambda+2\mu)\frac{\partial}{\partial x_k}U_{km} - 2\mu\frac{\partial}{\partial x_\gamma}U_{\gamma m}\}\, \Delta u_3\, dS \\ & - \int_S \mu\{\frac{\partial}{\partial x_3}U_{\alpha m} + \frac{\partial}{\partial x_\alpha}U_{3m}\}\, \Delta u_\alpha\, dS \end{aligned} \tag{3.1}$$

Upon substitution of the function U_{jm} from (2.9) and use of the relations (2.11) for the wave numbers the integral representation (3.1) takes the form

$$\begin{aligned} u_m^s(x_r) = & - \frac{i}{4\pi}\int_S \{(\frac{k^2}{h^2}\frac{\partial}{\partial x_m} - 2\delta_{m\gamma}\frac{\partial}{\partial x_\gamma})(kh_o(kr)) \\ & - \frac{\partial}{\partial x_m}(\frac{1}{h^2}\frac{\partial^2}{\partial x_k \partial x_k} - \frac{2}{k^2}\frac{\partial^2}{\partial x_\gamma \partial x_\gamma})(hh_o(hr) - kh_o(kr))\}\Delta u_3 dS \\ & - \frac{i}{4\pi}\int_S \{(\delta_{m\alpha}\frac{\partial}{\partial x_3} + \delta_{m3}\frac{\partial}{\partial x_\alpha})(kh_o(kr)) \\ & - \frac{2}{k^2}\frac{\partial^3}{\partial x_m \partial x_\alpha \partial x_3}(hh_o(hr) - kh_o(kr))\}\Delta u_\alpha\, dS \end{aligned} \tag{3.2}$$

The choice of the coordinates y_α in the plane of the crack is arbitrary, and it is therefore not to be expected that each of the components Δu_α will in it self represent any physical effect. In contrast Δu_3 uniquely represents the crack opening normal to its faces. A similar clear physical interpretation can be given the two scalars $\Delta u_{\alpha,\alpha}$ and $e_{\alpha\beta}\Delta u_{\alpha,\beta}$, where $e_{\alpha\beta}$ is the permutation symbol. $\Delta u_{\alpha,\alpha}$ is the difference in planar dilatation of points on opposing crack faces, while $e_{\alpha\beta}\Delta u_{\alpha,\beta}$ is the difference in rotation with respect to the normal to the crack faces.

A reformulation of the scattered displacement field (3.2) in terms of the discontinuities Δu_3, $\Delta u_{\alpha,\alpha}$ and $e_{\alpha\beta}\Delta u_{\alpha,\beta}$ can be obtained by resolving the kernels as sums of gradients and rotations and then using Gauss' theorem to effect an integration by parts. The boundary integrals vanish, because $\Delta u_\alpha = 0$ on the boundary. Although this procedure will accomplish the desired reformulation of (3.2), it is more illuminating first to separate the scattered field $u_m^s(x_r)$ into three types of waves corresponding to the well known P, SV and SH waves.

The separation of $u_m^s(x_r)$ is accomplished by representing the displacement field in terms of a scalar potential $\phi(x_r)$ and a vector potential $\psi_m(x_r)$. The vector potential may be constrained by one condition, and in the present context it is convenient to choose $\psi_3(x_r) = 0$. The potential representation then is [4]

$$u_\alpha = \frac{\partial\phi}{\partial x_\alpha} - e_{\alpha\beta}\frac{\partial\psi_\beta}{\partial x_3} \tag{3.3}$$

$$u_3 = \frac{\partial\phi}{\partial x_3} - e_{\alpha\beta}\frac{\partial\psi_\alpha}{\partial x_\beta} \tag{3.4}$$

The vector property of $\psi_\alpha(x_r)$ makes it difficult to separate transverse waves of different polarisation, and $\psi_\alpha(x_r)$ is therefore decomposed into a gradient and a rotation part.

$$\psi_\alpha = \frac{\partial\Lambda}{\partial x_\alpha} - e_{\alpha\beta}\frac{\partial\Psi}{\partial x_\beta} \tag{3.5}$$

This gives the displacement field $u_m^s(x_r)$ in terms of the three scalar potentials $\phi(x_r)$, $\Psi(x_r)$ and $\Lambda(x_r)$.

$$u_\alpha = \frac{\partial\phi}{\partial x_\alpha} - \frac{\partial^2\Psi}{\partial x_\alpha\partial x_3} - e_{\alpha\beta}\frac{\partial^2\Lambda}{\partial x_\beta\partial x_3} \tag{3.6}$$

$$u_3 = \frac{\partial\phi}{\partial x_3} + \frac{\partial^2\Psi}{\partial x_\gamma\partial x_\gamma} \tag{3.7}$$

The scalar potentials satisfy the differential equations

$$\left(\frac{\partial^2}{\partial x_m \partial x_m} + h^2\right)\phi = 0 \quad , \quad \left(\frac{\partial^2}{\partial x_m \partial x_m} + k^2\right)(\Psi, \Lambda) = 0 \tag{3.8}$$

and each of them is associated with one particular type of wave. The potential $\phi(x_r)$ is associated with longitudinal waves, or P waves, as demonstrated by the relation

$$\frac{\partial u_m}{\partial x_m} = \frac{\partial^2 \phi}{\partial x_m \partial x_m} = -h^2 \phi \tag{3.9}$$

The potentials $\Psi(x_r)$ and $\Lambda(x_r)$ are associated with transverse waves, or S waves, and it is seen from (3.7) that $\Lambda(x_r)$ does not contribute to the normal displacement component $u_3^s(x_r)$. Thus $\Lambda(x_r)$ represents SH waves, while $\Psi(x_r)$ represents SV waves.

Each type of waves can be expressed in terms of the discontinuities Δu_3, $\Delta u_{\alpha,\alpha}$ and $e_{\alpha\beta}\Delta u_{\alpha,\beta}$ by use of (3.2). Substitution of (3.2) into (3.9) followed by integration by parts gives

$$\phi(x_r) = -i\left(\frac{1}{2} + \frac{1}{k^2}\frac{\partial^2}{\partial x_\gamma \partial x_\gamma}\right)\frac{1}{2\pi}\int_S h h_o(hr)\Delta u_3(y_\gamma)\,dS$$

$$+ \frac{i}{k^2}\frac{\partial}{\partial x_3}\frac{1}{2\pi}\int_S h h_o(hr)\Delta u_{\alpha,\alpha}(y_\gamma)\,dS \tag{3.10}$$

Similarly from (3.7)

$$\frac{\partial^2}{\partial x_\gamma \partial x_\gamma}\Psi(x_r) = u_3(x_r) - \frac{\partial}{\partial x_3}\phi(x_r)$$

$$= \frac{i}{k^2}\frac{\partial^2}{\partial x_\gamma \partial x_\gamma}\frac{\partial}{\partial x_3}\frac{1}{2\pi}\int_S k h_o(kr)\Delta u_3(y_\gamma)\,dS$$

$$+ i\left(\frac{1}{2} + \frac{1}{k^2}\frac{\partial^2}{\partial x_\gamma \partial x_\gamma}\right)\frac{1}{2\pi}\int_S k h_o(kr)\Delta u_{\alpha,\alpha}(y_\gamma)\,dS \tag{3.11}$$

Finally the equation

$$e_{\alpha\beta}\frac{\partial u_\alpha}{\partial x_\beta} = -\frac{\partial}{\partial x_3}\left(\frac{\partial^2 \Lambda}{\partial x_\gamma \partial x_\gamma}\right) \tag{3.12}$$

can be integrated with respect to x_3 to give

$$\frac{\partial^2}{\partial x_\gamma \partial x_\gamma}\Lambda(x_r) = \frac{i}{4\pi}\int_S k h_o(kr) e_{\alpha\beta}\Delta u_{\alpha,\beta}(y_\gamma)\,dS \tag{3.13}$$

It follows from these relations that Δu_3 generates symmetric P waves and antisymmetric SV waves, $\Delta u_{\alpha,\alpha}$ generates antisymmetric P waves and symmetric SV waves, while $e_{\alpha\beta}\Delta u_{\alpha,\beta}$ alone generates SH waves. Symmetry is here with respect to the plane of the crack.

In order to obtain the potentials $\Psi(x_r)$ and $\Lambda(x_r)$ the differential equations (3.11) and (3.12) must be integrated. In effect this amounts to finding a function $F(r_o,x_3)$ of x_3 and

$$r_o = ((x_\gamma - y_\gamma)(x_\gamma - y_\gamma))^{\frac{1}{2}} \tag{3.14}$$

satisfying the differential equation

$$\frac{\partial^2}{\partial x_\gamma \partial x_\gamma} F = \frac{1}{r_o}\frac{\partial}{\partial r_o} r_o \frac{\partial}{\partial r_o} F = k h_o(kr) = \frac{e^{ikr}}{ir} \tag{3.15}$$

From (2.10) and (3.14)

$$r^2 = r_o^2 + x_3^2 \tag{3.16}$$

and thus for constant x_3

$$r\,dr = r_o\,dr_o \tag{3.17}$$

The differential equation (3.15) can then be integrated once to yield

$$r_o \frac{\partial}{\partial r_o} F = -i \int e^{ikr} \frac{r_o}{r} dr_o = -i \int e^{ikr} dr = -\frac{1}{k} e^{ikr} \tag{3.18}$$

The displacement field (3.6), (3.7) only makes use of $\Psi,_\gamma$ and $\Lambda,_\gamma$ and only the two dimensional gradient $F,_\gamma$ is therefore needed. From (3.18)

$$\frac{\partial F}{\partial x_\gamma} = \frac{\partial r_o}{\partial x_\gamma}\frac{\partial F}{\partial r_o} = \frac{x_\gamma - y_\gamma}{r_o}\frac{\partial F}{\partial r_o} = -\frac{x_\gamma - y_\gamma}{k r_o^2} e^{ikr} \tag{3.19}$$

With this result (3.11) and (3.13) can be integrated once to give

$$\frac{\partial}{\partial x_\gamma}\Psi(x_r) = \frac{i}{k^2}\frac{\partial}{\partial x_\gamma}\frac{\partial}{\partial x_3}\frac{1}{2\pi}\int_S k h_o(kr)\Delta u_3(y_\nu)\,dS \quad (+ \;....)$$

$$+ \frac{i}{2\pi} \int_S \{ - \frac{x_\gamma - y_\gamma}{2kr_o^2} e^{ikr} + \frac{1}{k} \frac{\partial}{\partial x_\gamma} h_o(kr) \} \Delta u_{\alpha,\alpha}(y_\nu) \, dS \tag{3.20}$$

and

$$\frac{\partial}{\partial x_\gamma} \Lambda(x_r) = - \frac{1}{4\pi} \int_S \frac{x_\gamma - y_\gamma}{kr_o^2} e^{ikr} e_{\alpha\beta} \Delta u_{\alpha,\beta}(y_\nu) \, dS \tag{3.21}$$

The displacement fields from P, SV and SH waves follow from the representation (3.6), (3.7) by substitution of (3.10), (3.20) and (3.21), respectively.

4 THE SCATTERED FAR FIELD

It is clear from the formulas (3.10), (3.11) and (3.13) that the potentials $\phi(x_r)$, $\Psi(x_r)$ and $\Lambda(x_r)$ are determined from integrals of the displacement discontinuities weighted with $h_o(hr)$ and $h_o(kr)$. At large distance from the crack the dependence of these weight functions on the integration variable y_γ may be extracted as an exponential factor, and a greatly simplified representation of the far field in terms of the two dimensional Fourier transform of the displacement discontinuities results.

Let the origin be inside the crack. Then for $|x_j| >> \max|y_\gamma|$

$$r = (x_j x_j - 2x_\gamma y_\gamma + y_\gamma y_\gamma)^{\frac{1}{2}} \sim x - \hat{x}_\gamma y_\gamma + \ldots. \tag{4.1}$$

where

$$x = |x_j| \quad , \quad \hat{x}_j = x_j / |x_j| \tag{4.2}$$

This relation gives the following leading terms

$$e^{ikr} \sim e^{ikx} e^{-ik\hat{x}_\gamma y_\gamma} \tag{4.3}$$

$$h_o(kr) \sim h_o(kx) \, e^{-ik\hat{x}_\gamma y_\gamma} \tag{4.4}$$

Similarly for the derivatives

$$\frac{\partial}{\partial x_m} e^{ikr} = ik\frac{\partial r}{\partial x_m} e^{ikr} \sim ik\hat{x}_m e^{ikx} e^{-ik\hat{x}_\gamma y_\gamma} \tag{4.5}$$

$$\frac{\partial}{\partial x_m} h_o(kr) \sim \hat{x}_m \frac{1}{x} e^{ikx} e^{-ik\hat{x}_\gamma y_\gamma} \tag{4.6}$$

It is now clear that the asymptotic form of the integrals of section 3 can be expressed in terms of the Fourier integrals

$$F\Delta u_3(\xi_\gamma) = \frac{1}{2\pi} \int_S \Delta u_3(y_\gamma)\, e^{-i\xi_\gamma y_\gamma}\, dS \tag{4.7}$$

$$F\Delta u_{\alpha,\alpha}(\xi_\gamma) = \frac{1}{2\pi} \int_S \Delta u_{\alpha,\alpha}(y_\gamma)\, e^{-i\xi_\gamma y_\gamma}\, dS \tag{4.8}$$

$$Fe_{\alpha\beta}\Delta u_{\alpha,\beta}(\xi_\gamma) = \frac{1}{2\pi} \int_S e_{\alpha\beta}\Delta u_{\alpha,\beta}(y_\gamma)\, e^{-i\xi_\gamma y_\gamma}\, dS \tag{4.9}$$

In fact, it follows directly from (3.6), (3.7) and (3.10) that the asymptotic form of the scattered P waves is

$$u_m^s(x_r) \sim \hat{x}_m \frac{1}{x} e^{ihx} \{ ih(\frac{h^2}{k^2} \hat{x}_\gamma \hat{x}_\gamma - \frac{1}{2}) F\Delta u_3(h\hat{x}_\gamma) - \frac{h^2}{k^2} \hat{x}_3 F\Delta u_{\alpha,\alpha}(h\hat{x}_\gamma) \} \tag{4.10}$$

The asymptotic P field is purely radial.

The description of the transverse waves is complicated by the need to account for their sense of polarisation. This is conveniently done by introducing the unit vectors at the point x_r in spherical coordinates with polar angle θ and azimuth ϕ. The appropriate unit vectors are easily found to be

$$n_m^\theta = (n_\alpha^\theta, n_3^\theta) = (\frac{\hat{x}_\alpha \hat{x}_3}{|\hat{x}_\gamma|}, -|\hat{x}_\gamma|) \tag{4.11}$$

$$n_m^\phi = (n_\alpha^\phi, 0) = (\frac{-e_{\alpha\beta}\hat{x}_\beta}{|\hat{x}_\gamma|}, 0) \tag{4.12}$$

With the definition (4.11) the asymptotic form of the SV waves follows from (3.6), (3.7) and (3.20) as

$$u_m^s(x_r) \sim n_m^\theta \frac{1}{x} e^{ikx} \{ ik\,\hat{x}_3 \,|\hat{x}_\gamma|\, F\Delta u_3(k\hat{x}_\gamma)$$

$$+ (\hat{x}_\gamma \hat{x}_\gamma - \tfrac{1}{2}) |\hat{x}_\gamma|^{-1} F\Delta u_{\alpha,\alpha}(k\hat{x}_\gamma) \} \qquad (4.13)$$

It is noted that the presence of the factor $|\hat{x}_\gamma|^{-1}$ does not lead to a singularity for $|\hat{x}_\gamma| = 0$, i.e. on the normal to the crack. Regularity is assured because the total two dimensional divergence of each crack face correspond to the area increase due to deformation. Thus

$$F\Delta u_{\alpha,\alpha}(0) = \frac{1}{2\pi} \int_S \Delta u_{\alpha,\alpha}(y_\gamma)\, dS = 0 \qquad (4.14)$$

The asymptotic form of the SH waves follows from (3.6) and (3.21).

$$u_m^s(x_r) \sim n_m^\phi \frac{1}{x} e^{ikx} \{ \tfrac{1}{2} \hat{x}_3 \,|\hat{x}_\gamma|^{-1}\, Fe_{\alpha\beta}\Delta u_{\alpha,\beta}(k\hat{x}_\gamma) \} \qquad (4.15)$$

Regularity of this expression is assured by the equality of the circulation of u_α around the crack boundary as evaluated from each crack face. Thus

$$Fe_{\alpha\beta}\Delta u_{\alpha,\beta}(0) = \frac{1}{2\pi} \int_S e_{\alpha\beta}\Delta u_{\alpha,\beta}(y_\gamma)\, dS = 0 \qquad (4.16)$$

The formulas (4.10), (4.13) and (4.15) give the angular dependence of P, SV and SH waves in the far field, and it is seen that the influence of the crack shape and the incident field are combined in the three Fourier transforms (4.7) - (4.9), each of which generates one or two definite types of waves. The wave amplitudes depend explicitly on the polar angle through

$$\hat{x}_3 = \cos\theta \quad , \quad |\hat{x}_\gamma| = \sin\theta \qquad (4.17)$$

5 INTEGRAL EQUATIONS

When the total stress vector on the crack faces vanishes, the scattered field satisfies the boundary conditions

$$\sigma^s_{3j}(x_\gamma) = - \sigma^i_{3j}(x_\gamma) \tag{5.1}$$

on the crack. The right hand side is assumed given, and the left hand side is defined as the limiting value of $\sigma^s_{3j}(x_\gamma, x_3)$ for $x_3 \to 0$. The stress vector follows from the displacement field by Hooke's law.

$$\sigma_{33} = (\lambda+2\mu)u_{m,m} - 2\mu u_{\gamma,\gamma} \tag{5.2}$$

$$\sigma_{3\beta} = \mu(u_{3,\beta} + u_{\beta,3}) \tag{5.3}$$

In order to obtain expressions for the stress vector in terms of the discontinuities Δu_3, $\Delta u_{\alpha,\alpha}$ and $e_{\alpha\beta}\Delta u_{\alpha,\beta}$ it is convenient to substitute the displacements from (3.6) and (3.7), whereby

$$\sigma^s_{33}/\mu = - k^2\phi - 2\frac{\partial^2}{\partial x_\gamma \partial x_\gamma}(\phi - \frac{\partial \Psi}{\partial x_3}) \tag{5.4}$$

$$\sigma^s_{3\beta}/\mu = \frac{\partial}{\partial x_\beta}(2\frac{\partial}{\partial x_3}\phi + (2\frac{\partial^2}{\partial x_\gamma \partial x_\gamma} + k^2)\Psi) - e_{\beta\gamma}\frac{\partial}{\partial x_\gamma}(\frac{\partial^2}{\partial x_3 \partial x_3}\Lambda) \tag{5.5}$$

The stress component $\sigma^s_{33}(x_r)$ follows from (5.4) by substitution of (3.10) and (3.11). In the limit $x_3 \to 0$ it becomes necessary to perform the integration before completing all the differentiations. The resulting equation is

$$\sigma^i_{33}(x_\gamma) = - \frac{\mu k^2}{4\pi}\int_S \frac{e^{ihr}}{r}\Delta u_3(y_\gamma)\, dS - \mu\frac{\partial^2}{\partial x_\alpha \partial x_\alpha}(1 + \frac{1}{k^2}\frac{\partial^2}{\partial x_\beta \partial x_\beta})\frac{1}{\pi}\int_S \frac{e^{ihr} - e^{ikr}}{r}\Delta u_3(y_\gamma)\, dS \tag{5.6}$$

It is notable, although not unexpected, that only the "opening mode" Δu_3 is excited by the normal stress σ^i_{33}.

For $x_3 = 0$ the differential equations (3.11) and (3.13) for Ψ and Λ can be integrated completely in terms of the exponential integral $E_1(\ )$, see e.g. [5]. From (3.18)

$$F(r_o,0) = \frac{1}{k}\int_{r_o}^{\infty} e^{ikr}\frac{dr}{r} = \frac{1}{k}E_1(-ikr_o) \tag{5.7}$$

With this result the integrated forms of (3.11) and (3.13) for $x_3 = 0$ are

$$\Psi(x_\gamma) = \frac{i}{4\pi k}\int_S \{E_1(-ikr) + 2h_o(kr)\}\,\Delta u_{\alpha,\alpha}(y_\gamma)\,dS \tag{5.8}$$

$$\Lambda(x_\gamma) = \frac{i}{4\pi k}\int_S E_1(-ikr)\,e_{\alpha\beta}\Delta u_{\alpha,\beta}(y_\gamma)\,dS \tag{5.9}$$

Substitution of these results together with (3.11) and (3.13) into (5.5) give the shear boundary condition

$$\sigma^i_{3\beta}(x_\gamma) = \frac{\partial}{\partial x_\beta}\left[-\frac{\mu}{4\pi}\int_S\{ik\,E_1(-ikr) + \frac{4}{k^2}\frac{\partial^2}{\partial x_3\partial x_3}\frac{e^{ihr} - e^{ikr}}{r}\}\Delta u_{\alpha,\alpha}(y_\gamma)\,dS\right]$$

$$+ e_{\beta\gamma}\frac{\partial}{\partial x_\gamma}\left[-\frac{\mu}{4\pi}\int_S\{E_1(-ikr) + h_o(kr)\}e_{\alpha\beta}\Delta u_{\alpha,\beta}(y_\gamma)\,dS\right] \tag{5.10}$$

This formulation keeps the integrals in scalar form. It is noted that although (5.10) is a resolution of $\sigma^i_{3\beta}(x_\gamma)$ into gradient and rotation parts, the resolution is not unique, and integration would introduce an arbitrary harmonic function and its conjugate.

6 SCATTERING CROSS SECTIONS

The present formulation of the scattering problem provides a convenient framework for the derivation of an expression for the scattering cross section in terms of a point property of the scattered far field. The original derivation of this result [6] made use of a power balance equation on a sphere, the radius of which increased to infinity. A later rederivation in [1] by a similar method included a spurious term. Here the result is derived for coplanar cracks without the use of a limit process by direct reformulation of the power integral over the crack faces.

The time average of the power in the scattered field is given by an integral over the crack faces.

$$\langle P\rangle = -\frac{1}{2}\int_S \mathrm{Re}\{i\omega\Delta u_j(y_\gamma)\sigma^i_{3j}(y_\gamma)\}\, dS \tag{6.1}$$

Let the shear stress be written as

$$\sigma^i_{3\alpha}(y_\gamma) = \frac{\partial}{\partial y_\alpha} S(y_\gamma) + e_{\alpha\beta}\frac{\partial}{\partial y_\beta} T(y_\gamma) \tag{6.2}$$

whereby (6.1) can be written as

$$\langle P\rangle = -\frac{1}{2}\int_S \mathrm{Re}\{ i\omega\,(\Delta u_3\bar{\sigma}^i_{33} - \Delta u_{\alpha,\alpha}\bar{S} - e_{\alpha\beta}\Delta u_{\alpha,\beta}\bar{T})\}\, dS \tag{6.3}$$

An incident plane wave with direction vector n_m can be expressed in terms of the potentials ϕ^i, Ψ^i and Λ^i,

$$h\phi^i = a\, e^{ihn_jx_j}\ , \quad k^2(\Psi^i,\Lambda^i) = (b,c)e^{ikn_jx_j} \tag{6.4}$$

The displacement field then is

$$u^i_\alpha(x_j) = ia\, n_\alpha\, e^{ihn_jx_j} + (b\, n_3n_\alpha + c\, n_3\, e_{\alpha\beta}n_\beta)e^{ikn_jx_j} \tag{6.5}$$

$$u^i_3(x_j) = ia\, n_3\, e^{ihn_jx_j} - b\, n_\alpha n_\alpha\, e^{ikn_jx_j} \tag{6.6}$$

and the incident stress in (6.3) is described in terms of

$$\bar{\sigma}^i_{33}(y_\gamma)/\mu = ah(2n_\alpha n_\alpha - \frac{k^2}{h^2})e^{-ihn_\gamma y_\gamma} + 2ibkn_3n_\alpha n_\alpha e^{-ikn_\gamma y_\gamma} \tag{6.7}$$

$$\bar{S}(y_\gamma)/\mu = -\,2ia\, n_3\, e^{-ihn_\gamma y_\gamma} - b(2n_\alpha n_\alpha - 1)e^{-ikn_\gamma y_\gamma} \tag{6.8}$$

$$\bar{T}(y_\gamma)/\mu = c\, n_3\, e^{-ikn_\gamma y_\gamma} \tag{6.9}$$

When these expressions are substituted into (6.3) the integral has the form of a Fourier transform, and the coefficients are seen to match those of the asymptotic far field, (4.10), (4.13) and (4.15). Thus the power integral takes the explicit form

$$\langle P\rangle = -\,2\pi\omega\mu\,\mathrm{Re}\{\frac{k^2}{h^2}\, a\, e^{-ihx}\, x\, n_m u^s_m(n_m x) + i\, b\, |n_\gamma| e^{-ikx}\, x\, n^\theta_m u^s_m(n_m x) - i\, c\, n_3 |n_\gamma| e^{-ikx}\, x\, n^\phi_m u^s_m(n_m x)\} \tag{6.10}$$

where n_m and n_m are unit vectors of a spherical coordinate system in accordance with (4.11) and (4.12). The factors in (6.10) are recognized as the conjugates of the components of the incident displacement field (6.5) and (6.6) in the forward direction. The factor x is removed by evaluating the asymptotic form of the scattered field at unit distance from origo. The time average of the scattered power then takes the form of a product of the incident and the asymptotic scattered fields in the forward direction.

$$\langle P\rangle = 2\pi\omega\mu\,\mathrm{Im}\{\,\bar{u}_m^i(n_j)u_m^s(n_j) + (\frac{k^2}{h^2} - 1)n_m\bar{u}_m^i(n_j)n_l u_l^s(n_j)\} \quad (6.11)$$

The scattering cross section with respect to an incident plane P or S wave is defined as the ratio of the scattered power to the energy flux of the incident wave through a unit area. The energy flux is

$$\langle P\rangle_P = \frac{\rho\omega^3}{2h}\,\bar{u}_l^i u_l^i \quad , \qquad \langle P\rangle_S = \frac{\rho\omega^3}{2k}\,\bar{u}_l^i u_l^i \quad (6.12)$$

for P and S waves, respectively. Thus the corresponding scattering cross sections are

$$\Sigma_P = \frac{4\pi}{h}\,\frac{\mathrm{Im}\{\,\bar{u}_m^i(n_j)u_m^s(n_j)\}}{\bar{u}_l^i u_l^i} \quad (6.13)$$

$$\Sigma_S = \frac{4\pi}{k}\,\frac{\mathrm{Im}\{\,\bar{u}_m^i(n_j)u_m^s(n_j)\}}{\bar{u}_l^i u_l^i} \quad (6.14)$$

This completes the conversion of the power integral over the crack faces to a property of the scattered far field in the forward dierction.

REFERENCES

1. J.E. Gubernatis, E. Domany and J.A. Krumhansl,"Formal aspects of the theory of the scattering of ultrasound by flaws in elastic materials", Journal of Applied Physics, Vol.48, pp. 2804-2811, 1977.

2. B. Budiansky and J.R. Rice, "An integral equation for dynamic elastic response of an isolated 3-D crack", Wave Motion, Vol. 1, pp.187-191, 1979.

3. V.D. Kupradze,"Dynamical problems in elasticity", in Progress in Solid Mechanics, Vol.3, Ed. I.N. Sneddon and R. Hill, North-Holland, Amsterdam, 1963.

4. J.D. Achenbach, Wave Propagation in Elastic Solids, North-Holland, Amsterdam, 1973.

5. M. Abramowitz and I. Stegun, Handbook of Mathematical Functions, Dover, New York, 1965.

6. P.J. Barratt and W.D. Collins,"The scattering cross section of an obstacle in an elastic solid for plane harmonic waves", Proceedings of the Cambridge Philosophical Society, Vol.61, pp.969-981, 1965.

BOUNDARY ELEMENT METHODS FOR SINGULARITY PROBLEMS IN FRACTURE MECHANICS

J.C. Mason*, A.P. Parker[I], R.N.L. Smith*, R.M. Thompson[‡]
Mathematics Branch* and Materials Branch[‡], RMCS,
Shrivenham, Swindon, Wiltshire, England.
Department of Mechanical and Civil Engineering[I],
North Staffordshire Polytechnic, Beaconside, Stafford, England.

The accurate modelling of crack-tip singularities is an important consideration in the solution of fracture mechanics problems, and this applies in particular when a boundary element method (BEM) is adopted, based on the formulation of the problem as a boundary integral equation (BIE). A survey is given here of a variety of the possible techniques available for treating or eliminating singularities in the BEM, including the use of special basis functions, special nodes, and special fundamental solutions. Particular emphasis is given to recent developments of special nodes for curved cracks using isoparametric elements, and to new "crack-implicit" methods based on special fundamental solutions for circular holes. Finally, numerical results are described, which have been obtained for a wide variety of fracture problems involving such aspects as complicated geometries, curved cracks, rotation and crack growth prediction.

1. THE DIRECT BEM

In implementing the BEM to model a mathematical problem, a number of decisions have to be made, which may vary considerably according to the circumstances. Broadly speaking there seem to be three main current approaches to BEM modelling. The first approach, which might be termed the "applied mathematics" approach, follows the early work of G.T. Symm and others in adopting a simple BIE model (eg constant or linear elements) in conjunction with accurate integration and singularity modelling. This is ideal for certain model problems with relatively simple geometries, and one may hence readily adapt and extend the BIE method to new problem areas, as has been done for example in the recent work of Kelmanson [1], [2]. The second approach, which might be called the "numerical methods in engineering" approach, follows C.A. Brebbia, J.C. Lachat and J.O. Watson and others in consistently adopting a reasonably accurate (but inexact) numerical model, based perhaps on quadratic elements, numerical quadrature, and the collocation method, and inspired by techniques from finite element methods (FEM). This approach is suitable for the development of general purpose computer software for a wide variety of geometries, and a recent example of its application may be found in Mason and Smith [3]. The third approach, which might be termed the "numerical analysis" approach, is that which has been studied for a number of years by W.L. Wendland and his co-workers. They adopt a model based on a Galerkin or Galerkin-collocation method, with the aim of including rigorous error estimation or analysis, and a discussion of their work is given in Wendland [4]. Clearly all three of these approaches have their virtues. However, since it is our aim to develop versatile software, we concentrate here on the second approach.

The main decisions in BEM modelling, relate to the following eight aspects: (i) the formulation of the BIE, (ii) the partitioning of the domain, (iii) the choice of elements and basic functions, (iv) the treatment of crack-tip and other singularities, (v) the method of numerical solution, (vi) the calculation of integrals, (vii) the solution of the resulting linear algebraic system, and (viii) the calculation of stress intensity factors (SIFs). In the present context, aspects (iii) and (iv) are the ones which we emphasise, although these in turn affect other aspects such as (ii) and (vi). In discussing our own chosen approach to the BEM we suppress many of the details, since these are more fully covered elsewhere.

1.1 Formulation

We adopt the "direct" BIE method based on Betti's equation in the tensor form

$$c_{lk}\, u_k + \int_\Gamma u_k\, T_{lk}\, d\Gamma = \int_\Gamma t_k\, U_{lk}\, d\Gamma . \quad (l, k = 1,2) \qquad (1)$$

Here Γ is the domain boundary, c_{lk} is a constant (equal to $\frac{1}{2}\delta_{lk}$ if the tangent plane to Γ is continuous at a point x), U_{lk} and T_{lk} are fundamental displacements and tractions (for a point force in an infinite domain - see [5]). Integration in (1) is performed as a point moves over all boundaries, the displacement u_k or traction t_k being specified.

1.2 Partitioning

Partitioning is carried out in the direct method (see Figure 1) so as to differentiate between pairs of crack surfaces, and this effectively means that "additional boundaries" (shown as broken lines in Figure 1) have to be inserted in the interior. Continuity conditions ($u^{(1)} = u^{(2)}$, $t^{(1)} = -t^{(2)}$) are imposed across partitions, but not of course across cracks.

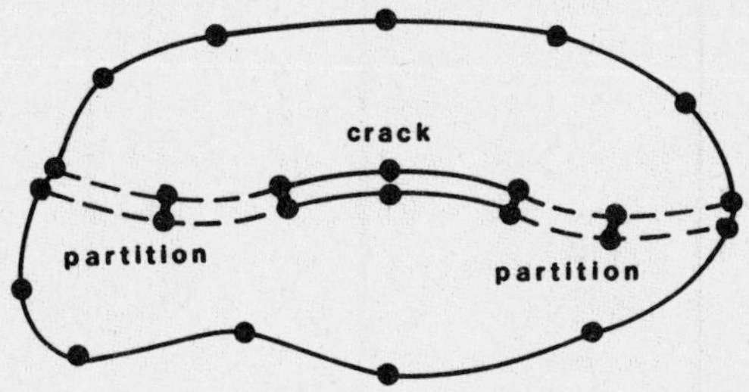

Figure 1 BOUNDARY ELEMENTS AND PARTITIONS

1.3 Elements and Basis Functions

The boundary is now divided up, in association with a corresponding set of basis functions, into a set of boundary elements. Typically m+1 equally spaced nodes are placed on each element, and an mth degree polynomial approximation is adopted. Isoparametric elements are used with curved boundary elements, and in the quadratic case, for example, this means that the cartesian coordinates x and y and the unknowns u and t are represented in terms of a parameter ξ in the forms

$$\begin{matrix} x \\ y \end{matrix} = a + b\xi + c\xi^2, \quad \begin{matrix} u \\ t \end{matrix} = d + e\xi + f\xi^2. \tag{2}$$

In this case three nodes are placed at the end points and mid point of each element, corresponding to parametric values $\xi = 0,1,\frac{1}{2}$, respectively. Shape functions M_1, M_2, M_3 are then introduced of the forms

$$M_1 = (2\xi-1)(\xi-1), \quad M_2 = -4\xi(\xi-1), \quad M_3 = \xi(2\xi-1) \tag{3}$$

so that x, y, u and t may be expressed conveniently as

$$x = x_1M_1 + x_2M_2 + x_3M_3, \text{ etc} \tag{4}$$

Here x_1, x_2, x_3 denote the values of x at $\xi = 0,\frac{1}{2},1$, and y_1, y_2, y_3, etc are all defined similarly.

1.4 Crack-Tips and Singularities

To model the crack-tips, either special nodes are introduced or special shape functions are adopted, and suitable techniques are discussed at length in §2 below.

1.5 Method of Solution

The key step in the BEM is that of converting the given problem (1), by way of the element model (2), into a routine numerical problem, namely the solution of a linear algebraic system. The method which we shall adopt here, for simplicity and convenience, is the collocation method, in which the modelled problem is solved exactly at all element nodes. This leads to a linear algebraic system

$$Hu - Gt = 0 \tag{5}$$

for the unknown displacements u and/or tractions t, which may be solved by a direct method such as Gauss elimination.

Alternatively, a Galerkin (or Galerkin-collocation) method might be adopted, as in the "numerical analysis approach" of Wendland [4], in which the left hand side of (5) is required to be orthogonal in a discrete sense to all element basis functions. This involves an additional integration over the boundary, but the extra expenditure can be worthwhile if realistic error analysis procedures are applicable.

1.6 Calculation of Integrals

Integration techniques are discussed in [3]. In two dimensional problems, Cauchy-singular integrals can be eliminated, integrals with logarithmic singularities can be treated by Gauss quadrature with a logarithmic weight function, and nicely behaved integrals can be dealt with by conventional Gauss quadrature. Gauss quadrature has been criticized by some workers in the literature as expensive. However, in the context of the overall model and accuracy which we adopt here, we have found a low order of integration (such as six) to be adequate and inexpensive.

1.7 Solution of Linear Systems

We adopt a Gauss elimination procedure, which we have to modify for the case of pure traction boundary conditions (see [3]). However, it is anticipated that more efficient procedures, which exploit special matrix structure, might become practicable in the near future.

1.8 Stress Intensity Factors

Coefficients of leading singular terms in crack-tip expansions of stresses are termed "stress intensity factors" (SIFs). Indeed this terminology is sometimes used colloquially for leading coefficients in singular expansions, regardless of whether or not these represent stresses in physical problems. In our approach, SIFs are typically evaluated by simple difference formulae based on nodal approximate solutions for u or t. Some care needs to be taken to ensure that these formulae are neither too crude nor too sophisticated. A good procedure in practice seems to be to adopt a relatively short crack-tip element and to use a simple difference formula, based for example on the two nodal values adjacent to the tip (see (16) below).

2. A SURVEY OF SINGULARITY MODELS

We now discuss five general types of methods for treating singularities, which we classify as (i) superposition, (ii) graded mesh, (iii) special basis functions, (iv) special nodes, and (v) special fundamental solutions. We note that a practical survey of finite element methods, with some emphasis on methods of types (i) and (iv), was given by Fawkes, Owen and Luxmoore [6].

2.1 Superposition

A typical superposition method solves a crack problem for a finite boundary geometry by combining (a) a classical solution for the same crack geometry, but for an infinite domain or for more elementary boundary conditions, and (b) a non-singular BEM (or FEM or other) solution for the residual problem. This approach in the context of fracture is sometimes referred to as Bueckner's principle (in recognition of the significant contribution of H.F. Bueckner). A number of examples of such an approach based on FEM are given in [6], and the BEM has been used successfully by M.A. Kelmanson in the treatment of viscous flows near corners [1] and singular slow flows [2].

The obvious advantage of this approach is that in effect the singularity is entirely removed. However, it has to be said that such an approach is only possible for a fairly limited class of model problems.

2.2 Graded Mesh

In principle, an elementary but effective general procedure for modelling singularities is to increase the number of elements in a well defined way in the vicinity of the singularity. Indeed sophisticated procedures are described by Johnson [7] and Fried and Yang [8] for grading element widths as the singularity approaches in such a way as to achieve precisely the same convergence rate as would be obtained in the absence of the singularity. In fact Johnson's procedure is based on the principle of achieving "equal errors per element in the energy norm".

A very much simpler procedure, which still increases accuracy substantially but does not attempt to achieve the full improvement ensured by Johnson's method, is to reduce appropriately the lengths of a few elements near the crack-tip. More specifically it is common to determine an optimal length for the (one) crack-tip element for a simple model problem, and then to adopt this element for a whole class of problems (see [3] for an example). Indeed it may well be worth adopting two or more levels of such simple grading, as in Smith and Mason [9] and §3.1 below.

2.3 Special Basis Functions

Another idea, which seems simple at first sight, is to retain a standard BEM mesh but to modify the basis functions on the crack-tip element (and possibly on adjacent elements). For example, Xanthis et al [10] adopt (instead of (3) above) the conforming singular shape functions

$$\hat{M}_1 = 1 - \hat{M}_2 - \hat{M}_3, \quad \hat{M}_2 = (2\xi)^{-\frac{1}{2}} M_2, \quad \hat{M}_3 = (\xi)^{-\frac{1}{2}} M_3, \qquad (6)$$

and essentially the same approach is proposed by Gomez Lera et al [11].

A similar approach (for FEM) is also adopted by Akin [12] based on the shape functions

$$\hat{M}_1 = 1 - \hat{M}_2 - \hat{M}_3, \quad \hat{M}_2 = (1 - M_1)^{-\frac{1}{2}} M_2, \quad \hat{M}_3 = (1 - M_1)^{-\frac{1}{2}} M_3, \quad (7)$$

More generally, we note that conforming shape functions may be chosen to treat an r^α singularity ($0<\alpha<1$) in the leading term (see [12]). The reader is also referred to the work of Okabe [13], who treats an r^α singularity by way of an element quadratic in r^α. In order to provide an accurate treatment of more complicated singular behaviours, especially in second and subsequent terms of an expansion, it appears to be necessary to adopt non-conforming elements. For example Kelmanson [1] adopts rather precise element approximations such as

$$u = a_0\xi^{\alpha_0} + a_1\xi^{\alpha_1} + a_2\xi^{\alpha_2} + \ldots + a_m\xi^{\alpha_m}$$

where α_0, α_1, α_2, ... are the true powers of ξ in a series expansion for u, since he has to solve a problem in which α_0, α_1, α_2, ... are essentially unrelated irrational numbers. Such an approach, necessary though it may be, does not fit very easily into general BEM procedures, and indeed it can only be applied when the appropriate values of $\alpha_0, \alpha_1, \alpha_2, \ldots, \alpha_m$ are available.

2.4 Special Nodes

In practice it is often most convenient to retain the standard basis functions, and to model the singularity by simply placing internal element nodes at special positions and adopting isoparametric elements. This approach, which has become extremely popular with engineers, is rooted in the "quarter-point" quadratic element proposed by Henshell and Shaw [14].

2.4.1 Straight Elements

Suppose that a crack-tip occurs at A, and that an isoparametric quadratic crack-tip element is placed along a straight boundary (or crack) y = 0. Then, in the quarter-point approach, nodes are placed at the ends x_1 and x_3 of the element (x_1 being at A) and at a position x_2 a quarter of the distance from x_1 to x_3. (See Figure 2.) The points x_1, x_2, x_3 are then identified with the parametric values $\xi = 0, \frac{1}{2}, 1$, respectively. Taking $x_1 = 0$, $x_3 = 1$, it follows that

$$x = \xi^2, \quad \text{and} \quad \xi = x^{\frac{1}{2}} = r^{\frac{1}{2}} \tag{8}$$

Thus the approximation (2) to the displacement u automatically includes the appropriate $r^{\frac{1}{2}}$ behaviour. However, in order to achieve a singularity of $r^{-\frac{1}{2}}$, which is proper for the traction t in the material, it is necessary to use the modified "traction-singular" quadratic element

$$t = \xi^{-1}(d + e\xi + f\xi^2) \tag{9}$$

(This is really a "special basis function", as discussed in §2.3, but it is a very simple example of this).

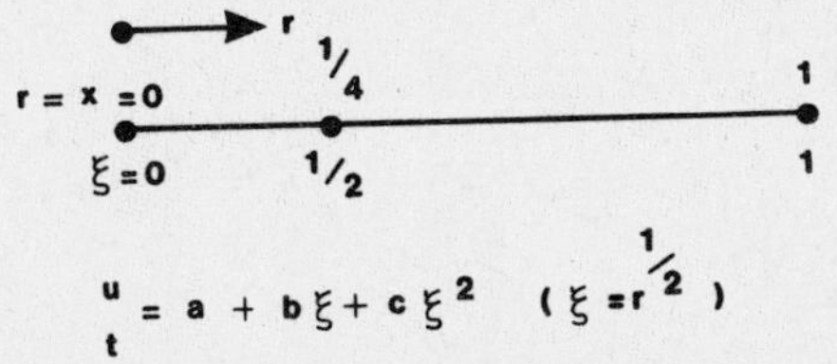

Figure 2 QUARTER POINT ELEMENT

Clearly the idea behind the quarter-point elements is easily extended to higher order elements. For example, in the cubic case two internal nodes may be placed at $x_2 = 1/9$ and $x_3 = 4/9$ (the nodes $x_1 = 0$ and $x_2 = 1$ being placed at the end points), and x_0, x_1, x_2, x_3 may be taken to correspond to $\xi = 0$, 1/3, 2/3. 1. The relationship (8) then holds again (the cubic form for x reducing to a quadratic), and a cubic approximation in ξ to u again includes an $r^{\frac{1}{2}}$ behaviour. However, we note that other choices of x_2 and x_3 are now also possible, and indeed it is sufficient that x should take the form (for arbitrary $c \neq 0$)

$$x = c\,\xi^2 + (1-c)\,\xi^3, \tag{10}$$

since ξ then has an infinite expansion in powers of $x^{\frac{1}{2}}$. The family of nodal positions corresponding to (10) is given by

$$x_2 = (2c+1)/27, \quad x_3 = 4(c+2)/27 \tag{11}$$

2.4.2 Curved Elements

Clearly the quarter-point method and its extensions, as discussed above, are only applicable to straight cracks or boundaries, and so it is necessary to modify the method to deal with singularities on curved boundaries or at curved cracks.

There is in fact no possible position for the internal node on a quadratic isoparametric curved element which exactly generalises the quarter-point for a straight element. However, if short crack-tip elements are adopted (as is our practice) then such elements are nearly straight. Mason and Smith [3], therefore place the internal node exactly or approximately a quarter of the distance along the curved element. This method, which we term the "quarter-point curved BEM", gives rather good results in practice, since it approximately reproduces the $r^{\frac{1}{2}}$ behaviour.

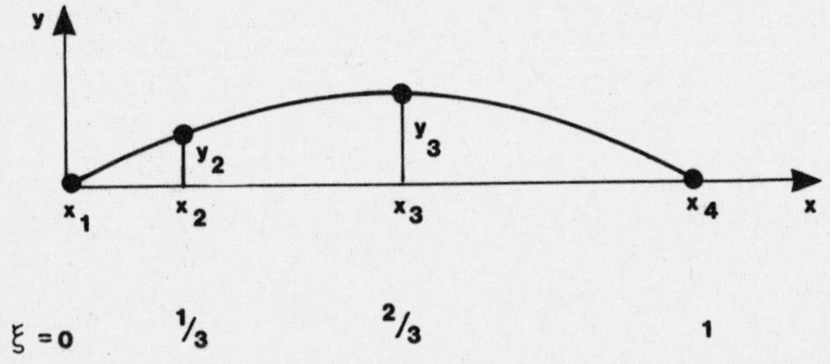

Figure 3 ISOPARAMETRIC CUBIC ELEMENT

However, by adopting isoparametric cubic or quartic elements, Mason [15] has been able to reproduce precisely the dominant singular behaviour at a curved crack-tip (or $r^{\frac{1}{2}}$ - type singularity). In the cubic case (Figure 3) four nodes (x_1,y_1), (x_2,y_2), (x_3,y_3), (x_4,y_4) (namely $\xi = 0,\ 1/3,\ 2/3,\ 1$) must be chosen for the element approximation

$$x = a_x + b_x\xi + c_x\xi^2 + d_x\xi^3,$$

$$y = a_y + b_y\xi + c_y\xi^2 + d_y\xi^3.$$

Taking $(x_1,y_1) = (0,0)$, $(x_4,y_4) = (1,0)$ the coefficients a_x and a_y vanish automatically. The coefficients b_x and b_y also vanish if x_2 and x_3 are determined to satisfy the nonlinear equation

$$y(x_3) - 2y(x_2) = 0, \quad \text{where } x_3 = 2x_2 + 2/9 \tag{12}$$

Here $y = y(x)$ is the equation of the curved boundary. A unique solution of (12) is guaranteed for a convex boundary, and a number of examples of nodal positions are given in [15] for various curves $y(x)$. For example, for a symmetrical parabolic boundary $y = Cx(1-x)$, internal nodes are placed at $x_2 = 0.1463$, $x_3 = 0.5148$.

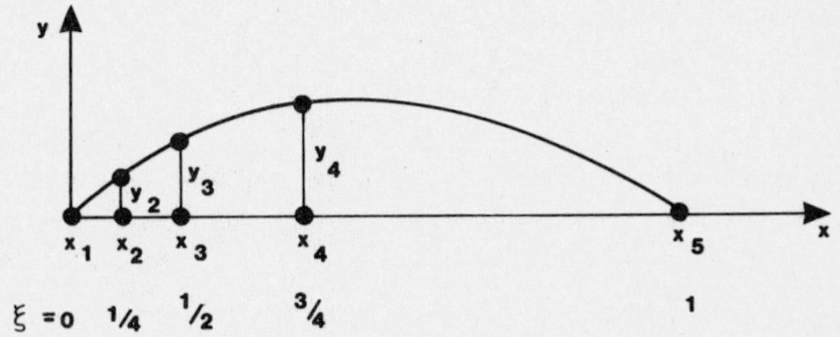

Figure 4 ISOPARAMETRIC QUARTIC ELEMENT

In the quartic case, with nodes $\xi = 0, 1/4, \frac{1}{2}, 3/4, 1$ (Figure 4), it is again necessary to ensure that a and b coefficients vanish in the element approximations

$$x = a_x + b_x\xi + c_x\xi^2 + d_x\xi^3 + e_x\xi^4$$

$$y = a_y + b_y\xi + c_y\xi^2 + d_y\xi^3 + e_y\xi^4$$

Since three internal nodes (x_2,y_2), (x_3,y_3), (x_4,y_4) are now required, the end points being $(x_1,y_1) = (0,0)$ and $(x_5,y_5) = (1,0)$, it is possible in principle to fix an (appropriate) direction for the tangent at the crack-tip (x_1,y_1). Three non-linear equations are then obtained for x_2, x_3, x_4; namely

$$12x_2 - 9x_3 + 4x_4 - 3/4 = 0, \quad 12y_2 - 9y_3 + 4y_4 = 0,$$
$$(9y_2 - 12y_3 + 7y_4) = C(9x_2 - 12x_3 + 7x_4 - 3/2) \tag{13}$$

where $y_1 = y(x_1)$ etc.

2.5 Special Fundamental Solutions

A method of rather a different character from those above, though having some of the flavour of superposition, has been adopted by Thompson [16] for simple crack (or hole) geometries. He uses, in place of the classical fundamental solutions U_{lk} and T_{lk} of §1.1, new fundamental solutions which implicitly take account of the crack (or hole). In the context of cracks we therefore refer to this approach as the "crack-implicit BEM", and in a more general context we refer to it as the "special fundamental solution method".

In Thompson's method, fundamental solutions for displacement and traction are defined in terms of complex stress functions $\phi(z)$ and $\psi(z)$ in the form

$$\phi(z) = \phi^o(z) + \phi^*(z), \quad \psi(z) = \psi^o(z) + \psi^*(z), \tag{14}$$

where ϕ^o and ψ^o are the standard stress functions for a point force in an infinite sheet, and ϕ^* and ψ^* are chosen to yield resultant zero tractions on the crack (or hole). We suppress the formulae here, since they are somewhat complicated, but details may be found in [16] and [17]. Essentially, classical formulae are already available for a (unit) circular hole, and the method of Muskhelishvili [18] may be applied to determine (14) by mapping the unit circle conformally into the required crack or hole. The classical mapping $z = \frac{1}{2}(w+w^{-1})$ is adopted for a straight crack along [-1,1].

The resulting new fundamental solutions U and T correspond to a stress-free crack or hole. The BEM may now be implemented and need only be applied to the remainder of the boundary. The advantages of Thompson's approach are that the boundary geometry is greatly simplified, that rather accurate results may be obtained at the crack, and that an element model simpler than usual (such as constant or linear elements) may often be adopted. However, there is bound to be a price to pay, and indeed the complexity of U and T makes the integration procedures more expensive in general problems.

3. NUMERICAL RESULTS

We now discuss some new and recent numerical results obtained for crack problems and stress intensity factors (SIFs) by our quarter-point curved BEM (QC-BEM) and by our crack-implicit BEM (CI-BEM) or (more generally) our special fundamental solution BEM (SFS-BEM).

3.1 Curved and Inclined Cracks

A variety of rectangular plate problems involving curved cracks (including circular arc and S-shaped cracks) have been successfully solved by the QC-BEM, and we refer the reader for details of these to Mason and Smith [4].

In order to cover two types of problems in one, we now illustrate the QC-BEM for a straight-crack problem, in which a centrally placed crack is inclined at angle α to the edges of a square plate under uniaxial tension (Figure 5). Let 2a and 2b denote the crack length and plate side, respectively, and suppose that the basic element mesh is that illustrated in Figure 5, with partitions indicated by dotted lines. Note that two levels of grading are used at the crack-tip, including an "optimal" crack-tip element of length c and an adjacent element of length 4c. The mesh is then subsequently refined by subdividing all elements other than those at the crack-tip.

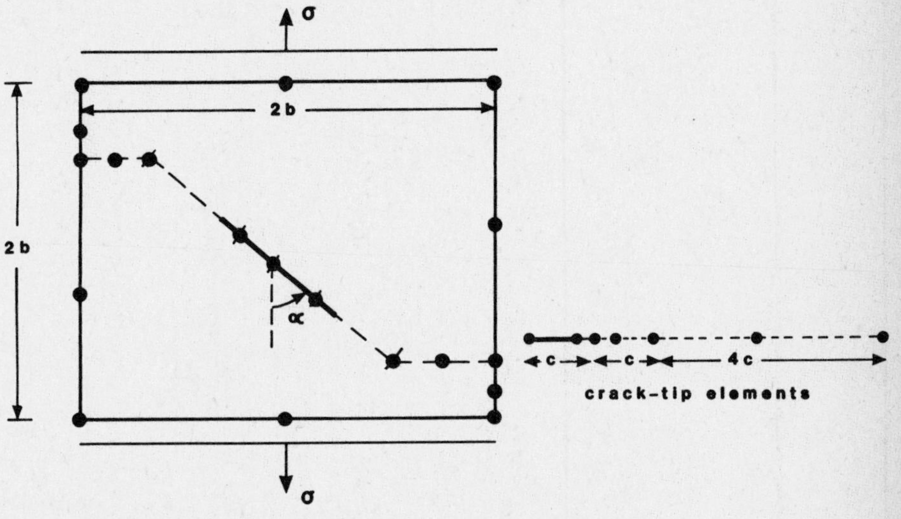

Figure 5 INCLINED CRACK UNDER UNIAXIAL TENSION-BEM MESH

The first terms in the expansions of u components u_x and u_y near the crack-tip, measured in radial coordinates (r,θ) relative to the crack-tip and its tangent, are

$$
\left.
\begin{aligned}
u_x \simeq u_x^{(1)} &= \mu^{-1}\,(r/2\pi)^{\frac{1}{2}} \left\{ \begin{aligned} &K_I \cos\tfrac{1}{2}\theta\,[\tfrac{1}{2}(\kappa-1) + \sin^2\tfrac{1}{2}\theta\,] \\ &+ K_{II}\sin\tfrac{1}{2}\theta\,[\tfrac{1}{2}(\kappa+1) + \cos^2\tfrac{1}{2}\theta\,] \end{aligned} \right\} \\
u_y \simeq u_y^{(1)} &= \mu^{-1}\,(r/2\pi)^{\frac{1}{2}} \left\{ \begin{aligned} &K_I \sin\tfrac{1}{2}\theta\,[\tfrac{1}{2}(\kappa+1) - \cos^2\tfrac{1}{2}\theta\,] \\ &+ K_{II}\cos\tfrac{1}{2}\theta\,[\tfrac{1}{2}(1-\kappa) + \sin^2\tfrac{1}{2}\theta\,] \end{aligned} \right\}
\end{aligned}
\right\} \quad (15)
$$

where μ = shear modulus, ν = Poisson's ratio, $\kappa = 3-4\nu$ or $(3-\nu)/(1+\nu)$ for plane strain or stress. Hence the SIFs K_I and K_{II} may be approximated by

$$
\frac{K_I}{u_{yC} - u_{yB}} \simeq \frac{K_{II}}{u_{xC} - u_{xB}} \simeq \left(\frac{2\pi}{r_B}\right)^{\frac{1}{2}} \cdot \frac{\mu}{\kappa+1} \qquad (16)
$$

where B and C are nodes adjacent to the crack-tip on opposite crack faces. Using just 36 elements and taking c = .1, values of K_I and K_{II} correct to within about one per cent were obtained in [9] for a variety of angles α. For example, for $\alpha = 60°$, the normalised SIFs K_I and K_{II} divided by $\sigma(\pi a)^{\frac{1}{2}}$, were determined to be:

.747 and .431 for a/b = .1; 1.073 and .619 for a/b = .6.

The former pair of values are consistent with known values of .75 and .433 for an infinite plate (a/b = 0). By way of comparison a corresponding FEM (see [9]) required 150 elements to achieve 2.5 per cent accuracy.

3.2 Secondary SIFs and Crack Growth Directions

Predictions of the direction of crack growth are commonly based on calculated values of the SIFs K_I and K_{II}. However, it was shown in [9] that these predictions can be rather inaccurate unless the second terms in the expansions for u_x and u_y are included, namely:

$$
\left.
\begin{aligned}
u_x &\simeq u_x^{(2)} = u_x^{(1)} + (8\mu)^{-1}[L_I(1+\kappa)(r\cos\theta+a) - L_{II}(1+\kappa)\,r\sin\theta] \\
u_y &\simeq u_y^{(2)} \simeq u_y^{(1)} + (8\mu)^{-1}[-L_I(3-\kappa)r\sin\theta + L_{II}(1+\kappa)(r\cos\theta+a)]
\end{aligned}
\right\} \quad (17)
$$

where L_I and L_{II} are constants which we call "secondary stress intensity factors". The latter may be approximated, in terms of displacements at the crack-tip A, by the formulae

$$L_I/u_{xA} \simeq L_{II}/u_{yA} \simeq 8\mu/[a(1+\kappa)] \qquad (18)$$

Using the 36 element mesh of §3.1 for the inclined crack problem, values of L_I, L_{II} correct to about one per cent were obtained in [9] for various values of α. For example, for $\alpha = 60°$, the normalised secondary SIFs, L_I and L_{II} divided by σ, were found to be -0.503 and 0.866 for a/b = .1; -0.993 and 0.956 for a/b = .6. The former pair of values are consistent with known values of -0.5 and 0.866 for an infinite plate (a/b = 0).

It was shown in [9] that, if a "maximum tensile stress" criterion were used to predict a crack growth direction θ, then the inclusion of L_I and L_{II} in the calculation would make a difference of up to ten per cent in θ. Moreover, only by including L_I and L_{II} could any distinction be made between the results for finite and infinite plates.

3.3 An Arc Crack in a Rotating Disc

It is clearly important to be able to treat the commonly occurring problem of cracks in rotating components (such as propellors, shafts and blades). A key model problem is that of a circular arc crack in a rotating disc, and Smith [19] has obtained consistent and accurate numerical results by a number of different methods. In particular his superposition approach involved solving both the classical problem of a rotating uncracked disc and the problem of a loaded stationary cracked disc (the loading being chosen to yield a stress-free crack on superposition). Adopting the basic mesh of Figure 6 in the latter problem and using a similar (quadratic) procedure to that of §1, results were obtained for a variety of values of θ and r/R which were accurate to within about 0.02. (Here r and R denote the radii of the crack and disc, respectively, and 2θ denotes the angle subtended by the arc). For example, for $\theta = 30°$ and r/R = 0.1, Smith obtained the values

$$K_I = 0.897, \quad K_{II} = 0.246$$

which agreed to with 0.01 with known values for an infinite plate (r/R = 0).

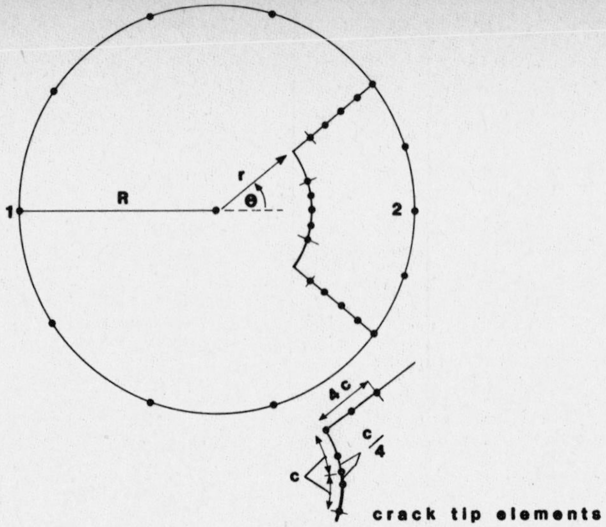

Figure 6 NODES FOR CIRCULAR ARC CRACK IN A DISC

3.4 Pressurised Thick Cylinder

Consider now the model problem (for the SFS-BEM) of a uniformly pressurised thick circular cylinder. By superposition this problem reduces to that of a stress-free circular hole in a circular cross-section (together with the classical problem of a pressurised circular hole in an infinite cross-section) (see Figure 7). By using the special fundamental solution for a circular hole, the SFS-BEM may be applied (on the exterior boundary only). To test this idea, we adopted an equally spaced mesh of n constant elements on one quarter of the circular boundary (exploiting double symmetry), and compared our numerical values for the stress concentration σ_{max}/P with the classical solution due to Lamé, namely

$$(R_1^2 + R_2^2)/(R_2^2 - R_1^2),$$

where R_1 and R_2 are the radii of the hole and exterior boundary, respectively. The results so obtained were found to be correct to within 0.5 per cent for n = 9 for all R_1/R_2 between .1 and .4, and full details may be found in [16]. In contrast, over 100 constant boundary elements were needed to achieve comparable accuracy by applying the BEM with classical fundamental solutions.

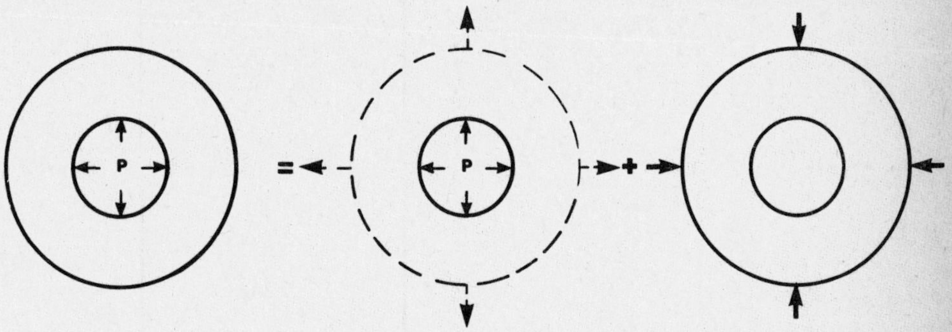

Figure 7 PRESSURISED THICK CYLINDER : SUPERPOSITION

3.5 Curved Cracks at Holes

A more complicated geometry, which occurs rather frequently in practice, is that of one or more cracks emanating from a circular hole. The QC-BEM is applicable to this general problem, and the CI-BEM is applicable in the special case of a straight crack. (We remark that there appears to be little point in adopting the special fundamental solution for a circular hole, since this does not eliminate the crack. In contrast the CI-BEM "erases" the straight crack, leaving only a circular hole and the exterior geometry).

A wide variety of problems of this type have been discussed by us in some detail in [17], and consistent results were obtained by both QC-BEM and CI-BEM for straight cracks at holes. As expected the CI-BEM was advantageous for straight cracks, and the QC-BEM gave accurate results for curved cracks. By way of an illustration of the CI-BEM, new numerical values of SIFs are given in Figure 8 for the problem of an infinite array of stress-free cracks at pressurised circular holes, and these results are believed to be correct to within about one per cent (.01 for values less than 1). The boundary conditions on the (dotted) lines of symmetry have in this case been taken to be

$$u_x = \text{constant}, \quad \int \sigma_x \, dy = 0.$$

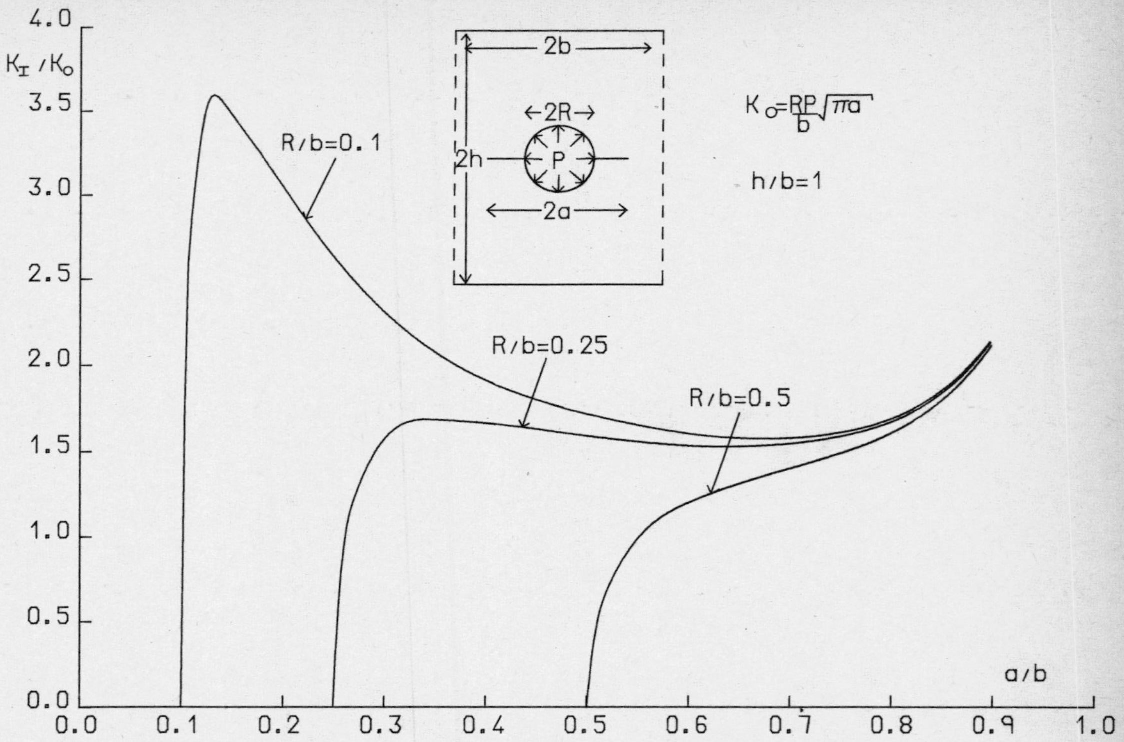

Figure 8. ARRAY PRESSURE IN HOLE

4. REFERENCES

[1] M A Kelmanson, Modified Integral Solution of Viscous Flows Near Sharp Corners. Computer and Fluids (1983) (in press).

[2] M A Kelmanson, An Integral Equation Method for the Solution of Singular Slow Flow Problems. J. Comp. Phys. 51 (1983) (in press).

[3] J C Mason and R N L Smith, Boundary Integral Equation Methods for a Variety of Curved Crack Problems. In: "Integral Equations and Their Numerical Treatment", C T H Baker and G F Miller (Ed), Academic Press, London 1982, pp 239-252

[4] W L Wendland, Boundary Element Methods and Their Asymptotic Convergence. In: "Theoretical Acoustics and Numerical Techniques" P Filippi (Ed), Lecture Notes in Physics, Springer-Verlag, Berlin, 1983 (in press).

[5] G E Blandford, A R Ingraffen, and J A Liggett, "Two-Dimensional Stress Intensity Factor Computations Using the Boundary Element Method". Int. J. Num. Meth. Engg. 17 (1981), 387-404.

[6] A J Fawkes, D R J Owen and A R Luxmoore. An Assessment of Crack-Tip Singularity Models for Use With Isoparametric Elements. Engg. Frac. Mech. 11 (1979), 143-159.

[7] A R Johnson, On the Accuracy of Polynomial Finite Elements for Crack Problems. Int. J. Num. Meth. Engg. 17 (1981), 1835-1842.

[8] I Fried and S K Yang, Best Finite Elements Distribution Around a Singularity. AIAA Journal. 10 (1972), 1244-1246.

[9] R N L Smith and J C Mason, Prediction of Crack Growth Direction Using the Boundary Element Method. In: "Boundary Elements" (Proc. 5th Int Conf.) C A Brebbia, T Futagami, and M Taneka (Ed), Springer-Verlag, Berlin, 1983, pp 495-503.

[10] L S Xanthis, M J M Bernal, and C Atkinson, The Treatment of Singularities in the Calculation of Stress Intensity Factors Using the Boundary Integral Equation Method. Comp. Meth. Appl. Mech. Engg. (to appear).

[11] S Gomez Lera, E Paris and E Alarcon, Treatment of Singularities in 2-D Domains Using BIEM. Appl. Math. Modelling 6 (1982), 111-118.

[12] J E Akin, The Generation of Elements With Singularities. Int. J. Num. Meth. Engg. 10 (1976), 1249-1259.

[13] M Okabe, Fundamental Theory of the Semi-Radial Singularity Mapping With Applications to Fracture Mechanics. Comp. Meth. Appl. Mech. Engg. 26 (1981), 53-73.

[14] R D Henshell and K G Shaw, Crack-Tip Finite Elements are Unnecessary. Int. J. Num. Meth. Engg. 9 (1975), 495-507.

[15] J C Mason, Isoparametric Curved Boundary Element Nodes for $r^{\frac{1}{2}}$-Type Singularities. In preparation, 1984.

[16] R M Thompson, The Application of the Boundary Integral Equation Method to Some Potential and Elasticity Problems. Tech. Note M/S 13, Materials and Structures Branch, RMCS Shrivenham, April 1983.

[17] J C Mason, R N L Smith, and R M Thompson, Stress Intensity Factor Evaluation by Boundary Element Methods for Cracks at Holes. Proceedings, 3rd International Conference on Numerical Methods in Fracture Mechanics, Swansea 1984. (To appear).

[18] N I Muskhelishvili "Some Basic Problems of Mathematical Theory of Elasticity", Noordhof, Groningen (Holland), 1953.

[19] R N L Smith, Stress Intensity Factors for an Arc Crack in a Rotating Disc. Engg. Frac. Mech. (1983), (submitted for publication).

SUR L'APPROXIMATION DES SOLUTIONS DU PROBLEME DE DIRICHLET DANS UN OUVERT AVEC COINS

M.A. MOUSSAOUI
Institut de Mathématiques . USTHB
B.P. 9 Dar El Beida . Algérie .

INTRODUCTION

Dans l'approximation des solutions d'un problème elliptique dans un ouvert régulier , l'ordre de l'erreur d'approximation dépend des éléments finisfi--nis et du maillage utilisés , mais aussi de la régularité des solutions . Ce dernier paramètre apparait comme une obstruction dans la situation où, l'ouvert dans lequel on traite le problème présente des singularités , ou lorsque les données aux limites presentent des discontinuités. En effet lorsque par exemple , et c'est le cas qui nous interessera dans ce travail,l'ou--vert est un polygône plan , la régularité de la solution d'un problème elli--ptique ,ne dépend pas uniquement de la régularité des données mais également de la géométrie de ce polygône . Prenons pour illustrer cela l'exemple du pro-blème de Dirichlet pour l'équation de Laplace dans un polygône Ω de fron--tière Γ :

$$\Delta u = f \quad \text{dans } \Omega \quad (1)$$

$$u = o \quad \text{sur } \Gamma \quad (2)$$

Pour f donnée dans,par exemple dans $L^2(\Omega)$, il est bien connu que ce problè--me admet une unique solution faible (ou variationnelle) dans $H_o^1(\Omega)$. Cette solution peut ne pas être dans $H^2(\Omega)$, comme ce serait toujours le cas si l'ouvert était régulier . Par suite la mise en oeuvre de la méthode d'éléments finis classique ne donne pas une erreur d'approximation optimale . En fait dans beaucoup de situations , comme celle à laquelle nous nous interesserons par la suite , la solution se décompose en deux parties , l'une régulière (que l'on notera plus loin u_o) et l'autre formée d'une somme finie de fonc--tions "singulières" dont la régularité est liée à la géométie de l'ouvert. Pour préciser ces idées considérons le problème de Dirichlet (1) , (2) dans le domaine plan en "forme de L" disposé de telle sorte que son angle rentrant soit porté par les axes de coordonnées comme indiqué sur la figure 1 . Pour une donnée f dans $L^2(\Omega)$ la solution variationnelle u de ce problème se dé--compose d'après [1] de la manière suivante :

$u = u_o + \lambda u_s$ avec u_o dans $H^2(\Omega)$ $H_o^1(\Omega)$ et u_s s'écrivant au voisina--ge de l'origine et en coordonnées polaires :

$u_s(r,\theta) = r^{2/3}\sin 2/3\theta$.

On peut vérifier que u_s est dans $H^s(\Omega)$ pour tout $s < 5/3$ mais pas pour $5/3$. Par ailleurs , le coefficient λ dépend linéairement de f , mais de manière globale . Comme on le verra par la suite λ peut être non nul même pour f très régulière par exemple indéfiniment dérivable à support compact .

Considérons à présent un schéma d'approximation de u par la méthode d'éléments finis classique d'ordre un introduite par exemple dans [2] . Posant $V = H_0^1(\Omega)$ u est alors l'unique élément de V tel que :

$$a(u,v) = \int_\Omega \nabla u . \nabla v \, dx = \int_\Omega fv \, dx \qquad \forall v \in V$$

Soit V_h un sous espace de dimension finie de V . Appelons u_h l'unique élément de V_h qui vérifie :

$$a(u_h,v) = \int_\Omega fv \, dx \quad \text{pour tout } v \text{ dans } V_h .$$

Munissant V de la norme induite par la forme bilinéaire a , il est alors bien connu que l'on a l'inégalité suivante :

$$\|u - u_h\| \leqslant \inf \|u - v\| \quad \text{pour } v \text{ parcourant } V_h$$

et que si la solution u est dans $H^2(\Omega)$ alors

$$\|u - u_h\| \leqslant C_1 h$$
$$|u - u_h| \leqslant C_2 h^2$$

où h désigne le diamètre des éléments de la triangulation adoptée , $\|.\|$ la norme dans V et $|.|$ la norme L^2 . Les constantes C_1 et C_2 sont indépendan--tes de h mais dépendent entre autres de la norme L^2 des dérivées secondes de la solution u . Ce que nous pouvons obtenir dans le cas d'un ouvert polygônal ce sont les estimations suivantes :

$$\|u - u_h\| \leqslant Ch^{1/2}$$
$$|u - u_h| \leqslant Ch$$

Nous avons donc une perte dans l'ordre de convergence. Pour pallier à cet in-convénient , plusieurs méthodes de calcul ont été proposées dont nous cite--rons entre autres :

a) Adjonctions des singularités à l'espace d'approximation V_h cf [3] , [4].

b) Utilisation des espaces de Sobolev avec poids et maillage raffiné au voi--sinage des sommets adapté aux singularités cf [5] , [10]

c) Méthode de séparation de l'ouvert en sous domaines cf [6]

Toutes ces méthodes permettent d'obtenir un ordre de convergence equivalent à celui obtenu dans le cas d'un ouvert régulier .

Nous nous proposons ici de developper une autre méthode basée essentiellement sur le problème adjoint . Cette méthode semble particulièrement intéressante dans les situations où l'on doit résoudre un grand nombre de fois le même problème avec des seconds membres différents et lorsque seule est désirée la connaissance de la valeur des coefficients des singularités et non les solutions approchées en elles mêmes . Nous le ferons en détail sur le cas modèle donné precedemment et nous signalerons quelques extensions possibles.

QUELQUES RESULTATS THEORIQUES DANS LE CAS MODELE

Fixons tout d'abord quelques notations : Ω désignera l'ouvert ci-dessous (figure 1). Sa frontière Γ étant supposée être la réunion de Γ_1 formée des deux côtés issus de l'origine et Γ_0 son complémentaire . Dans toute la sui--te on posera k = 2/3 et (r , θ) désigneront les coordonnées polaires à par--tir de l'origine . On définit alors dans Ω la fonction :

$$S^*(r,\theta) = r^{-k}\sin k\theta + \bar{S}^*(r,\theta)$$

où $\bar{S}^*(r,\theta)$ est l'unique élément de $H^1(\Omega)$ solution de :

$$\Delta \bar{S}^* = 0 \quad \text{dans } \Omega$$

$$\bar{S}^* = - r^{-k}\sin k\theta \quad \text{sur } \Gamma$$

La fonction $r^{-k}\sin k\theta$ est nulle sur Γ_1 et régulière sur Γ_0 . L'existence de $\bar{S}^*$ est alors bien connue . Il est aisé de verifier que :

i) S^* est un élément de $L^2(\Omega)$ non identiquement nul

ii) S^* est harmonique et donc régulière à l'intérieur de Ω

iii) S^* est nulle sur Γ

Remarquons par ailleurs que S^* est dans $L^p(\Omega)$ pour tout $p < 2/k$.

Posons $a = |S|$ et définissons à présent la fonction S comme l'unique solu--tion dans $H_0^1(\Omega)$ de l'équation :

$$\Delta S = S^*$$

S representera en quelque sorte la singularité modèle pour le problème de Dichlet (1) , (2) . Nous avons alors les deux lemmes suivants dûs à [1] :

Lemme 2.1

Soient f donnée dans $L^2(\Omega)$ et u la solution variationnelle de (1),(2)

$$u \in H^2(\Omega) \Longleftrightarrow (f,S^*) = 0$$

Ici (.,.) désigne le produit scalaire dans $L^2(\Omega)$.

Lemme 2.2

Pour f donnée dans $L^2(\Omega)$, la solution u du problème (1),(2) s'écrit

$$u = \lambda\, r^k \sin k\theta + u_0 \quad \text{avec } u_0 \text{ dans } H^2(\Omega)$$

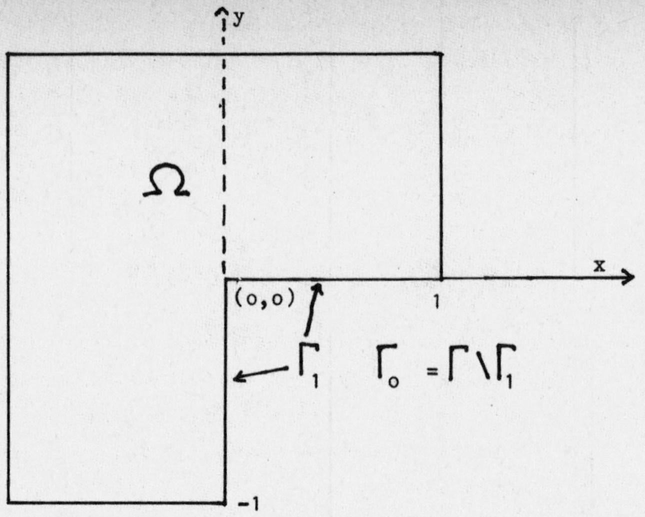

figure 1

Ces deux lemmes nous permettent d'énoncer la

Proposition 2.3

La solution S de l'équation $\Delta S = S^*$ s'écrit :

i) $S = \lambda r^k \sin k\theta + \overline{S}$ avec $\overline{S}$ dans $H^2(\Omega)$ et

ii) $\lambda = (-1/\pi)a^2$

D'après le lemme 2.2 , il reste à démontrer l'égalité ii):
Remarquons d'abord que S^* étant dans un espace L^p avec $p > 2$, la fonction $\overline{S}$ est dans $W^{2,p}(\Omega)$ avec $p > 2$ (convenablement choisi) . Elle est donc dans $C^1(\overline{\Omega})$ d'après les théorèmes d'injection de Sobolev .
Pour $t > o$ suffisamment petit on définit :

$$D_t = \{(r,\theta) \in \Omega \ ; \ o < r < t\} \ C_t = \{(r,\theta) \in \Omega ; r = t\} \text{ et } \Omega_t = \Omega \setminus \overline{D}_t$$

Nous avons alors :

$$(S^*,S^*)_t = (\Delta S, S^*)_t = \int_{C_t} (S^* \frac{\partial S}{\partial n} - S \frac{\partial S^*}{\partial n}) \, d\sigma$$

où $(,)_t$ désigne le produit scalaire dans $L^2(\Omega_t)$. En décomposant S et S en leurs parties régulières et singulières et en utilisant la régularité de S on obtient le résultat désiré en faisant tendre t vers zero .
A partir de ces résultats nous avons alors la

Proposition 2.4

Soit f donnée dans $L^2(\Omega)$. La solution u du problème (1),(2) s'écrit

$u = u_o + \lambda S$ avec u_o dans $H^2(\Omega)$ et

$\lambda = (1/a^2).(f,S^*)$

Remarque 2.2 : d'après la proposition précedente on peut aussi écrire :

$u = (-1/\pi) r^k \sin k\theta \ + \ u_1$ avec u_1 dans $H^2(\Omega)$.

Preuve On décompose f sous la forme :

$f = (f, \frac{S^*}{a}) \frac{S^*}{a} + f_o$ et on applique Δ^{-1} aux deux membres de cette égalité. Le résultat est alors une conséquence du lemme 2.1

Remarque 2.3 La relation qui donne le coefficient de la singularité montre bien que la régularité de la solution ne dépend pas uniquement de la régularité de f, en ce sens que ce coefficient peut être non nul même si f est très régulière .

Avant de clore ce paragraphe donnons un résultat analogue concernant le problème non homogène :

$$\Delta u = 0 \text{ dans } \quad (3)$$

$$u = g \quad \text{sur} \quad (4)$$

avec g dans $H^{3/2}(\Gamma)$ que l'on supposera (quitte à modifier u par une constante) nulle en (o,o) . On a alors comme precedemment la

Proposition 2.5

La solution u du problème (3) , (4) s'écrit sous la forme :

$u = u_o + \lambda S$ avec u_o dans $H^2(\Omega)$ et

$$\lambda = (1/a^2) \int_\Gamma g \frac{\partial S^*}{\partial n} d$$

n désignant la normale à Γ orientée vers l'extérieur de Ω.

Pour obteneir ce résultat il suffit d'écrire dans Ω_t la formule de Green puis de faire tendre t vers zero .

APPROXIMATION NUMERIQUE DANS CE CAS MODELE

1. Approximation du coefficient de la singularité.

Elle repose sur la proposition 2.4 . Ainsi pour f donnée dans $L^2(\Omega)$ la so-lution u du problème (1) , (2) s'écrit $u = u_o + \lambda S$ avec u_o régulière et λ donné par la relation :

$$a^2 \lambda = (f , S^*)$$

ou encore de manière plus explicite :

$$u = (-1/\pi).(f,S^*) \, r^k \sin k\theta \ + u_1 \quad \text{avec } u_1 \text{ dans } H^2(\Omega) .$$

Ainsi la connaissance de S^* permet de determiner le coefficient de la sin-gularité sans avoir à calculer la solution approchée du problème.

La première étape consistera pour nous à calculer une approximation de S^*.

Rappelons que l'on a :

$$S^* = r^{-k} \sin k\theta \ + \ \bar{S}^*$$

$$\Delta \bar{S}^* = 0 \quad \text{dans } \Omega$$

$$\bar{S}^* = g = -r^{-k}\sin k\theta \quad \text{sur } \Gamma$$

Introduisons une méthode d'éléments finis . On notera $V = H_o^1(\Omega)$ et P_1 l'espace des polynômes de degré inférieur ou égal à 1 .

On considère une triangulation régulière , T_h , de Ω formée de triangles K avec pour chaque K un diamètre de l'ordre de h . Posons :

$$W_h(\Omega) = \left\{ v \in H^1(\Omega) \; ; \; v|_K \in P_1 \; , \; \forall K \in T_h \; , \; v = g \text{ aux sommets frontière} \right\}$$

$$V_h = \left\{ v \in H^1(\Omega) \; ; \; v|_K \in P_1 \; , \; \forall K \in T_h \; , \; v = 0 \text{ aux sommets frontière} \right\}$$

Une approximation de $\bar{S}^*$ est l'unique solution de :

$$\int_\Omega \nabla \bar{S}_h^* . \nabla v_h \, dx = 0 \qquad \forall v_h \in V_h$$

$$\bar{S}_h^* \in W_h$$

Utilisant alors les résultats classiques cf [2], et l'interpolation on a

Proposition 3.1

Il existe une constante C ,indépendante de h,telle que:

$$|\bar{S}^* - \bar{S}_h^*| \leq Ch$$

(Ici et dans la suite $|\,.\,|$ désignera la norme L^2 et $\|\,.\,\|$ celle de V) .

Remarque 3.1 On obtient en fait une erreur en $O(h^s)$ avec $s < 2k$

Remarque 3.2 On peut obtenir une meilleure approximation en utilisant la méthode introduite dans [7] . Cette méthode necessite la résolution de deux problèmes de Dirichlet et donne une erreur en $O(h^s)$ avec $s < 1+k$.

On a alorsla proposition

Proposition 3.2

Posant $S_h^* = r^{-k}\sin k\theta + \bar{S}_h^*$ et $\lambda_h = (-1/\pi).(f,S_h^*)$ on a :

$$|\lambda - \lambda_h| \leq (1/\pi) \, |f| \, |\bar{S}^* - \bar{S}_h^*|$$

C'est une conséquence directe de la formule établie precedemment . Ainsi l'approximation du coefficient λ est du même ordre que celui de l'approximation de $\bar{S}^*$ en norme L^2.

Par ailleurs le calcul de S_h^* est fait independamment de f , et peut être exécuté une fois pour toutes .

2. Approximation de la solution u .

Pour f donnée dans $L^2(\Omega)$ on cherche une approximation de u solution de (1),(2) . Nous savons par la proposition 2.4 que l'on peut écrire u:

$u = \lambda r^k \sin k\theta + w$ avec w dans $H^2(\Omega)$.

En posant $g = -r^k \sin k\theta$ sur Γ, w est alors solution du problème :

$\Delta w = f$ dans Ω (1)

$w = \lambda g$ sur Γ (2)'

Par la première partie de ce paragraphe nous avons une approximation λ_h de λ. On cherchera alors , par la méthode d'éléments finis introduite prece--demment une approximation de w , en discrétisant le problème (1) ,(2)' où λ est remplacé par λ_h , soit w_h . L'approximation de u sera alors :

$$u_h = w_h + \lambda_h r^k \sin k\theta \ .$$

En utilisant les mêmes arguments que dans [7] nous avons la

Proposition 3.3

Il existe une constante C , indépendante de h , telle que :

$$\| u - u_h \| \leq Ch$$

3. Utilisation de la proposition 2.5

Nous voudrions pour terminer signaler une utilisation possible de la propo--sition 2.5 . Pour simplifier plaçons nous dans le cas modèle précédent. Appelons $\tilde{\Omega}$ le carré $]-1,1[\times]-1,1[$ qui contient l'ouvert Ω. Pour deter--miner le coefficient λ de la singularité dans le problème (1) , (2) on com--mence par résoudre par une méthode d'éléments finis le problème :

$\Delta U = F$ dans $\tilde{\Omega}$

$U = 0$ sur le bord $\tilde{\Gamma}$ de $\tilde{\Omega}$

où F est un prolongement quelconque de f à $\tilde{\Omega}$.

En posant $G = U$ sur $\tilde{\Gamma}$ alors u peut s'écrire sous la forme :

$u = U + G(o,o) + w$ avec w solution de

$\Delta w = 0$ dans Ω

$w = G - G(o,o) = g$ sur Γ

Le coefficient λ est alors donné par la proposition 2.5 car U est réguliè--re et l'on peut montrer que l'on a une approximation de λ avec une erreur en O(h) en posant

$$\lambda_h = (1/\pi) \int_\Gamma g(\sigma) . \frac{\partial S_h^*}{\partial n} d\sigma$$

où g est la restriction à Γ de l'approximation de U dans $\tilde{\Omega}$. Cela suppose bien entendu S_h^* déjà connu .

Extensions de la méthode

. Cette méthode peut être étendue à d'autres types de conditions aux limi--tes , pour le même opérateur .

. Elle est utilisée par Abdelatif - Zaïr [8] pour le système de l'elasti--cité linéaire dans un domaine plan fissuré .

. Il est peut être possible d'améliorer les ordres d'erreur en utilisant les résultats de [9] .

. Enfin nous voudrions avant de terminer signaler qu'une méthode similai-a été developpée parallèlement dans [11] à partir des formules de repré--sentation des coefficients des singularités contenues dans [12] .

Bibliographie

[1] Grisvard : Alternative de Fredholm relative au problème de Dirichlet dans un polygône ou un polyèdre .1ère partie. Bol. UMI 14 - 5 - 1972 .

[2] Ciarlet : The finite element method for elliptic problems . North Holland 1977 .

[3] Strang - Fix: An analysis of the f.e.m. Prentice Hall New York 1972

[4] Lelièvre : Utilisation des fonctions singulières dans la méthode des éléments finis . Thèse de spécialité 1977 Marseille .

[5] Raugel : Résolution numérique des problèmes elliptiques dans des domai--nes avec coins . Thèse de troisième cycle . 1978 Rennes .

[6] Bardos - Céa - Grisvard : Calcul numérique des solutions singulières de problèmes aux limites . Séminaire Université de Nice 1977 .

[7] Bellout - Moussaoui : Remarques sur l'approximation numérique du problè--me de Dirichlet dans un ouvert avec coins . CSTN - CROA - L5 - 1981

[8] Abdelatif - Zaïr : Approximation des coefficients d'intensité de con--traintes dans un domaine plan fissuré . A paraître . Alger 1984 .

[9] Schatz - Wahlbin : Maximum norm estimates in the finite element method on plane polygonal domains . Math. of Comp. (32) 1978 .

[10] Babuska : Finite element method for domains with corners . Computing 6 264 - 273 (1970) .

[11] Blum - Dobrowolski : On finite element methods for elliptic equations on domains with corners , Computing , 28 , 53 - 63 (1982) .

[12] Maz'ja - Planemevskij : Coefficients in the asymptotics of the soluti--ons of elliptic boundary value problems in a cone , Sem. LOMI 52, 110-127 (1975) (= J.Sov.Math. 9, 750 - 764 (1978) .

THE TREATMENT OF SINGULARITIES IN ORTHONORMALIZATION METHODS FOR NUMERICAL CONFORMAL MAPPING

N. Papamichael
Department of Mathematics and Statistics
Brunel Universiy
Uxbridge, Middlesex, UB8 3PH, U.K.

1. INTRODUCTION

This paper is a report of some recent advances concerning the treatment of singularities in certain methods for numerical conformal mapping. The methods considered are orthonormalization methods for determining approximations to the following three conformal maps:

CM1: The mapping of a domain interior to a closed Jordan curve onto the interior of the unit disc.

CM2: The mapping of a domain exterior to a closed Jordan curve onto the exterior of the unit disc.

CM3: The mapping of a doubly-connected domain, bounded by two closed Jordan curves, onto a circular annulus.

The above three conformal maps are of considerable practical interest and have important applications in, for example, fluid mechanics, electrostatics and stress analysis. Several such applications to specific physical problems are noted in the survey article by Laura [10]. A more general application, which has received considerable interest recently, concerns the computer generation of orthogonal curvilinear meshes for the finite difference solution of partial differential equations; see e.g. Thompson et al [24].

2. THE CONFORMAL MAPPING PROBLEMS

Let $\partial\Omega$ be a closed piecewise analytic Jordan curve in the complex z-plane, and assume that the origin 0 lies in $\mathrm{Int}(\partial\Omega)$. Then, the two problems associated with the conformal maps CM1 and CM2 can be stated as follows:

Problem P1. To determine the function

$$w = f(z) , \tag{2.1}$$

which maps $\Omega_I = \mathrm{Int}(\partial\Omega)$ one-one conformally onto the unit disc

$$D_I = \{w : |w| < 1\} , \tag{2.2}$$

so that

$$f(0) = 0 \quad \text{and} \quad f'(0) > 0 . \tag{2.3}$$

Problem P2. To determine the function

$$w = g(z) , \tag{2.4}$$

which maps $\Omega_E = \mathrm{Ext}(\partial\Omega)$ one-one conformally onto the exterior of the unit disc

$$D_E = \{w : |w| > 1\} , \tag{2.5}$$

so that

$$g(\infty) = \infty \quad \text{and} \quad \lim_{z\to\infty} g'(z) > 0 . \tag{2.6}$$

The above two problems can be related to each other by means of the transformation

$$z \longrightarrow z^{-1} . \tag{2.7}$$

This simple inversion transforms $\partial\Omega$ onto a Jordan curve $\partial\hat{\Omega}$ and maps conformally Ω_I onto $\hat{\Omega}_E = \mathrm{Ext}(\partial\hat{\Omega})$ and Ω_E onto $\hat{\Omega}_I = \mathrm{Int}(\partial\hat{\Omega})$. Therefore, if $\hat{f}$ and $\hat{g}$ are the interior and exterior mapping functions associated respectively with $\hat{\Omega}_I$ and $\hat{\Omega}_E$ then

$$g(z) = \left\{\hat{f}(z^{-1})\right\}^{-1} \quad \text{and} \quad f(z) = \left\{\hat{g}(z^{-1})\right\}^{-1} . \tag{2.8}$$

There are two important domain functionals associated with problems P1 and P2. These are respectively the *conformal radius*

$$R = 1/f'(0) \tag{2.9}$$

of Ω_I at 0, and the *capacity*

$$c = \lim_{z\to\infty} \left\{g'(z)\right\}^{-1} \tag{2.10}$$

of the curve $\partial\Omega$. That is, R is the radius of the disc which is the image of Ω_I under the normalized mapping

$$w = f(z)/f'(0) \ , \tag{2.11}$$

and c is the reciprocal of the conformal radius of the interior domain $\hat{\Omega}_I$. More precisely, it follows from (2.8) that if $\hat{R}$ and $\hat{c}$ are respectively the conformal radius of $\hat{\Omega}_I$ and the capacity of $\partial\hat{\Omega}$ then

$$c = 1/\hat{R} \quad \text{and} \quad R = 1/\hat{c} \ . \tag{2.12}$$

Let now Ω_D be a finite doubly-connected domain with boundary $\partial\Omega_D = \partial\Omega_1 \cup \partial\Omega_2$, where $\partial\Omega_i$; $i = 1,2$ are piecewise analytic Jordan curves. We assume that $\partial\Omega_1$ and $\partial\Omega_2$ are respectively the inner and outer components of $\partial\Omega_D$ and that the origin 0 lies in the "hole" of Ω_D, i.e. $0 \in \mathrm{Int}(\partial\Omega_1)$. Then, the problem associated with the conformal map CM3 can be stated as follows.

Problem P3. To determine the function

$$w = F(z) \tag{2.13}$$

which maps Ω_D one-one conformally onto a circular annulus

$$A(r_1,r_2) = \{w : r_1 < |w| < r_2\} \tag{2.14}$$

so that

$$F(\zeta_1) = r_1 \ , \tag{2.15}$$

where ζ_1 is some fixed point on $\partial\Omega_1$ and r_1 is a prescribed number.

The condition (2.15) determines uniquely the radius r_2 of the outer circle and ensures that $\partial\Omega_1$ and $\partial\Omega_2$ are mapped respectively onto the two circles $|w| = r_1$ and $|w| = r_2$. The ratio

$$M = r_2/r_1 \ , \tag{2.16}$$

of the two radii of $A(r_1,r_2)$ is an important domain functional known as the *conformal modulus* of Ω_D. This ratio determines completely the conformal equivalence class of Ω_D, i.e. two doubly-connected domains can be mapped conformally onto each other if and only if they have the same conformal modulus. Furthermore, the value of M is of special significance in many practial applications; see e.g. [13] and [8,p.346]. See also Kantorovich and Krylov [9,p.362] who state that *"the finding of M is one of the difficult problems of the theory of conformal transformations".*

3. THE NUMERICAL METHODS

The methods considered here are the well-known Bergman kernel method, for the solution of problems P1 and P2, and another, lesser known, orthonormalization method for the solution of problem P3. We shall refer to these two methods as the BKM and ONM respectively.

3.1 The BKM for the solution of problem P1.

The BKM is an orthonormalization method based on the properties of the so-called Bergman kernel function $K(.,0)$ of the interior domain Ω_I. The theory of this method is treated extensively in the literature; see e.g. [1,6,7,14].

Let $L_2(\Omega_I)$ be the Hilbert space of all square integrable analytic functions in Ω_I, and denote the inner product of $L_2(\Omega_I)$ by $(.,.)_{\Omega_I}$, i.e.

$$(u,v)_{\Omega_I} = \iint_{\Omega_I} u(z)\overline{v(z)}dS_z \ . \tag{3.1}$$

Then the kernel $K(.,0)$ has the reproducing property

$$(\eta,K(.,0))_{\Omega_I} = \eta(0) \ , \quad \forall \ \eta \in L_2(\Omega_I) \ , \tag{3.2}$$

and is related to the mapping function f of problem P1 by means of

$$K(z,0) = \frac{1}{\pi} f'(0)f'(z) \ ; \tag{3.3}$$

see e.g. [7,pp.36-39]. In the BKM the approximation to f is obtained from (3.3), after first approximating the kernel $K(.,0)$ by a finite Fourier series sum. More precisely, if $\{\eta_j\}$ is a complete set of $L_2(\Omega_I)$ then the details of the method are as follows.

The set $\left\{\eta_j\right\}_{j=1}^{n}$ is orthonormalized, by means of the Cram-Schmidt process, to give the orthonormal set $\left\{\eta_j^*\right\}_{j=1}^{n}$. Then, because of (3.2),

$$\begin{aligned} K_n(z,0) &= \sum_{j=1}^{n} (K(.,0),\eta_j^*)_{\Omega_I} \eta_j^*(z) \\ &= \sum_{j=1}^{n} \overline{\eta_j^*(0)}\eta_j^*(z) \end{aligned} \tag{3.4}$$

is the nth partial Fourier series sum of the kernel function and hence, from (3.3),

$$f_n(z) = \left\{\frac{\pi}{K(0,0)}\right\}^{1/2} \int_0^z K_n(t,0)\,dt \tag{3.5}$$

is the nth BKM approximation to the mapping function f of problem P1. Also, from (2.9),

$$R_n = 1/\left\{\pi K_n(0,0)\right\}^{1/2} \tag{3.6}$$

is the nth BKM approximation to the conformal radius of Ω_I at 0.

Clearly,

$$\lim_{n\to\infty} \| K_n(.,0) - K(.,0) \|_{\Omega_I} = 0 \tag{3.7}$$

and, in the space $L_2(\Omega_I)$, this norm convergence implies that $K_n(z,0) \to K(z,0)$ uniformly in every compact subset of Ω_I. Hence, we have that $f_n(z) \to f(z)$ uniformly in every compact subset of Ω_I.

3.2 The BKM for the solution of problem P2.

Let $\partial\hat{\Omega}$ be the image of the curve $\partial\Omega$ under the inversion (2.7) and, as in Section 2, let $\hat{f}$ and $\hat{R}$ be respectively the interior mapping function and the conformal radius associated with $\hat{\Omega}_I = \text{Int}(\partial\hat{\Omega})$. Also, let $\hat{f}_n$ and $\hat{R}_n$ be the nth BKM approximations to $\hat{f}$ and $\hat{R}$ obtained as described in Section 3.1, by applying the BKM to the domain $\hat{\Omega}_I$. Then, the BKM approximations to the mapping function g, solving problem P2, and to the capacity c of $\partial\Omega$ are obtained from the approximations $\hat{f}_n$ and $\hat{R}_n$, by using the relations (2.8) and (2.12). That is, the two approximations are respectively

$$g_n(z) = \left\{\hat{f}_n(z^{-1})\right\}^{-1}, \tag{3.8}$$

and

$$c_n = 1/\hat{R}_n . \tag{3.9}$$

3.3 The ONM for the solution of problem P3.

The ONM may be regarded as the generalization of the BKM to the mapping of doubly-connected domains. The method emerges easily from the theory contained in [6,p.249; 1,p.102; 14,p.373].

Let Ω_D be the doubly-connected domain of problem P3 and observe that the function F solving this problem can be expressed as

$$F(z) = z \exp A(z) \ , \qquad (3.10)$$

where

$$A(z) = \log F(z) - \log z \qquad (3.11)$$

is analytic and single-valued in Ω_D. Also, let $L_2^S(\Omega_D)$ denote the closed subspace of all functions in $L_2(\Omega_D)$ which also possess a single-valued indefinite integral in Ω_D. Finally, let

$$\begin{aligned} H(z) &= A'(z) \\ &= F'(z)/F(z) - 1/z \ , \end{aligned} \qquad (3.12)$$

and assume that the components $\partial\Omega_1$ and $\partial\Omega_2$ of $\partial\Omega_D$ are analytic curves. Then, it is shown in [6,p.250] that, for every function $\eta \in L_2^S(\Omega_D)$ which is also analytic on $\partial\Omega_D$

$$(\eta, H)_{\Omega_D} = i \int_{\partial\Omega_D} \eta(z) \log|z| dz \ , \qquad (3.13)$$

and that

$$\log M = \left\{ \frac{1}{i} \int_{\partial\Omega_D} \frac{1}{z} \log|z| dz - \| H \|^2_{\Omega_D} \right\} \Big/ 2\pi \ , \qquad (3.14)$$

where M is the conformal modulus of Ω_D. For our applications, it is important to observe that (3.13) and (3.14) hold under somewhat weaker requirements than those stated in [6]. In particular, it can be shown easily that the two results hold when $\partial\Omega_1$ and $\partial\Omega_2$ are piecewise analytic and when, in (3.13), η is continuous on $\partial\Omega_D$ except for a finite number of boundary singularities of the form $(z - z_0)^\alpha$, $z_0 \in \partial\Omega_D$, $\alpha > -1/2$.

As was previously remarked, the ONM can be regarded as the generalization of the BKM to the solution of problem P3. In fact, the main details of the two methods are the same, except that in the ONM the function H replaces the kernel K(.,0), property (3.13) plays the role of the reproducing property (3.2), and the set $\left\{\eta_j^*\right\}_{j=1}^n$ is formed by orthonormalizing the first n functions of a set $\{\eta_j\}$ which forms a basis in the space $L_2^S(\Omega_D)$. That is, the function H is approximated by the Fourier series sum

$$H_\eta(z) = \sum_{j=1}^{n} (H, \eta_j^*)_{\Omega_D} \eta_j^*(z) \ , \qquad (3.15)$$

where the Fourier coefficients are known by means of (3.13). Then, because of (3.12) and (3.14), the nth ONM approximations to the mapping function F and to the modulus M of Ω_D are given respectively by

$$F_n(z) = \frac{r_1}{\zeta_1} \exp\left\{ \int_{\zeta_1}^{z} H_n(t)\,dt \right\} , \tag{3.16}$$

and

$$M_n = \exp\left\{\left(\frac{1}{i} \int_{\partial\Omega_D} \frac{1}{z} \log z\,dz - \| H_n \|^2_{\Omega_D} \right) \Big/ 2\pi \right\} . \tag{3.17}$$

As in the case of the BKM, we have that $F_n(z) \to F(z)$ uniformly in every compact subset of Ω_D.

4. SINGULARITIES - CHOICE OF BASIS

The BKM and ONM have the important advantage that they approximate the mapping functions f and F, of problems P1 and P3, by explicit formulae of the type

$$f_n(z) = \sum_{j=1}^{n} a_j \mu_j(z) , \tag{4.1}$$

and

$$F_n(z) = \frac{r_1}{\zeta_1} \exp\left\{ \sum_{j=1}^{n} a_j \mu_j(z) \right\} , \tag{4.2}$$

where, in each case, the μ_j are integrals of the basis functions η_j. (Of course, the same remark applies to the approximation g_n, of the mapping function g, which is determined by means of (3.8), from the BKM approximation to the interior mapping function $\hat{f}$.) Unfortunately, the determination of the coefficients a_j in (4.1) and (4.2) involves the use of the Cram-Schmidt process and, as is well-known, this process can be extremely unstable. For this reason, the success of the methods depends strongly on the speed of convergence of the approximating series, and this in turn depends on the set of basis functions used.

A computationally convenient basis for use with the BKM is the *monomial set*

$$\eta_j(z) = z^{j-1} \; ; \quad j = 1,2,\ldots , \tag{4.3}$$

which is complete in the space $L_2(\Omega_I)$. Unfortunately, the convergence of the resulting approximating series is often extremely slow and, because of the instability, the use of the basis set (4.3) does not always lead to approximations of acceptable accuracy. The same applies to the use of the *monomial set*

$$\eta_{2j-1}(z) = z^{j-1} \ , \quad \eta_{2j}(z) = 1/z^{j+1} \ ; \quad j = 1,2,\ldots \ , \tag{4.4}$$

in connection with the ONM. That is, (4.4) forms a basis in the space $L_2^S(\Omega_D)$ but, in many cases, the convergence of the corresponding ONM approximations is very slow. As might be expected, the slow convergence associated with the use of the basis sets (4.3) and (4.4) is due to the presence of singularities of the mapping function f and of the function H, defined by (3.12), in the complements of the domains Ω_I and Ω_D respectively.

The purpose of this section is to describe how available information about the singularities of the functions f and H can be used to improve the speed of convergence of the BKM and ONM approximations. This involves the use of *augmented basis sets*, formed by introducing into the sets (4.3) and (4.4) *singular functions* that reflect the main singular behaviour of f and H in compl(Ω_I) and compl(Ω_D) respectively. Such basis sets have been used successfully, in connection with the BKM, the ONM and two closely related variational methods in [12,15,16,17]; see also [5].

The singular functions needed for constructing the augmented basis sets are determined by considering the *corner singularities* of f on $\partial\Omega$ and of H on $\partial\Omega_D$, and also the *pole* or *pole-type singularities* that the two functions may have in the complements of $\overline{\Omega}_I$ and $\overline{\Omega}_D$.

4.1 Corner singularities

Any boundary singularities of the functions f and H are *corner singularities*, similar to those that arise in the study of elliptic boundary value problems. The asymptotic form of these singularities can be determined by using the results of Lehman [11], concerning the development of the mapping function at an analytic corner.

We consider first problem P1, and assume that part of the boundary of the domain Ω_I consists of two analytic arcs which meet at a point z_0 and form there a corner of interior angle $\alpha\pi$. Then, depending on whether α is rational or irrational, the results of [11] lead to the following two asymptotic expansions.

(i) If $\alpha = p/q$, with p and q relatively prime, then, as $z \to z_0$,

$$f(z) - f(z_0) = \sum_{k,l,m} B_{k,\ell,m}(z - z_0)^{k+\ell/\alpha}(\mathrm{Log}(z - z_0))^m , \tag{4.5}$$

where k,ℓ and m run over all integers $k \geq 0$, $1 \leq \ell \leq p$, $0 \leq m \leq k/q$ and where $B_{0,1,0} \neq 0$. Also, the terms in (4.5) are ordered so that the term corresponding to $B_{k,\ell,m}$ precedes the term corresponding to $B_{k',\ell',m'}$ if either $k + \ell/\alpha < k' + \ell'/\alpha$ or $k + \ell/\alpha = k' + \ell'/\alpha$ and $m > m'$.

ii) If α is irrational then, as $z \rightarrow z_0$,

$$f(z) - f(z_0) = \sum_{k,\ell} B_{k,\ell}(z - z_0)^{k+\ell/\alpha} , \tag{4.6}$$

where now k and ℓ run over all integers $k \geq 0$, $\ell \geq 1$ and where $B_{0,1} \neq 0$.

The expansions (4.5) and (4.6) simplify considerably when the two arms of the corner z_0 are both straight lines or circular arcs. Then, it follows from the Schwarz-Christoffel formula that, as $z \rightarrow z_0$,

$$f(z) - f(z_0) = \sum_{\ell=1}^{\infty} B_\ell(z - z_0)^{\ell/\alpha} , \quad B_1 \neq 0 ; \tag{4.7}$$

see e.g. [4,p.170] and [14,p.189].

It follows from the above that the dominant term in the asymptotic expansion of f is always $(z - z_0)^{1/\alpha}$. This reflects the geometric property that, under the mapping f, the angle $\alpha\pi$ at $z_0 \in \partial\Omega$ is transformed onto an angle π at the point $f(z_0)$ on the unit circle. Therefore, when $1/\alpha$ is not an integer, a branch point singularity always occurs at the corner z_0. Furthermore, because of the logarithmic terms in (4.5), a branch point singularity might occur even when $1/\alpha$ is an integer. This means, in particular, that the use of preliminary transformations, which is frequently proposed as a method for rectifying corners, does not necessarily remove corner singularities.

The slow convergence due to a branch point singularity at a corner z_0 can be overcome, as proposed in [12,15], by using a basis set that reflects the singular behaviour at z_0 of the kernel function $K(.,0)$. More precisely, since $K(.,0)$ is related to the derivative of f by means of (3.3), the basis set is constructed by introducing into the monomial set (4.3) the derivatives of the first few singular terms of the appropriate asymptotic series (4.5), (4.6) or (4.7). That is, the BKM singular basis functions for dealing with corner singularities are of the form

$$\eta(z) = \frac{d}{dz}\left\{ (z - z_0)^r \right\} ; \quad r = k + \ell/\alpha \quad \text{or} \quad r = \ell/\alpha , \tag{4.8a}$$

and

$$\eta(z) = \frac{d}{dz}\left\{(z - z_o)^{k+\ell/\alpha}(\mathrm{Log}(z - z_o))^m\right\} . \tag{4.8b}$$

In the case of problem P2, a corner of exterior angle $\alpha\pi$ at $z_0 \in \partial\Omega$ is transformed, under the inversion (2.7), into a corner of interior angle $\alpha\pi$ at the point $1/z_0 \in \partial\hat{\Omega}$. Therefore, since the approximation to the mapping function g is determined, as described in Section 3.2, from the BKM approximation to the interior mapping function $\hat{f}$, the details for constructing an augmented basis for dealing with corner singularities are the same as for the interior problem; see [16].

We consider next the problem P3, and let $z_0 \in \partial\Omega_D$ be a corner of interior angle $\alpha\pi$, where by interior angle we mean interior to the domain Ω_D. Then, the question regarding the choice of basis functions, for dealing with the singularities of the function H at z_0, can again be answered by using the results of [11]. However, in this case, the form of the singular functions used for augmenting the set (4.4) depends on whether z_0 lies on the inner or outer component of $\partial\Omega_D$. That is, the ONM singular basis functions are of the form (4.8) when z_0 is a corner on the outer boundary $\partial\Omega_2$, and of the form

$$\eta(z) = \frac{d}{dz}\left\{\left(\frac{1}{z} - \frac{1}{z_0}\right)^r\right\} \; ; \quad r = k + \ell/\alpha \quad \text{or} \quad r = \ell/\alpha \; , \tag{4.9a}$$

and

$$\eta(z) = \frac{d}{dz}\left\{\left(\frac{1}{z} - \frac{1}{z_0}\right)^{k+\ell/\alpha}\left(\mathrm{Log}\left(\frac{1}{z} - \frac{1}{z_0}\right)\right)^m\right\} , \tag{4.9b}$$

when z_0 is a corner of the inner boundary $\partial\Omega_1$; see [17]. (Oberve that the form of the functions (4.9) ensures that these singular functions have single-valued indefinite integrals in Ω_D.)

4.2 *Pole-type singularities*

Apart from corner singularities the functions f and H may also have serious singularities, off the boundary, in

$$\mathrm{compl}(\overline{\Omega}_1) = \mathrm{Ext}(\partial\Omega) ,$$

and

$$\mathrm{compl}(\overline{\Omega}_D) = \mathrm{Int}(\partial\Omega_1) \cup \mathrm{Ext}(\partial\Omega_2) ,$$

respectively. We shall first outline a method, which has been used recantly in [18], for determining the dominant singularities of the mapping function f in $\mathrm{Ext}(\partial\Omega)$, i.e. the singularities of the analytic extension of f which are closest to $\partial\Omega$.

With the notation of problem P1, we let Γ be an analytic arc of $\partial\Omega$ with analytic parametric equation

$$z = \tau(s) \; ; \qquad s_1 < s < s_2 \, , \tag{4.10}$$

and assume that the function

$$z = \tau(\zeta) \, , \tag{4.11}$$

of the complex variable $\zeta = s + it$, is one-one analytic in some simply-connected domain G containing the straight line

$$L = \{\zeta : \zeta = s + it, \; s_1 < s < s_2, \; t = 0\} \, . \tag{4.12}$$

We also assume that G has a symmetric partition with respect to L, so that

$$G = G_1 \cup L \cup G_2 \, ,$$

where G_2 is the mirror image of G_1 in the straight line L, and where the image of G_1 under the transformation (4.11) is contained within Ω_I. More precisely, we assume that (4.11) maps conformally G onto a domain $\Omega_1 \cup \Gamma \cup \Omega_2$ so that the straight line L and the domains G_i; $i = 1,2$ are mapped respectively onto the arc Γ and the domains $\Omega_1 \subseteq \Omega_I$ and $\Omega_2 \subset \mathrm{Ext}(\partial\Omega)$. Then, the function

$$\phi(z) = \begin{cases} f(z) & z \in \Omega_1 \cup \Gamma \, , \\ 1/\overline{f(I(z))} \, , & z \in \Omega_2 \, , \end{cases} \tag{4.13a}$$

where

$$I(z) = \tau\{\overline{\tau^{[-1]}(z)}\} \, , \tag{4.13b}$$

is meromorphic in Ω_2 and defines the analytic continuation of f across Γ into Ω_2. This analytic extension of f is a particular case of the *symmetry principle of analytic arcs*, and the points z, I(z) are called *symmetric points with respect to the arc* Γ; see e.g. [22,p.102].

It follows from the above that the singularities of f in $\overline{\Omega}_2$, i.e. the singularities of the analytic extension ϕ, can be determined by considering the function (4.13). For example, it can be shown easily that if $0 \in \Omega_1$ then f has a simple pole at the symmetric point of the origin with respect to Γ, i.e. at the point

$$z_0 = I(0) \in \Omega_2 \, . \tag{4.14}$$

Although simple poles are the most frequently occurring singularities, other types of singularities may also occur. For example, when $0 \in \partial\Omega_1/\Gamma$ then f may have a double pole at $z_0 = I(0)$ or a branch point of the form

$$(z - z_0)^{-1/2} \ . \tag{4.15}$$

For simplicity, we shall refer to all singularities of the analytic extension of f as *"pole singularities of f"*.

Full details regarding the determination of the pole singularities of f, by a method based on the analysis outlined above, and several specific examples can be found in [18]. Here, we only make some additional remarks concerning the special case where Γ is a straight line or a circular arc. In this case, (4.13) leads to the results predicted by the well-known *Schwarz reflection principle*, i.e. if $0 \in \Omega_1 \cup \partial\Omega_1/\Gamma$ then f has a simple pole at the symmetric point $z_0 = I(0)$, where now z_0 coincides with the mirror image of 0 in the straight line or with the geometric inverse of 0 with respect to the circular arc. Therefore, the determination of the dominant poles of f is particularly simple in the case where $\partial\Omega$ consists of straight line segments and circular arcs. In fact, this is the only geometry for which Levin et al [12] and Papamichael and Kokkinos [15] were able to determine the precise location and nature of the pole singularities of f.

The procedure for treating pole singularities is exactly the same as that used in the case of singular corners. That is, the BKM basis set is formed by introducing into the monomial set (4.3) singular functions that relfect the dominant poles of f. For example, the singular function corresponding to a simple pole at z_0 is

$$\eta(z) = \frac{d}{dz}\left\{\frac{z}{z - z_0}\right\} \ ; \tag{4.16}$$

see [12,15,18].

In the case of problem P2, the dominant pole singularities of the mapping function f, associated with the interior domain $\hat{\Omega}_I$, can again be determined by using the procedure of [18]. However, because of the intermediate transformation (2.7), such pole singularities do not occur as frequently as in the case of problem P1. This is illustrated by the following two results, which can be established easily either from (4.13) or, as in [16,p.193], directly from the Schwarz reflection principle:

(i) If the original boundary $\partial\Omega$ is a polygon then $\hat{f}$ has no pole singularities.

(ii) If $\partial\Omega$ consists of straight line segments and circular arcs then the only poles of $\hat{f}$ are due to the circular arcs. More precisely, a pole occurs only if the centre of a circular arc is in Int($\partial\Omega$) and does not coincide with the origin of the z-plane. If $z_0 \in$ Int($\partial\Omega$) is such a centre then f has a simple pole at the point $\hat{z}_0 = 1/z_0$ Ext($\partial\hat{\Omega}$).

In the case of problem P3, the situation regarding the behaviour of the function H in compl($\overline{\Omega}_D$) is much more involved. In fact Papamichael and Kokkinos [17], who first considered the numerical implementation of the ONM, were unable to provide any information concerning the singularities of the analytic extension of H. However, the problem has also been studied recently in [19], where it is shown that, in many cases, H has singularities in compl($\overline{\Omega}_D$), at the so-called *common symmetric points* with respect to the two boundary components $\partial\Omega_1$ and $\partial\Omega_2$.

Let Γ_j; j = 1,2 be two analytic arcs of $\partial\Omega_j$, j = 1,2 respectively. Also, let $I_j(z)$; j = 1,2 be the two functions corresponding to (4.13b), which define respectively pairs of symmetric points $(z, I_j(z))$; j = 1,2 with respect to the arcs Γ_j; j = 1,2. Then, the points

$$z_1 \in \text{Int}(\partial\Omega_1) \quad \text{and} \quad z_2 \in \text{Ext}(\partial\Omega_2) \tag{4.17}$$

are said to be *common symmetric points with respect to* Γ_1 *and* Γ_2 if

$$z_1 = I_j(z_2) \quad \text{and} \quad z_2 = I_j(z_1) \ ; \quad j = 1,2 \ , \tag{4.18}$$

i.e. if z_1 and z_2 are both fixed points of the two composite functions

$$S_1 = I_1 \circ I_2 \quad \text{and} \quad S_2 = I_2 \circ I_1 \ . \tag{4.19}$$

As an example, consider the case where the inner boundary $\partial\Omega_1$ is the circle

$$\partial\Omega_1 = \{z : |z| = a, \ a < 1\} \ , \tag{4.20}$$

and the outer boundary $\partial\Omega_2$ is a concentric n-sided regular polygon with

$$\ell = \{z : z = 1 + iy, \ |y| \leq \tan(\pi/n)\} \ , \tag{4.21}$$

as one of its sides. Then, with $\Gamma_1 = \partial\Omega_1$ and $\Gamma_2 = \ell$ we have that

$$I_1(z) = a^2/\overline{z} \ , \quad I_2(z) = 2 - \overline{z} \ , \tag{4.22}$$

and hence

$$S_1(z) = a^2/(2-z) \ , \quad S_2(z) = 2 - a^2/z \ . \tag{4.23}$$

Therefore, in this particular case, the common symmetric points with respect to Γ_1 and Γ_2 are

$$z_1 = 1 - (1-a^2)^{1/2} \quad \text{and} \quad z_2 = 1 - (1-a^2)^{1/2} \tag{4.24}$$

More precisely, in this case, there are n pairs of common symmetric points associated with the circle $\partial\Omega_1$ and each of the n sides of the polygon. These points are respectively

$$z_1^{(j)} = z_1\omega_n^{j-1} \quad \text{and} \quad z_2^{(j)} = z_2\omega_n^{j-1} \; ; \quad j = 1,2,\ldots,n \; , \tag{4.25}$$

where z_1 and z_2 are defined by (4.24) and

$$\omega_n = \exp\{2\pi i/n\} \; .$$

Of course there are geometries for which no common symmetric points exist, However, as the above example illustrates, in many cases the common symmetric points $z_1 \in \text{Int}(\partial\Omega_1)$ and $z_2 \in \text{Ext}(\partial\Omega_2)$ can be determined easily from the composite functions (4.19). In such cases, an analysis based on the repeated application of the Schwarz reflection principle show that, under certain conditions, the points z_1 and z_2 are singular points of the function H. The full details of this analysis can be found in [19], where it is also shown that, for the purposes of the ONM, the singular behaviour of H at z_1 and z_2 may be reflected approximately by introducing into the monomial set (4.4) the two singular functions

$$\eta_1(z) = 1/(z - z_1) - 1/z \; , \tag{4.26a}$$

and

$$\eta_2(z) = 1/(z - z_2) \; . \tag{4.26b}$$

5. COMPUTATIONAL DETAILS AND GENERAL REMARKS

5.1 Evaluation of inner products

Let G be a domain of finite connectivity bounded by closed piecewise analytic Jordan curves and, as before, let $(.,.)_G$ denote the inner product of the Hilbert space $L_2(G)$. Then, it is well-known that the *Green's formula*

$$(u,v')_G = \frac{1}{2i}\int_{\partial G} u(z)\overline{v(z)}dz \tag{5.1}$$

holds, for any functions $u,v \in L_2(G)$ which also saitsfy certain continuity conditions on the boundary G; see e.g. [1,p.96;7,p.18;14,p.241]. In particular, it can be shown easily that (5.1) holds when u and v are continuous on ∂G except for a finite number of branch point singularities of the form $(z-z_0)^{\alpha}, z_0 \in \partial\Omega, -1/2 < \alpha < 0$. Since all the basis functions considered in Section 4 satisfy this boundary continuity requirement, it follows that (5.1) can be used to express as contour integrals all the inner products needed for the orthonormalization of the BKM and ONM basis sets $\{\eta_j\}$. For example the inner products needed for the BKM and ONM solutions of problems P1 and P3 can be expressed respectively as

$$(\eta_r,\eta_s)_{\Omega_1} = \frac{1}{2i}\int_{\partial\Omega}\eta_r(z)\overline{\mu_s(z)}dz \tag{5.2}$$

and

$$(\eta_r,\eta_s)_{\Omega_D} = \frac{1}{2i}\int_{\partial\Omega_1\cup\partial\Omega_2}\eta_r(z)\overline{\mu_s(z)}dz \tag{5.3}$$

where, in each case, $\mu_s' = \eta_s$. Similarly, the inner products $(\eta_r,\eta_s)_{\hat{\Omega}_I}$, needed for determining the BKM approximation $\hat{f}_n$ in the case of problem P2, can be expressed as contour integrals over the original boundary $\partial\Omega$. That is, by using (2.7),

$$(\eta_r,\eta_s)_{\hat{\Omega}_I} = \frac{1}{2i}\int_{\partial\Omega}\frac{1}{z^2}\,\eta_r(\tfrac{1}{z})\overline{\mu_s(\tfrac{1}{z})}dz\ , \tag{5.4}$$

where the path of integration is in the positive sence with respect to $\mathrm{Ext}(\partial\Omega)$.

In [12,15,16,17], all the inner products (η_r,η_s) are computed from (5.2), (5.3) or (5.4) by Gauss-Legendre quadrature. Similarly, in the case of the ONM, each inner product $(\eta_j^*,H)_{\Omega_D}$ is computed by applying to the contour integral

$$(\eta_j^*,H)_{\Omega_D} = i\int_{\partial\Omega_D}\eta_j^*(z)\log|z|dz \tag{5.5}$$

the Gaussian rule used for the evaluation of (5.3). Of course, when performing the quadrature care must be taken to deal with integrand singularities that occur, when due to the presence of a corner at z_0, the basis set contains singular functions of the form (4.8) or (4.9). However, the effect of such integrand singularities can either be reduced sufficiently or, in many cases, removed completely, by using a suitable parametrization of the boundary; see [12,15,16,17] for further details

5.2 *Error estimates*

In both the BKM and ONM, *the principle of maximum modulus* leads to a reliable method for determining an estimate of the maximum error in the modulus of the approximate conformal maps. For example, in the cases of problems P1 and P2, this error estimate is given respectively by

$$E_n = \max_j \left| 1 - |f_n(z_j)| \right| , \qquad (5.6)$$

and

$$E_n = \max_j \left| 1 - |g_n(1/z_j)| \right| , \qquad (5.7)$$

where, in each case, $\{z_j\}$ is a set of *boundary test points* on $\partial\Omega$. Similarly, in the case of problem P3 the error estimate is given by

$$E_n = \max\left\{ \max_j \left| r_1 - |F_n(z_{1,j})| \right| , \max_j \left| r_1 M_n - |F_n(z_{2,j})| \right| \right\} , \qquad (5.8)$$

where $\{z_{1,j}\}$ and $\{z_{2,j}\}$ are two sets of boundary test points on $\partial\Omega_1$ and $\partial\Omega_2$ respectively.

Each numerical method should be programmed so that it computes recursively a sequence of successive approximate conformal maps. Then the comparison of successive error estimates E_n provides a simple termination criterion, for determining the *optimum number* n = Nopt of basis functions which gives maximum accuracy in the sense described in [12,p.178] and [17,§5]. Of course, the computable error estimates, defined by (5.6), (5.7) and (5.8), the optimum number Nopt of basis functions and the corresponding best approximations depend critically on both the convergence and stability properties of the numerical method used.

5.3 *Rotational symmetry*

Let G be a domain of finite connectivity and let

$$\omega_M = \exp\{2\pi i/M\} ,$$

where M is a positive integer. Then, G is said to have *M-fold, $M \geq 2$, rotational symmetry* about the origin if $z \in G$ implies that $\omega_M z \in G$. When the domain under consideration has such rotational symmetry then the number of basis functions used in the BKM and ONM can be reduced considerably. For example, if the domain Ω_I of problem P1 is M-fold symmetric then the monomial basis set (4.3) can be replaced by

$$z^{iM} ; \quad j = 0,1,2,\ldots . \qquad (5.9)$$

Similarly, if the doubly-connected domain Ω_D of problem P3 is M-fold symmetric then the ONM monomial set (4.4) can be replaced by

$$z^{jM-1} \; ; \quad j = \pm 1, \pm 2, \pm 3, \ldots \; . \tag{5.10}$$

Also, in both cases, the M singular basis functions corresponding to pole or corner singularities at the M points $z_0^{(1)}$, $z_0^{(2)}, \ldots, z_0^{(M)}$, defined by

$$z_0^{(j+1)} = \omega_M z_0^{(j)} \; ; \quad j = 1(1)M-1 \; , \tag{5.11}$$

can be combined into a single function, see [12,15-17].

5.4 Stability

The stability properties of the BKM and the ONM are studied in detail in [20] where, in particular, a geometrical characterization of the degree of instability in the orthonormalization of the monomial sets (4.3) and (4.4) is established, by using certain well-known results of Carleman [3]; see also [6;p.136]. The significance of this characterization is that it provides theoretical justification for some intuitively apparent results, concerning the relation between the stability properties of the method and the geometry of the domain under consideration. For example, we have the following results, in connection with the use of the basis set (4.3) for the BKM solution of problem P1:

(i) Best stability occurs when $\partial\Omega$ is *"nearly circular"*. Conversely, the orthonormalization process is seriously unstable when Ω_I is a *"thin domain"*.

(ii) For the purposes of stability, the origin 0 should be positioned so that its maximum distance from $\partial\Omega$ is as small as possible.

Regarding the use of augmented basis sets, it is shown in [20] that the introduction of singular basis functions can, in some cases, lead to a deterioration of the stability of the process. This occurs when singular functions of the form (4.6) or (4.26) are used to reflect pole-type singularities at points far from the boundary.

5.5 Convergence

As was previously remarked both the BKM and ONM approximations converge uniformly in every compact subset of the domain under consideration. Furthermore, the two books of Gaier [6,7] contain a number of results which establish the uniform convergence in the closure of the domain, of the BKM and ONM approximations corresponding to the monomial basis sets

(4.3) and (4.4). These results also give information about the rate of convergence of the approximations, but they are established under rather restrictive boundary smoothness requirements. There is also an important result due to Simonenko [23], which establishes the uniform convergence in $\Omega_I \cup \partial\Omega$ of the BKM polynomial approximations to the mapping function f of problem P1. This result holds for any piecewise analytic boundary $\partial\Omega$, but unfortunately it does not give any information about the rate of convergence of the approximations. Finally, the convergence properties of the BKM and ONM are investigated in [20] where, in particular, the results of [6,7] are used to provide some theoretical explanation of the improvement in accuracy which is achieved when the sets (4.3) and (4.4) are augmented by the introduction of singular functions.

5.6 Numerical results

Many practical examples, illustrating the very considerable improvement in accuracy which is achieved when the basis sets (4.3) and (4.4) are augmented by introducing appropriate singular functions can be found in references [12,15-20]. Here, we merely present the numerical results of three examples taken respectively from [12,Ex5], [16,Ex3.3] and [17,5.2].

(i) *Interior of thin ellipse:* Problem P1 with

$$\Omega_I = \{(x,y) : x^2/100 + y^2 < 1\} .$$

The numerical difficulty here is due to the two simple poles that the mapping function f has at the points

$$z_0 = \pm 20/\sqrt{99i} = 2.0101 .$$

The BKM with monomial basis gives:

$$\text{Nopt} = 11 \quad \text{and} \quad E_{11} = 9.4 \times 10^{-2} ,$$

whereas the monomial basis with one singular function reflecting the two pole singularities of f gives:

$$\text{Nopt} = 11 \quad \text{and} \quad E_{11} = 3.3 \times 10^{-10} .$$

(ii) *Exterior of equilateral triangle:* Problem P2 with Ω_E taken to be the exterior of the equilateral triangle whose corners are the points $A = (\sqrt{3},-1)$, $B = (-\sqrt{3},-1)$ and $C = (2,0)$.

The difficulty here is due to the branch point singularities at the three *re-entrant corners* A, B and C. (With the notation of Section 4.1.

at each of these corners $\alpha = 5/3$).

The BKM with monomial basis gives:

$$\text{Nopt} = 20 \ , \quad E_{20} = 2.7 \times 10^{-1} \ , \quad c_{20} = 1.399 \ ,$$

where, as in Section 3.2, we use c_n to denote the nth BKM approximation to $\text{cap}\,\partial\Omega$.

The BKM with an augmented basis, including four singular functions for dealing with the branch point singularities at A, B and C, gives:

$$\text{Nopt} = 13 \ , \quad E_{13} = 3.2 \times 10^{-5} \ , \quad c_{13} = 1.460\ 998\ 57 \ .$$

The exact value of $\text{cap}\,\partial\Omega$, obtained by using the formula of Pólya and Szegö [21,p.256], is:

$$c = 1.460\ 998\ 49 \ .$$

(iii) *Square frame:* Problem P3 with

$$\Omega_D = \{(x,y) : |x| < 1, |y| < 1\} \cap \{(x,y) : |x| > 0.5, |y| > 0.5\} \ .$$

The difficulty here is due to the branch point singularities at the four re-entrant corners of the inner square.

The ONM with monomial basis gives:

$$\text{Nopt} = 30 \ , \quad E_{30} = 4.3 \times 10^{-2} \ , \quad M_{30} = 1.857 \ ,$$

where, as in Section 3.3, M_n denotes the nth approximation to the conformal modulus of Ω_D.

The ONM with an augmented basis, including four singular functions for dealing with the branch point singularities, gives:

$$\text{Nopt} = 24 \ , \quad E_{24} = 5.0 \times 10^{-8} \ , \quad M_{24} = 1.847\ 709\ 011\ 22 \ .$$

The exact value of M obtained by using the formula of Bowman [2,p.104], is:

$$M = 1.847\ 709\ 011\ 24 \ .$$

REFERENCES

1. S. Bergman, The kernel function and conformal mapping. Math. Surveys 5 (A.M.S., Providence, R.I., 2nd ed., 1970).

2. F. Bowman, Introduction to Elliptic Functions (English University Press, London, 1950).

3. T. Carleman, Über die Approximation analytischer Functionen durch lineare Aggregate von vorgegebenen Potenzen, Ark. Mat. Astr. Fys. 17 (1923) 1-30.

4. E.T. Copson, Partial Differential Equations (Cambridge University Press, London, 1975).

5. W. Eidel, Konforme Abbildung mehrfach zusammenhängender Gebiete durch Lösung von Variationsproblemen, Diplomarbeit Giessen, 1979.

6. D. Gaier, Konstruktive Methoden der Konformen Abbildung (Springer Verlag, Berlin, 1964).

7. D. Gaier, Vorlesengen über Approximation im Komplexen (Bïrkhauser Verlag, Basel, 1980).

8. P. Henrici, Applied and Computational Complex Analysis, Vol.1 (Wiley, New York, 1974).

9. L.V. Kantorovich and V.I. Krylov, Approximate Methods of Higher Analysis (Wiley, New York, 1964).

10. P.A.A. Laura, A survey of modern applications of the method of conformal mapping, Revista De La Union Mathematica Argentina 27 (1975) 167-179.

11. R.S. Lehman, Development of the mapping function at an analytic corner, Pacific J. Math. 7 (1957) 1437-1449.

12. D. Levin, N. Papamichael and A? Sideridis, The Bergman kernel method for the numerical conformal mapping of simply-connected domains, J. Inst. Math. Appl. 22 (1978) 171-187.

13. G.K. Lewis, Flow and load parameters of hydrostatic oil bearing for several port shapes, J. Mech. Engrg. Sci. 8 (1966) 173-184.

14. Z. Nehari, Conformal mapping (McGraw-Hill, New York, 1952).

15. N. Papamichael and C.A. Kokkinos, Two numerical methods for the conformal mapping of simply-connected domains, Comput. Meths. Appl. Mech. Engrg. 28 (1981) 285-307.

16. N. Papamichael and C.A. Kokkinos, Numerical conformal mapping of exterior domains, Comput. Meths. Appl. Mech. Engrg. 31 (1982) 189-203.

17. N. Papamichael and C.A. Kokkinos, The use of singular functions for the approximate conformal mapping of doubly-connected domains, Tech. Rep. TR/01/82, Dept. of Maths., Brunel University 1982. (To appear in SIAM J. Sci. Stat. Comput.).

18. N. Papamichael, M.K. Warby and D.M. Hough, The determination of the poles of the mapping function and their use in numerical conformal mapping, J. Comp. Appl. Maths. 9 (1983) 155-166.

19. N. Papamichael and M.K. Warby, Pole-type-singularities and the numerical conformal mapping of doubly-connected domains, J. Comp. Appl. Maths. (1984, in press).

20. N. Papamichael and M.K. Warby, Stability and convergence properties of kernel function methods for numerical conformal mapping, Tech. Rept., Dept. of Maths. and Stats., Brunel Univeriesty. In preparation.

21. G. Pólya and G. Szegö, Isoperimetric Inequalities in Mathematical Physics (Princeton University Press, Princeton, N.J., 1951).

22. G. Sansone and J. Gerretsen, Lectures on the Theory of Functions of a Complex Variable, Vol.II (Walters-Noordhoff, Groningen, 1969).

23. I.B. Simonenko, On the convergence of Bieberbach polynomials in the case of a Lipschitz domain, Math. USSR Izvestija, 13 (1979) 166-174.

24. J.F. Thompson, Z.U.A. Warsi and C.W. Masti, Boundary fitted coordinate systems for numerical solution of partial differential equations. J. Comp. Phys. 47 (1982) 1-108.

CALCULATION OF POTENTIAL IN A SECTOR

J. Barkley Rosser

ABSTRACT

About 25 years ago Wasow and Lehman got an asymptotic series for a harmonic function in a sector. This showed that, in various important cases, the first derivative of the function becomes infinite as one approaches the vertex of the sector. For this reason, finite difference or finite element methods of a fixed mesh give poor accuracy near the vertex. The asymptotic series of Wasow and Lehman cannot be used for calculations near the vertex since the derivations of the series gave no means to compute the numerical values of the coefficients of the series.

In this talk means are provided for computing the numerical values of the coefficients. Moreover, by suitably pairing some terms of the series, the resulting series of terms and pairs turns out to be convergent. It is therefore quite suitable for calculating values of the harmonic function near the vertex.

AMS (MOS) Classification: 3104, 31A05, 35J25

Key Words and Phrases: Boundary value problems, harmonic functions and computation of same

CALCULATION OF POTENTIAL IN A SECTOR

J. Barkley Rosser
Mathematics Research Center
University of Wisconsin-Madison
Madison, Wisconsin 53705/USA

0. Background. One can have a harmonic function, $u(x,y)$, in a domain for which the boundary has a cusp. In Wasow [1957] and Lehman [1959] the asymptotic behavior of $u(x,y)$ inside the cusp is studied. This study showed that, in many important cases, the first partial derivatives of $u(x,y)$ become unbounded as one approaches the tip of the cusp. For this reason, finite differences or finite elements of a fixed mesh give poor approximations near the tip. However, the series given by Wasow and Lehman also cannot be used for calculations near the tip because they are only asymptotic series, besides which the derivations of the series gave no means to determine the numerical values of the coefficients of the series.

In this paper we specialize to the case where the sides of the cusp are straight lines. Thereby we can provide means for determining the numerical values of the coefficients of the series. Moreover, by suitably pairing some terms of the series, the resulting series of terms and pairs turns out to be convergent. It is therefore quite suitable for calculating values of $u(x,y)$, expecially near the tip.

We confine our attention to the region near the tip. Specifically, we deal with a sector of a circle of radius A cut out by two radii separated by an angle $\alpha\pi$. The vertex of the sector serves as the tip of the cusp. Furthermore, whereas Wasow and Lehman dealt only with the Dirichlet case, in which values of $u(x,y)$ are prescribed on the sides of the cusp (now the radii of our sector), we shall consider also the Neumann case, in which values of the normal derivative are prescribed along the radii.

1. Auxiliary results. We give some results that we shall have need of.

Let Ω be a sector, with boundary Γ consisting of two radii and a circular arc. We shall so orient our system of coordinates that, with rectangular and polar coordinates related by the familiar

$$z = x + iy = re^{i\theta} , \qquad [1.1]$$

Ω is given by

$$0 < r < A, \quad 0 < \theta < \alpha\pi \ . \tag{1.2}$$

We write $u(x,y)$ or $u(r;\theta)$ for the value of a real function which is assumed to be harmonic inside Ω. Then, with z given by (1.1), we see by Theorem III on page 345 of Kellogg [1953] that there is a function u which is analytic inside Ω, such that inside Ω

$$u(x,y) = u(r;\theta) = \mathcal{R}u(z) \ . \tag{1.3}$$

Throughout this report, we shall assume the boundary conditions (whether Dirichlet or Neumann) to be very smooth except at the vertex. Along the arc it will be assumed that the boundary conditions are given by a continuous function with a continuous first derivative and a second derivative existing almost everywhere and of bounded variation. On the bounding radii, existence of even more derivatives may be required, according to what case we are dealing with (Dirichlet or Neumann). Details will be specified at the appropriate point.

Our requirement of smoothness is less restrictive than might appear at first sight. If one has boundary conditions with simple discontinuities of either the boundary values or their first four derivatives, these discontinuities can be removed. See pp. 221-222 of Milne [1953].

We shall occasionally use other coordinates, in which ξ and η stand for distances from the origin on two orthogonal axes:

$$\zeta = \xi + i\eta = \rho e^{i\phi} \ . \tag{1.4}$$

The quantities $r(\xi,\eta)$ and $s(\xi,\eta)$, which will be used frequently, are defined by

$$r(\xi,\eta) = \frac{\eta}{\pi} \int_{-\infty}^{\infty} \frac{g(t)dt}{(\xi - t)^2 + \eta^2} \quad \text{for} \quad \eta \neq 0 \ , \tag{1.5}$$

$$r(\xi,0) = g(\xi) \ , \tag{1.6}$$

$$s(\xi,\eta) = \frac{1}{2\pi} \int_{-\infty}^{\infty} g(t)\log\{(\xi - t)^2 + \eta^2\}dt \ . \tag{1.7}$$

Note that if g is such that $s(\xi,\eta)$ exists, then for $\eta > 0$ or $\eta < 0$

$$r(\xi,\eta) = \frac{\partial}{\partial\eta} s(\xi,\eta) \ . \tag{1.8}$$

<u>Definition 1.1</u>. We say that f is finitely integrable if and only if for each finite C and D with $C < D$ the integral

$$\int_C^D |f(x)|dx$$

exists.

Definition 1.2. We say that f is absolutely integrable on the real axis if and only if

$$\int_{-\infty}^{\infty} |f(x)|dx$$

exists. In such case, it is common to write $f \in L^1(\mathbf{R})$.

Convention 1.1. If f is measurable and if there is a positive M such that

$$|f(x)| \leq M$$

except for a set of measure zero, we shall write $f \in L^{\infty}(\mathbf{R})$.

Theorem 1.1. Let $g(t) = f(t) + h(t)$, where $f \in L^1(\mathbf{R})$ and $h \in L^{\infty}(\mathbf{R})$. Then $r(\xi,\eta)$ is harmonic in each of the domains $\eta > 0$ and $\eta < 0$ separately. If $g(t)$ is continuous at $t = a$, then $r(\xi,\eta)$ is continuous at $(a,0)$, PROVIDED continuity is taken only with respect to the region $\eta \geq 0$.

NOTE. If one approaches $(a,0)$ through negative values of η, $r(\xi,\eta)$ will have the limit $-g(a)$.

This theorem is similar to results in Evans [1927]. By the methods of Evans [1927] one can extend the results therein to give exactly our theorem.

Definition 1.3. A real function, f, is said to be quite reasonable if and only if it is finitely integrable, and at each point a there is a constant K_a such that in some neighborhood of $x = a$

$$|f(x)| \leq \frac{K_a}{|x - a||\log(x - a)^2|^3} . \qquad [1.9]$$

If f is quite reasonable, we call it a QR-function.

Theorem 1.2. Assume that g is a QR-function and that $t(\log t^2)^3 g(t)$ is bounded for all sufficiently large $|t|$. Then the function $s(\xi,\eta)$ is continuous for all ξ and η. Moreover, $s(\xi,\eta)$ is harmonic in each of the domains $\eta > 0$ and $\eta < 0$ separately.

Like Theorem 1.1, this can be proved by the methods of Evans [1927] from results therein.

Corollary. In the interval $E < t < F$, let $k(t)$ be discontinuous only at a set of points of measure zero, and let $k(t)$

be bounded for $E < t < F$. If we define

$$t(\xi,\eta) = \frac{1}{2\pi}\int_E^F k(t)\log\{(\xi - t)^2 + \eta^2\}dt$$

for all ξ and η, then $t(\xi,\eta)$ is continuous for all ξ and η.

Proof. In $s(\xi,\eta)$, define $g(t)$ to be $k(t)$ at those values of t with $E < t < F$ at which $k(t)$ is defined, and take $g(t) = 0$ at all other values of t.

Theorem 1.3. In the interval $E < t < F$, let $g(t)$ and its first n derivatives be continuous, and let the $(n+1)$-st derivative be bounded and continuous except at a set of points of measure zero. Let also $g(t) = f(t) + h(t)$, where $f \in L^1(\mathbf{R})$ and $h \in L^\infty(\mathbf{R})$. Then for $\eta > 0$ each partial derivative of $r(\xi,\eta)$ of order $\leq n$ has a continuous extension to $\eta = 0$ for $E < \xi < F$.

Proof. See the proof of the next theorem.

Theorem 1.4. In the interval $E < t < F$, let $g(t)$ and its first $n-1$ derivatives be continuous, and let the n-th derivative be bounded and continuous except at a set of points of measure zero. Let also g be a finitely integrable function such that $t(\log t^2)^3 g(t)$ is bounded for all sufficiently large $|t|$. Then for $\eta > 0$ each partial derivative of $s(\xi,\eta)$ of order $\leq n$ has a continuous extension to $\eta = 0$ for $E < \xi < F$.

Proof. We prove by induction on N that both Theorem 1.3 and Theorem 1.4 hold for $n \leq N$. First take $N = 0$. To establish Theorem 1.3 for $n = 0$, use Theorem 1.1. To establish Theorem 1.4 for $n = 0$, we have to prove that $s(\xi,\eta)$ itself has a continuous extension to $\eta = 0$ for $E < \xi < F$; this must be done under only the hypothesis that g is a finitely integrable function such that $t(\log t^2)^3 g(t)$ is bounded for all sufficiently large $|t|$ and $g(t)$ is bounded and continuous for $E < t < F$ except at a set of points of measure zero. Write

$$s(\xi,\eta) = \frac{1}{2\pi}\int_{-\infty}^E g(t)\log\{(\xi - t)^2 + \eta^2\}dt$$

$$+ \frac{1}{2\pi}\int_E^F g(t)\log\{(\xi - t)^2 + \eta^2\}dt$$

$$+ \frac{1}{2\pi}\int_F^\infty g(t)\log\{(\xi - t)^2 + \eta^2\}dt .$$

Differentiation under the integral sign shows that the first integral is a function which is harmonic except for $\xi \leq E$. The third integral is a function which is harmonic except for $F < \xi$. By the corollary to Theorem 1.2, the second integral is continuous for all ξ and η.

So assume that we have established the desired result for N and let $n < N+1$. If $n < N$, we invoke the hypothesis of the induction. So let $n = N+1$. Consider a partial derivative of $r(\xi,\eta)$ of order $N+1$.

Case 1. The partial derivative is the N-th derivative of

$$\frac{\partial}{\partial\xi} r(\xi,\eta) = \frac{-2\eta}{\pi} \int_{-\infty}^{E} \frac{g(t)(\xi - t)dt}{\{(\xi - t)^2 + \eta^2\}^2}$$

$$- \frac{2\eta}{\pi} \int_{E}^{F} \frac{g(t)(\xi - t)dt}{\{(\xi - t)^2 + \eta^2\}^2}$$

$$- \frac{2\eta}{\pi} \int_{F}^{\infty} \frac{g(t)(\xi - t)dt}{\{(\xi - t)^2 + \eta^2\}^2} .$$

In the first and third integrals, bring the η under the integral sign. Then differentiation under the integral sign shows that both integrals are functions harmonic for $E < \xi < F$ and all η. So they and their derivatives of all orders are continuous.

Integrate by parts in the second integral. We get

$$\left[- \frac{\eta}{\pi} \frac{g(t)}{(\xi - t)^2 + \eta^2}\right]_E^F + \frac{\eta}{\pi} \int_E^F \frac{g'(t)dt}{(\xi - t)^2 + \eta^2} .$$

The first term is harmonic for $E < \xi < F$ and all η. Define

$$g^*(t) = \begin{cases} g'(t) & E < t < F \\ 0 & \text{otherwise} . \end{cases} \qquad [1.10]$$

By hypothesis, Theorem 1.3 holds if $n < N$. Applying it with $g^*(t)$ in place of $g(t)$ gives the desired conclusion, since $g^* \in L^\infty(\mathbf{R})$ because $g'(t)$ is bounded for $E < \xi < F$ because its derivative is bounded.

Case 2. The partial derivative is the N-th derivative of

$$\frac{\partial}{\partial\eta} r(\xi,\eta) = \int_{-\infty}^{E} \phi(t)dt + \int_{E}^{F} \phi(t)dt + \int_{F}^{\infty} \phi(t)dt ,$$

where temporarily we take

$$\phi(t) = \frac{1}{\pi} \frac{g(t)\{(\xi - t)^2 - \eta^2\}}{\{(\xi - t)^2 + \eta^2\}^2} .$$

As in Case 1, the first and third integrals pose no problem. Integrating by parts twice in the second integral gives

$$- \frac{1}{\pi} \int_E^F g''(t)\log\{(\xi - t)^2 + \eta^2\}dt$$

plus a function which is harmonic for $E < \xi < F$ and all η. Define $g^*(t)$ to be $g''(t)$ at those values of t with $E < t < F$ at which $g''(t)$ exists and take $g^*(t) = 0$ at all other values of t. By hypothesis, Theorem 1.4 holds if $n < N$. Applying it with $g^*(t)$ in place of $g(t)$ gives the desired conclusion.

Consider a partial derivative of $s(\xi,\eta)$ of order $N + 1$.

Case 1. The partial derivative is the N-th derivative of

$$\frac{\partial}{\partial \xi} s(\xi,\eta) = \frac{1}{\pi} \int_{-\infty}^{E} \frac{g(t)(\xi - t)dt}{(\xi - t)^2 + \eta^2}$$

$$+ \frac{1}{\pi} \int_{E}^{F} \frac{g(t)(\xi - t)dt}{(\xi - t)^2 + \eta^2}$$

$$+ \frac{1}{\pi} \int_{F}^{\infty} \frac{g(t)(\xi - t)dt}{(\xi - t)^2 + \eta^2} .$$

By integrating by parts in the second integral, we are reduced to

$$\frac{1}{2\pi} \int_E^F g'(t)\log\{(\xi - t)^2 + \eta^2\}dt$$

plus a harmonic function. The usual trick reduces this to considering Theorem 1.4 for $n < N$.

Case 2. The partial derivative is the N-th derivative of

$$\frac{\partial}{\partial \eta} s(\xi,\eta) = r(\xi,\eta) ,$$

and we apply Theorem 1.3 for $n < N$.

2. <u>The first auxiliary function</u>. We determine $u(x,y)$ as the sum of two auxiliary functions:

$$u(x,y) = u^{(s)}(x,y) + u^{(r)}(x,y) . \qquad [2.1]$$

We use $u^{(s)}$ to handle the boundary conditions along the radii, after which $u^{(r)}$ can easily handle the boundary conditions along the arc.

What we require of $u^{(s)}$ is that it should satisfy the boundary conditions along the radii. What it does for $r > A$ is clearly quite irrelevant in (2.1). So we are free to assign its behavior at will for $r > A$. This we do in a way that will be helpful when we get around to defining $u^{(r)}$. Details will be given at the appropriate places. In particular, we always invent an extension of the boundary conditions along the radii, out to $r = C$, where $C > A$. So, given boundary conditions out to $r = C$ on each radius, let us see how to define $u^{(s)}$ to satisfy such conditions.

We first describe the D-technique; "D" stands for "Dirichlet".

From the origin, project straight lines at angles $\theta = \beta$ and $\theta = \gamma$, with $\beta < \gamma$. Prescribe boundary conditions:

$$u^{(s)}(r;\theta) = f_\beta(r) \qquad \theta = \beta, \quad 0 < r < C \tag{2.2}$$

$$u^{(s)}(r;\theta) = f_\gamma(r) \qquad \theta = \gamma, \quad 0 < r < C\ . \tag{2.3}$$

In accordance with our smoothness criteria, we assume that $f_\beta(r)$ and $f_\gamma(r)$ are each continuous for $0 < r < C$.

We undertake to find a function $u^{(s)}(r;\theta)$ which is harmonic for $\beta < \theta < \gamma$ and $0 < r < C$, is continuous up to the lines $\theta = \beta$ and $\theta = \gamma$, and satisfies (2.2) and (2.3) along these lines. As we said earlier, we do not care what $u^{(s)}(r;\theta)$ does for $r > A$, and therefore for $r > C$. We define $f_\beta(r) = f_\gamma(r) = 0$ for $r > C$.

We define

$$\mu = \pi/(\gamma - \beta)\ . \tag{2.4}$$

We take

$$\zeta = e^{-i\mu\beta}z^\mu\ , \tag{2.5}$$

and invoke (1.4). This will entail

$$\rho = r^\mu \tag{2.6}$$

$$\phi = \mu(\theta - \beta)\ , \tag{2.7}$$

in view of (1.1). We will shortly define a $g(t)$, in terms of which we will set

$$v(\xi,\eta) = r(\xi,\eta)\ ; \tag{2.8}$$

recall (1.5) and (1.6). We will take $g(t) = f(t) + h(t)$, where $f \in L^1(\mathbf{R})$ and $h \in L^\infty(\mathbf{R})$. So, by Theorem 1.1, $v(\xi,\eta)$ is harmonic for $\eta > 0$, and if $g(t)$ is continuous at $t = a$, then $v(\xi,\eta)$ is continuous at $(a,0)$ if we approach $(a,0)$ through points with non-negative η. Also, by (1.6),

$$v(a,0) = g(a)\ . \tag{2.9}$$

Consider the semi-circular area, S, bounded by the circle $\rho = C^\mu$ and with $\eta > 0$. If we apply the inverse of the transformation (2.5),

$$z = e^{i\beta}\zeta^{1/\mu} , \qquad [2.10]$$

then the interior of S will go into the interior of the sector bounded by $\theta = \beta$, $\theta = \gamma$, and $r = C$. This transformation will be conformal from S to the sector. So, as $v(\xi,\eta)$ is harmonic inside S, it will go into a harmonic function inside the sector. We take this to be $u^{(s)}(x,y)$. That is

$$u^{(s)}(x,y) = v(\xi,\eta) , \qquad [2.11]$$

where (x,y) and (ξ,η) are related by (2.10). The values in the ζ-plane with $\eta = 0$ and $\xi > 0$ go into values along $\theta = \beta$, and the values with $\eta = 0$ and $\xi < 0$ go into values along $\theta = \gamma$. Hence, to satisfy (2.2) and (2.3), we define:

$$g(t) = \begin{cases} 0 & t < -C^{\mu} \\ f_{\gamma}((-t)^{1/\mu}) & -C^{\mu} < t < 0 \\ 0 & t = 0 \\ f_{\beta}(t^{1/\mu}) & 0 < t < C^{\mu} \\ 0 & t > C^{\mu} . \end{cases} \qquad [2.12]$$

Since f_{β} and f_{γ} are continuous, we see easily that $g \in L^1(\mathbf{R})$ and that (2.2) and (2.3) are satisfied.

In some cases, we will need to know that the second derivative of $u^{(s)}(x,y)$ along the arc $r = A$ is of bounded variation for $\beta < \theta < \gamma$. Suppose f_{β} and f_{γ} have continuous first and second derivatives and a third derivative which is bounded and continuous except at a set of points of measure zero for $0 < r < C$. By (2.12), the same will hold for $g(t)$ for $-C^{\mu} < t < 0$ and also for $0 < t < C^{\mu}$. Therefore, by Theorem 1.3, the first and second partial derivatives of $v(\xi,\eta)$ will have continuous extensions to $\eta = 0$ for $-C^{\mu} < \xi < 0$ and also for $0 < \xi < C^{\mu}$. Transforming to $u^{(s)}$ by (2.10) and (2.11) tells us that the first and second derivatives of $u^{(s)}$ have continuous extensions to the lines $\theta = \beta$ and $\theta = \gamma$. Between $\theta = \beta$ and $\theta = \gamma$, $u^{(s)}$ is harmonic, and has continuous derivatives of all orders. So the only way a second derivative could have an infinite variation would be in the limit, as one approaches $\theta = \beta$ or $\theta = \gamma$. But this cannot be, as the second derivative has a continuous extension as θ approaches β or γ from inside the angle.

In other cases, we will need to know that the normal derivative of $u^{(s)}(x,y)$ along the arc $r = A$ has a second derivative which is of bounded variation for $\beta < \theta < \gamma$. If f_β and f_γ have continuous derivatives up to the third and a fourth derivative which is bounded and continuous except at a set of points of measure zero for $0 < r < C$, we can go through an analogous argument.

For the $u^{(s)}(x,y)$ that we need for (2.1), we take $\beta = 0$ and $\gamma = \alpha\pi$. Then $\mu = 1/\alpha$. Also, we need only satisfy the boundary conditions out as far as $r = A$. However, it is no trouble to imagine them extended out to $r = C$, giving us the existence of derivatives inside an interval, as specified in Theorem 1.3.

We now describe the N-technique; "N" stands for "Neumann".

Again, we take the angle between $\theta = \beta$ and $\theta = \gamma$. We prescribe boundary conditions:

$$\frac{\partial}{\partial n} u^{(s)}(r;\theta) = g_\beta(r) \qquad \theta = \beta, \quad 0 < r < C \tag{2.13}$$

$$\frac{\partial}{\partial n} u^{(s)}(r;\theta) = g_\gamma(r) \qquad \theta = \gamma, \quad 0 < r < C\ . \tag{2.14}$$

Here $\partial/\partial n$ means the normal derivative OUTWARD. We assume that $g_\beta(r)$ and $g_\gamma(r)$ are each continuous for $0 < r < C$.

We undertake to find a function $u^{(s)}(r;\theta)$ which is harmonic for $\beta < \theta < \gamma$ and $0 < r < C$, and has first derivatives normal to the lines $\theta = \beta$ and $\theta = \gamma$ which are continuous up to the lines $\theta = \beta$ and $\theta = \gamma$, and which satisfies (2.13) and (2.14) along these lines. We define $g_\beta(r) = g_\gamma(r) = 0$ for $r > C$.

We take μ and ζ as in the D-technique. For a suitable $g(t)$ we will set

$$v(\xi,\eta) = s(\xi,\eta)\ ; \tag{2.15}$$

recall (1.7). We will arrange that g is a QR-function and that $t(\log t^2)^3 g(t)$ is bounded for all sufficiently large $|t|$. So, by Theorem 1.2, $v(\xi,\eta)$ is continuous for all ξ and η. Moreover, $v(\xi,\eta)$ is harmonic in each of the domains $\eta > 0$ and $\eta < 0$ separately. In particular, $v(\xi,\eta)$ is harmonic inside the semicircle, S, that we referred to earlier.

We apply (2.10), which takes $v(\xi,\eta)$ into a function which is harmonic inside the sector bounded by $\theta = \beta$, $\theta = \gamma$, and $r = C$. We take this to be $u^{(s)}(x,y)$; that is, we invoke (2.11) again. We define:

$$g(t) = \begin{cases} 0 & t \le -C^{\mu} \\ \frac{1}{\mu}(-t)^{(1/\mu)-1} g_{\gamma}((-t)^{1/\mu}) & -C^{\mu} < t < 0 \\ 0 & t = 0 \\ -\frac{1}{\mu}(t)^{(1/\mu)-1} g_{\beta}((t)^{1/\mu}) & 0 < t < C^{\mu} \\ 0 & t \ge C^{\mu} . \end{cases} \qquad [2.16]$$

If we assume that $g_{\beta}(t)$ and $g_{\gamma}(t)$ are bounded for $0 < r < C$, then g is a QR-function. Also, obviously $t(\log t^2)^3 g(t)$ is bounded for $|t| \ge C^{\mu}$. Hence it remains to show that (2.13) and (2.14) are satisfied.

Take first (2.13). We rotate the z-plane so that the line that was $\theta = \beta$ becomes the right half of the real axis. Then the relation between ζ and the new z is given by

$$\zeta = z^{\mu} . \qquad [2.17]$$

That is,

$$\xi + i\eta = (x + iy)^{\mu} = r^{\mu} e^{i\theta\mu} .$$

Hence

$$\frac{\partial \xi}{\partial y} + i \frac{\partial \eta}{\partial y} = i\mu(x + iy)^{\mu-1} = i\mu r^{\mu-1} e^{i\theta(\mu-1)} .$$

Equating real and imaginary parts gives

$$\frac{\partial \xi}{\partial y} = -\mu r^{\mu-1} \sin(\mu - 1)\theta \qquad [2.18]$$

$$\frac{\partial \eta}{\partial y} = \mu r^{\mu-1} \cos(\mu - 1)\theta . \qquad [2.19]$$

Choose an x_0 with $0 < x_0 < C$. Take the point $x_0 + iy$, and let $y \to 0$, so that we approach the point $(x_0, 0)$ on the real axis. Then $r \to x_0$, and $\theta \to 0$. Also, in the ζ plane, we are approaching the point $(x_0^{\mu}, 0)$. That is, $\xi \to x_0^{\mu}$ and $\eta \to 0$. By (2.11) and (2.15)

$$\frac{\partial}{\partial y} u^{(s)}(x,y) = \frac{\partial}{\partial \xi} s(\xi,\eta) \frac{\partial \xi}{\partial y} + \frac{\partial}{\partial \eta} s(\xi,\eta) \frac{\partial \eta}{\partial y} . \qquad [2.20]$$

By (1.8),

$$\frac{\partial}{\partial \eta} s(\xi,\eta) = r(\xi,\eta) . \qquad [2.21]$$

But $g(t)$ is continuous at $t = x_0^{\mu}$. Then, by Theorem 1.1, $r(\xi,\eta)$ is continuous at $\xi = x_0^{\mu}$, $\eta = 0$, and by (1.6) takes the value $g(x_0^{\mu})$ at that point. Hence, by (2.21) and (2.16)

$$\frac{\partial}{\partial\eta}\, s(\xi,\eta) \to -\frac{1}{\mu}\,(x_0)^{1-\mu} g_\beta(x_0) . \qquad [2.22]$$

Since $r \to x_0$ and $\theta \to 0$, we see by (2.19) that

$$\frac{\partial\eta}{\partial y} \to \mu(x_0)^{\mu-1} .$$

Hence, by (2.22)

$$\frac{\partial}{\partial\eta}\, s(\xi,\eta)\, \frac{\partial\eta}{\partial y} \to -g_\beta(x_0) . \qquad [2.23]$$

As $\theta \to 0$, $\sin(\mu - 1)\theta$ approaches equality with $(\mu - 1)\theta$. By (2.17), $\rho = r^\mu$ and $\phi = \mu\theta$. As $\eta = \rho \sin \phi$, we get for θ near 0 that η is near equality with $x_0^\mu \mu\theta$. So, by (2.18), $\partial\xi/\partial y$ is near equality with $-\mu x_0^{\mu-1}(\mu - 1)\theta$, which is $-((\mu - 1)/x_0)(x_0^\mu \mu\theta)$. That is, $\partial\xi/\partial y$ approximates $-((\mu - 1) / x_0)\eta$. Let us show that $\eta(\partial s(\xi,\eta)/\partial\xi) \to 0$, which will tell us that

$$\frac{\partial}{\partial\xi}\, s(\xi,\eta)\, \frac{\partial\xi}{\partial y} \to 0 . \qquad [2.24]$$

By (1.7)

$$\frac{\partial}{\partial\xi}\, s(\xi,\eta) = \frac{1}{\pi} \int_{-\infty}^{\infty} \frac{g(t)(\xi - t)dt}{(\xi - t)^2 + \eta^2} .$$

Hence for $\eta > 0$

$$\left|\eta \frac{\partial}{\partial\xi}\, s(\xi,\eta)\right| < \frac{\eta}{\pi} \int_{-\infty}^{\infty} \frac{|g(t)||x_0^\mu - t|dt}{(\xi - t)^2 + \eta^2} + |\xi - x_0^\mu|\, \frac{\eta}{\pi} \int_{-\infty}^{\infty} \frac{|g(t)|dt}{(\xi - t)^2 + \eta^2} .$$

As $|g(t)|$ is continuous at $t = x_0^\mu$,

$$\frac{\eta}{\pi} \int_{-\infty}^{\infty} \frac{|g(t)|dt}{(\xi - t)^2 + \eta^2} \to |g(x_0^\mu)|$$

by Theorem 1.1. As $\xi \to x_0^\mu$, we get

$$|\xi - x_0^\mu|\, \frac{\eta}{\pi} \int_{-\infty}^{\infty} \frac{|g(t)|dt}{(\xi - t)^2 + \eta^2} \to 0 .$$

Also $|g(t)||x_0^\mu - t|$ is continuous at $t = x_0^\mu$. So, by Theorem 1.1

$$\frac{\eta}{\pi} \int_{-\infty}^{\infty} \frac{|g(t)||x_0^\mu - t|dt}{(\xi - t)^2 + \eta^2} \to \left[|g(t)||x_0^\mu - t|\right]_{t=x_0^\mu} = 0 .$$

Hence, by (2.20), (2.23), and (2.24), we get

$$\frac{\partial}{\partial y} u^{(s)}(x,y) = -g_\beta(x_0)$$

at $x = x_0$, $y = 0$. But $\partial/\partial y$ is the inward normal. So the outward normal is $g_\beta(x_0)$, satisfying (2.13).

The treatment for the line $\theta = \gamma$ is similar, except that this time $\partial/\partial y$ is the outward normal.

In various cases, we will need to know that $u^{(s)}(x,y)$ or its normal derivative along the arc $r = A$ has a second derivative which is of bounded variation for $\beta \leq \theta \leq \gamma$. By Theorem 1.4, this can be achieved by suitable assumptions about the derivatives of g_β and g_γ.

As in the D-technique, we take $\beta = 0$ and $\gamma = \alpha\pi$. Then $\mu = 1/\alpha$. Also, the extension out to $r = C$ is no trouble, and gives us the existence of derivatives inside an interval, as specified in Theorem 1.4.

We have two more cases to consider, namely Dirichlet conditions on one radius and Neumann conditions on the other. Let us prescribe

$$u^{(s)}(r;\theta) = f_0(r) \qquad \theta = 0, \quad 0 < r < C \tag{2.25}$$

$$\frac{\partial}{\partial n} u^{(s)}(r;\theta) = g_\alpha(r) \qquad \theta = \alpha\pi, \quad 0 < r < C\ . \tag{2.26}$$

First we use the D-technique to define a $u_1(r;\theta)$ satisfying

$$u_1(r;\theta) = f_0(r) \qquad \theta = 0, \qquad 0 < r < C$$

$$u_1(r;\theta) = f_0(r) \qquad \theta = 2\alpha\pi, \quad 0 < r < C\ .$$

By symmetry

$$\frac{\partial}{\partial n} u_1(r;\theta) = 0 \qquad \theta = \alpha\pi, \quad 0 < r < C\ .$$

Now use the N-technique to define a $u_2(r;\theta)$ satisfying

$$\frac{\partial}{\partial n} u_2(r;\theta) = -g_\alpha(r) \qquad \theta = -\alpha\pi, \quad 0 < r < C$$

$$\frac{\partial}{\partial n} u_2(r;\theta) = \ g_\alpha(r) \qquad \theta = \ \alpha\pi, \quad 0 < r < C\ .$$

By antisymmetry

$$u_2(r;\theta) = 0 \qquad \theta = 0, \quad 0 < r < C\ .$$

Clearly $u_1(r;\theta) + u_2(r;\theta)$ will serve as our $u^{(s)}(r;\theta)$. We will have $\mu = 1/2\alpha$.

If we prescribe Neumann conditions for $\theta = 0$ and Dirichlet conditions for $\theta = \alpha\pi$, the procedure is similar.

3. <u>The Second Auxiliary Function</u>. What do we require for $u^{(r)}(x,y)$? Recall (2.1). Since $u^{(s)}(x,y)$ satisfies the same boundary conditions as $u(x,y)$ along the radii, $u^{(r)}(x,y)$ must satisfy homogeneous boundary conditions along the radii. That is, if $u(x,y)$ satisfies Dirichlet conditions along a radius, then $u^{(r)}(x,y)$ must be zero along that radius. If $u(x,y)$ satisfies Neumann conditions along a radius, then the normal derivative of $u^{(r)}(x,y)$ must be zero along that radius.

This produces four cases to consider; Dirichlet conditions on each radius, or Neumann conditions on each radius, or Dirichlet on one radius and Neumann on the other radius. Besides this, the conditions on the arc $r = A$ may be either Dirichlet or Neumann. So we have eight cases to treat. Fortunately, the treatments are fairly similar, though differing in details. We shall do a couple of cases fairly carefully, and then sketch the rest.

Consider first Dirichlet conditions all around. We take

$$u^{(r)}(r;\theta) = \sum_{m=1}^{\infty} D_m r^{m/\alpha} \sin \frac{\theta m}{\alpha} . \qquad [3.1]$$

Clearly, the series is harmonic inside its radius of convergence. Also, it has the sum 0 for both $\theta = 0$ and $\theta = \alpha\pi$. If we can choose the D_m's so that the sum equals $u^{(r)}(r;\theta)$ along the arc $r = A$, $0 \le \theta \le \alpha\pi$, then $u^{(r)}(r;\theta)$ and the series take the same values around the entire perimeter of the sector. As $u^{(r)}(r;\theta)$ and the series are both harmonic inside the sector, (3.1) must hold inside the sector, as well as around the perimeter.

Let $j(\theta)$ be the value that $u^{(r)}(r;\theta)$ must take along the arc $r = A$. By (2.1)

$$j(\theta) = u(A;\theta) - u^{(s)}(A;\theta) \qquad 0 \le \theta \le \alpha\pi . \qquad [3.2]$$

Because $u^{(s)}(x,y)$ satisfies the same boundary conditions as $u(x,y)$ along the radii, we have

$$j(0) = j(\alpha\pi) = 0 .$$

Among our smoothness conditions was that, along the arc $r = A$, $u(r;\theta)$ is continuous with a continuous first derivative and a second derivative existing almost everywhere and of bounded variation. Recall that we can insure that the same smoothness conditions hold for $u^{(s)}(x,y)$ by imposing suitable requirements on the boundary values of $u^{(s)}(x,y)$ along the radii and on some of the derivatives of said boundary values. Let us assume that these requirements have been imposed. Then by (3.2), we conclude that $j(\theta)$ is continuous with a continuous first derivative and a second derivative existing almost everywhere and of bounded variation.

By (3.1) and (3.2),

$$j(\theta) = \sum_{m=1}^{\infty} D_m A^{m/\alpha} \sin \frac{\theta m}{\alpha}, \quad 0 < \theta < \alpha\pi . \tag{3.3}$$

The right side of (3.3) is a wholly standard Fourier series in terms of θ/α. So, by the standard theory of Fourier series, we have

$$D_n = \frac{2}{\pi\alpha A^{n/\alpha}} \int_0^{\alpha\pi} j(\theta) \sin \frac{\theta n}{\alpha} \, d\theta . \tag{3.4}$$

Since $j(0) = j(\alpha\pi) = 0$, we can integrate by parts twice in this, and get

$$D_n = \frac{-2\alpha}{\pi n^2 A^{n/\alpha}} \int_0^{\alpha\pi} j''(\theta) \sin \frac{\theta n}{\alpha} \, d\theta . \tag{3.5}$$

But $j''(\theta)$ is continuous almost everywhere, and is of bounded variation. So, by the result on p. 172 of Whittaker and Watson [1946], the absolute value of the integral on the right of (3.5) goes to zero at least as fast as $1/n$. Hence $D_n A^{n/\alpha}$ goes to zero at least as fast as $1/n^3$. Therefore, we have very fast absolute convergence in (3.1) for $|r| < A$. Indeed, we may truncate the series on the right of (3.1) and still have very good accuracy. Consequently, we may calculate as many of the D_n's as needed quite quickly by means of the Fast Fourier Transform (see Bergland [1969]).

Note that by (3.1), $u^{(r)}(r;\theta)$ is the imaginary part of a power series in $z^{1/\alpha}$ with real coefficients.

So much for the case of Dirichlet conditions all around. Let us turn to the case of Dirichlet conditions on the radii, but Neumann conditions on the arc. Say

$$\frac{\partial}{\partial r} u^{(r)}(r;\theta) = k(\theta), \quad r = A, \quad 0 < \theta < \alpha\pi . \tag{3.6}$$

So the partial with respect to r of the series on the right of (3.1) must equal $k(\theta)$ when $r = A$. That is,

$$k(\theta) = \sum_{m=1}^{\infty} \frac{m}{\alpha} D_m A^{(m/\alpha)-1} \sin \frac{\theta m}{\alpha} . \tag{3.7}$$

As $u(x,y)$ and $u^{(s)}(x,y)$ satisfy the same boundary conditions along the radii, their derivatives with respect to r must be the same. By (2.1), $k(\theta)$ is the difference of these derivatives. Hence

$$k(0) = k(\alpha\pi) = 0 .$$

So we can reason as in the previous case, and conclude that

$$\frac{n}{\alpha} D_n A^{(n/\alpha)-1}$$

goes to zero at least as fast as $1/n^3$. This tells us that we have rapid absolute convergence, and can truncate the series, and can calculate the needed D_n's quickly by means of the Fast Fourier Transform.

Let us turn to the case of Neumann conditions on the radii. We replace (3.1) by

$$u^{(r)}(r;\theta) = \sum_{m=0}^{\infty} D_m r^{m/\alpha} \cos \frac{\theta m}{\alpha} . \qquad [3.8]$$

This has the normal derivative, $\partial/\partial\theta$, equal to zero on each radius. To handle Dirichlet conditions on the arc, let $j(\theta)$ again be the value that $u^{(r)}(r;\theta)$ must take along the arc. Instead of $j(0) = j(\alpha\pi) = 0$ we will have

$$j'(0) = j'(\alpha\pi) = 0 . \qquad [3.9]$$

Instead of (3.4), we will have

$$D_0 = \frac{1}{\pi\alpha} \int_0^{\alpha\pi} j(\theta) d\theta \qquad [3.10]$$

$$D_n = \frac{2}{\pi\alpha A^{n/\alpha}} \int_0^{\alpha\pi} j(\theta) \cos \frac{\theta n}{\alpha} d\theta . \qquad [3.11]$$

Because of (3.9), we can integrate by parts twice in (3.11) and conclude that $D_n A^{n/\alpha}$ goes to zero at least as fast as $1/n^3$ because $j(\theta)$ has a second derivative of bounded variation. If it happens that $j(\theta)$ has a third derivative of bounded variation, we can integrate by parts yet one more time in (3.11). Then we can conclude that $D_n A^{n/\alpha}$ goes to zero at least as fast as $1/n^4$, which will give even faster calculation of the D_n's by means of the Fast Fourier Transform.

Note that by (3.8), $u^{(r)}(r;\theta)$ is this time the real part of a power series in $z^{1/\alpha}$ with real coefficients.

To handle Neumann conditions on the arc, we take $k(\theta)$ as in (3.6). We recall that the integral of the normal derivative all around the sector must be zero; see Theorem I on p. 212 of Kellogg [1953]. Analogous to (3.10) we would have

$$D_0 = \frac{1}{\pi\alpha} \int_0^{\alpha\pi} k(\theta) d\theta .$$

But since the normal derivative of $u^{(r)}(x,y)$ is zero all along both radii, we can conclude that $D_0 = 0$. As $\partial u^{(r)}(r;\theta)/\partial\theta = 0$ all along each radius, we must have

$$\frac{\partial^2 u^{(r)}(r;\theta)}{\partial r \partial \theta} = 0 .$$

This entails

$$\frac{\partial^2 u^{(r)}(r;\theta)}{\partial \theta \partial r} = 0$$

at $r = A$, $\theta = 0$. So, taking the partial derivative with respect to θ on both sides of (3.6) gives $k'(\theta) = 0$ at $r = A$, $\theta = 0$. Similarly we get $k'(\theta) = 0$ at $r = A$, $\theta = \alpha\pi$. So we can integrate by parts twice in the analog of (3.11), to conclude that

$$\frac{n}{\alpha} D_n A^{(n/\alpha)-1}$$

goes to zero at least as fast as $1/n^3$. If $k(\theta)$ has a third derivative of bounded variation, we can integrate by parts one more time, and do even better.

We still have to handle the cases where there are Dirichlet conditions on one radius and Neumann conditions on the other.

First let us have Dirichlet conditions on $\theta = 0$ and Neumann conditions on $\theta = \alpha\pi$. Reflect the sector antisymmetrically about the real axis. That is, put $-u^{(r)}(x,y)$ at the point $(x,-y)$. It does not matter whether we have Dirichlet or Neumann conditions along the arc. If Dirichlet, we have values $j(\theta)$ along the arc. But $j(0) = 0$, so that the values along the combination of reflected and original arc will be continuous and have a continuous first derivative at $\theta = 0$. Naturally the second derivative still has bounded variation. So we have twice as large a sector, with Neumann conditions on each bounding radius. We solve this as explained above. By antisymmetry, the solution will have the value zero along the real axis. So we throw away what is below the real axis. If we have Neumann conditions along the arc, we repeat the previous seven sentences with "Neumann" replacing "Dirichlet" and k replacing j, with such other changes as these entail.

This time, it will turn out that $u^{(r)}(r;\theta)$ is the real part of a power series in $z^{1/2\alpha}$ with real coefficients.

The remaining cases are just symmetric to the two we just did with respect to the line $\theta = \alpha\pi/2$.

4. Computational Details. As we noted in the previous section, the series that are used to define $u^{(r)}(x,y)$ converge with reasonable rapidity. We can truncate them and use the Fast Fourier Transform to calculate the coefficients, and can proceed without difficulty. However, $u^{(s)}(x,y)$ is defined in terms of $r(\xi,\eta)$ or $s(\xi,\eta)$; see (2.8) and (2.11), or (2.15) and (2.11). Fortunately, by (2.12) or (2.16), the integrals involved go only from $-C^{\mu}$ to C^{μ}. So we can appeal to techniques for numerical quadrature.

In some cases, we may have $\eta = 0$, as when Neumann conditions are prescribed along a radius, and we wish a value for $u^{(s)}(x,y)$ along that radius. Then $s(\xi,0)$ is given by

$$s(\xi,0) = \frac{1}{2\pi} \int_{-C^{\mu}}^{C^{\mu}} g(t)\log(\xi - t)^2 dt \ . \qquad [4.1]$$

For a point on the radius, we will get ξ with $-C^{\mu} \leq \xi \leq C^{\mu}$. If $-C^{\mu} < \xi < C^{\mu}$, we break the integral into the sum of

$$\frac{1}{2\pi} \int_{-C^{\mu}}^{\xi} g(t)\log(\xi - t)^2 dt \qquad [4.2]$$

and

$$\frac{1}{2\pi} \int_{\xi}^{C^{\mu}} g(t)\log(\xi - t)^2 dt \ . \qquad [4.3]$$

In the first, substitute $t = \xi - e^{u}$ and get

$$\frac{1}{\pi} \int_{-\infty}^{K_1} ug(\xi - e^{u})e^{u} du \qquad [4.4]$$

where $K_1 = \log(C^{\mu} + \xi)$. In the second, substitute $t = \xi + e^{u}$ and get

$$\frac{1}{\pi} \int_{-\infty}^{K_2} ug(\xi + e^{u})e^{u} du \qquad [4.5]$$

where $K_2 = \log(C^{\mu} - \xi)$. There is no difficulty about evaluating either (4.4) or (4.5) by a numerical quadrature, since e^{u} becomes so small as $u \to -\infty$ that we may replace the lower limit of integration by a reasonable negative number, say -20 or -30. If $\xi = -C^{\mu}$, then (4.2) is zero, and we take the limit of (4.5) as $\xi \to -C^{\mu}$. Similarly if $\xi = C^{\mu}$.

We can have trouble if η is close to zero. Specifically, suppose η_0 is very close to 0, and we wish the value of $r(\xi_0,\eta_0)$

or $s(\xi_0, \eta_0)$. If $g(\xi_0)$ is of normal size, then the values of

$$\frac{g(t)}{(\xi_0 - t)^2 + \eta_0^2} \qquad [4.6]$$

that occur in the integrand of $r(\xi_0, \eta_\xi)$ will be very large and changing rapidly for t near ξ_0. Thus there will be considerable cancellation errors if one attempts to calculate $r(\xi_0, \eta_0)$ by numerical quadrature, and it will be difficult to get an approximation to good accuracy. However, suppose $g(t)$ has three derivatives in the neighborhood of $t = \xi_0$. Then we can write

$$g(t) = g(\xi_0) + (t - \xi_0)g'(\xi_0) + \frac{(t - \xi_0)^2}{2} g''(\xi_0) + R \, , \qquad [4.7]$$

where R will be quite small for t near ξ_0. If we substitute this for $g(t)$ in (4.6), we will get three terms that can be integrated explicitly plus a term that can be handled by numerical quadrature, since its integrand will remain of modest size throughout.

This suggests that if we wish an approximation for $r(\xi_0, \eta_0)$, with η_0 very small, we choose a B of reasonable size and break the integral into four parts, for example one from $-C^\mu$ to $\xi_0 - B$, one from $\xi_0 - B$ to ξ_0, one from ξ_0 to $\xi_0 + B$, and one from $\xi_0 + B$ to C^μ. The first and last of these can be handled well by numerical quadrature, and for the two intermediate ones we substitute (4.7) into (4.6).

Note that

$$\frac{\eta_0}{\pi} \int_{\xi_0}^{\xi_0+B} \frac{(t - \xi_0)^n dt}{(\xi_0 - t)^2 + \eta_0^2} = \frac{1}{\pi} I \int_0^B \frac{u^n du}{u - i\eta_0} \, . \qquad [4.8]$$

$$\frac{\eta_0}{\pi} \int_{\xi_0-B}^{\xi_0} \frac{(t - \xi_0)^n dt}{(\xi_0 - t)^2 + \eta_0^2} = \frac{(-1)^{n+1}}{\pi} I \int_0^B \frac{u^n du}{u + i\eta_0} \, . \qquad [4.9]$$

We can handle $s(\xi_0, \eta_0)$ similarly. Define

$$G(u) = \int_0^u g(t)dt \, . \qquad [4.10]$$

Then if $g(t)$ has some derivatives at $t = \xi_0$, $G(u)$ will have one more. Note that

$$\frac{1}{2\pi}\int_E^F g(t)\log\{(\xi - t)^2 + \eta^2\}dt \qquad [4.11]$$

$$= \frac{1}{2\pi}\left[G(t)\log\{(\xi - t)^2 + \eta^2\}\right]_E^F - \frac{1}{\pi}\int_E^F \frac{G(t)(t - \xi)dt}{(\xi - t)^2 + \eta^2} .$$

So we substitute an expansion like (4.7) into the final integral of (4.11). Observe that

$$\frac{1}{\pi}\int_{\xi_0}^{\xi_0+B} \frac{(t - \xi_0)^{n+1}dt}{(\xi_0 - t)^2 + \eta_0^2} = \frac{1}{\pi}\, \mathcal{R}\int_0^B \frac{u^n du}{u - i\eta_0} \qquad [4.12]$$

$$\frac{1}{\pi}\int_{\xi_0-B}^{\xi_0} \frac{(t - \xi_0)^{n+1}dt}{(\xi_0 - t)^2 + \eta_0^2} = \frac{(-1)^{n+1}}{\pi}\, \mathcal{R}\int_0^B \frac{u^n du}{u + i\eta_0} . \qquad [4.13]$$

So we are led to seek an evaluation for $F(\tau,\lambda)$, where

$$F(\tau,\lambda) = \frac{1}{\pi}\int_0^B \frac{t^\tau dt}{t - \lambda} . \qquad [4.14]$$

For $\arg\lambda = 0$, we define $F(\tau,\lambda)$ by taking the limit as $\arg\lambda$ approaches 0 through positive values, and for $\arg\lambda = 2\pi$, we define $F(\tau,\lambda)$ by taking the limit as $\arg\lambda$ approaches 2π through smaller values.

If τ is a non-negative integer, $F(\tau,\lambda)$ is easily evaluated. We write

$$F(\tau,\lambda) = \frac{1}{\pi}\int_0^B \frac{t^\tau - \lambda^\tau}{t - \lambda}\,dt - \frac{\lambda^\tau}{\pi}\int_0^B \frac{dt}{\lambda - t}$$

$$= \frac{1}{\pi}\int_0^B \left\{\sum_{n=0}^{\tau-1} t^{\tau-1-n}\lambda^n\right\}dt + \frac{\lambda^\tau}{\pi}\left[\log(\lambda - t)\right]_0^B .$$

So we get

$$F(\tau,\lambda) = \frac{1}{\pi}\sum_{n=0}^{\tau-1} \frac{B^{\tau-n}\lambda^n}{\tau - n} + \frac{\lambda^\tau}{\pi}\log(\lambda - B) - \frac{\lambda^\tau}{\pi}\log\lambda . \qquad [4.15]$$

This is valid for $0 < |\lambda| < B$ and for $0 < \arg\lambda < 2\pi$. By continuity, it holds also for $\arg\lambda = 0$ and for $\arg\lambda = 2\pi$.

We will need also an evaluation for $F(\tau,\lambda)$ when τ is not an integer.

Theorem 4.1. Let τ not be an integer and let $-1 < \tau$. Let $0 \leq \arg\lambda \leq 2\pi$ and $0 < |\lambda| < B$. Then

$$F(\tau,\lambda) = -\frac{e^{-\pi i\tau}\lambda^{\tau}}{\sin \pi\tau} + \frac{B^{\tau}}{\pi}\sum_{n=0}^{\infty}\frac{(\lambda/B)^n}{\tau - n} . \qquad [4.16]$$

Proof. Clearly it suffices to prove (4.16) for $0 < \arg\lambda < 2\pi$ and $0 < |\lambda| < B$, since we can then deduce (4.16) for $\arg\lambda = 0$ and $\arg\lambda = 2\pi$ by taking limits. So let $0 < \arg\lambda < 2\pi$ and $0 < |\lambda| < B$. Consider the contour in the t-plane specified as follows. Start at the origin, and proceed right along the real axis to $t = B$, keeping $\arg t = 0$; then proceed counterclockwise around the circle of radius B with center at the origin back to $t = B$, but now with $\arg t = 2\pi$; keeping $\arg t = 2\pi$ proceed left along the real axis to the origin. By Cauchy's integral theorem, we have

$$\frac{1}{2\pi i}\oint \frac{t^{\tau}dt}{t - \lambda} = \lambda^{\tau} ,$$

where the integration is around the indicated contour.

Alternatively, let us write the integral as the sum of three integrals along the three parts of the contour. This gives

$$\frac{1}{2\pi i}\oint \frac{t^{\tau}dt}{t - \lambda} = \frac{1}{2\pi i}\int_0^B \frac{t^{\tau}dt}{t - \lambda} + \frac{B^{\tau+1}}{2\pi}\int_0^{2\pi} \frac{e^{i(\tau+1)\theta}d\theta}{Be^{i\theta} - \lambda} + \frac{e^{2\pi i\tau}}{2\pi i}\int_B^0 \frac{t^{\tau}dt}{t - \lambda} .$$

So we conclude

$$\lambda^{\tau} = \frac{1 - e^{2\pi i\tau}}{2i} F(\tau,\lambda) + \frac{B^{\tau}}{2\pi}\int_0^{2\pi} \frac{e^{i\tau\theta}d\theta}{1 - (\lambda/Be^{i\theta})} .$$

As $|\lambda| < B$, we can expand the integrand into an infinite series and integrate term-by-term. Thus we get

$$\lambda^{\tau} = \frac{1 - e^{2\pi i\tau}}{2i} F(\tau,\lambda) + \frac{B^{\tau}(e^{2\pi i\tau} - 1)}{2\pi i}\sum_{n=0}^{\infty}\frac{(\lambda/B)^n}{\tau - n} .$$

From this, (4.16) follows.

We turn to the case where the bounding conditions have a complete Maclaurin expansion at $r = 0$. As before, we have a multiplicity of cases, since we can have either a Dirichlet condition or a Neumann condition on each radius. We shall give a sample of each on the radius $\theta = 0$. Similar developments will hold for the radius $\theta = \alpha\pi$. The reader can put them together to fit the case he is dealing with.

We start with a Dirichlet condition. Let us require

$$u^{(s)}(r;\theta) = f_0(r) \qquad \theta = 0, \quad 0 < r < C$$

where

$$f_0(r) = \sum_{k=0}^{\infty} f_k r^k \qquad [4.17]$$

holds for $|r| < D$, with the f_k all real. By (2.8), (2.11), and (2.12), one portion of $u^{(s)}(x,y)$ will be

$$\frac{\eta}{\pi} \int_0^B \frac{f_0(t^{1/\mu})dt}{(\xi - t)^2 + \eta^2} . \qquad [4.18]$$

According to whether we have a Dirichlet or Neumann condition on $\theta = \alpha\pi$, we will have $\mu = 1/\alpha$ or $\mu = 1/2\alpha$. Let us proceed as though $\mu = 1/\alpha$, since the other situation can easily be handled by putting 2α for α. So (4.18) becomes

$$\frac{\eta}{\pi} \int_0^B \frac{f_0(t^{\alpha})dt}{(\xi - t)^2 + \eta^2} .$$

Suppose we have chosen B so that $B^{\alpha} < D$. So

$$f_0(t^{\alpha}) = \sum_{k=0}^{\infty} f_k t^{k\alpha} \qquad 0 < t < B .$$

Then (4.18) equals

$$I \frac{1}{\pi} \sum_{k=0}^{\infty} f_k \int_0^B \frac{t^{k\alpha}}{t - \zeta} . \qquad [4.19]$$

By (4.14), (4.19) becomes

$$I \sum_{k=0}^{\infty} f_k F(k\alpha, \zeta) . \qquad [4.20]$$

Turn now to a Neumann condition. Let us require

$$\frac{\partial}{\partial n} u^{(s)}(r;\theta) = g_0(r) \quad \theta = 0, \quad 0 < r < C$$

where

$$g_0(r) = \sum_{k=0}^{\infty} g_k r^k \qquad [4.21]$$

holds for $|r| < D$, with the g_k all real. By (2.15), (2.11), and (2.16), one portion of $u^{(s)}(x,y)$ will be

$$\frac{1}{2\pi} \int_0^B - \frac{1}{\mu} t^{(1/\mu)-1} g_0((t)^{1/\mu}) \log\{(\xi - t)^2 + \eta^2\} dt . \qquad [4.22]$$

As before, μ may be $1/\alpha$ or $1/2\alpha$, but we can proceed as though $\mu = 1/\alpha$. So (4.22) becomes

$$\frac{1}{2\pi}\int_0^B -\alpha t^{\alpha-1} g_0(t^\alpha)\log\{(\xi - t)^2 + \eta^2\}dt .$$

Integrating by parts and recalling (4.10) gives

$$[-\frac{1}{2\pi} G(t^\alpha)\log\{(\xi - t)^2 + \eta^2\}]_0^B + \frac{1}{\pi}\int_0^B \frac{G(t^\alpha)(t - \xi)dt}{(\xi - t)^2 + \eta^2} .$$

As before, choose B so that $B^\alpha < D$. Then, by (4.21)

$$G(t^\alpha) = \sum_{k=1}^{\infty} \frac{1}{k} g_{k-1} t^{k\alpha} \qquad 0 < t < B .$$

Then (4.22) equals

$$[-\frac{1}{2\pi} G(t^\alpha)\log\{(\xi - t)^2 + \eta^2\}]_0^B + R \sum_{k=1}^{\infty} \frac{g_{k-1}}{\pi k} \int_0^B \frac{t^{k\alpha}dt}{t - \zeta} . \qquad [4.23]$$

By (4.14), (4.23) becomes

$$[-\frac{1}{2\pi} G(t^\alpha)\log\{(\xi - t)^2 + \eta^2\}]_0^B + R \sum_{k=1}^{\infty} \frac{g_{k-1}}{k} F(k\alpha,\zeta) . \qquad [4.24]$$

It is tempting to substitute (4.16) into (4.20) and interchange the order of summation. This would give

$$I \sum_{k=0}^{\infty} \frac{-f_k e^{-\pi i k\alpha}\zeta^{k\alpha}}{\sin \pi k\alpha} + I \frac{1}{\pi} \sum_{n=0}^{\infty} (\frac{\zeta}{B})^n \sum_{k=0}^{\infty} \frac{f_k B^{k\alpha}}{k\alpha - n} . \qquad [4.25]$$

If α is rational, this will not work, since occasionally we would have $k\alpha$ equal to an integer (as when $k = 0$); then both the terms

$$\frac{-f_k e^{-\pi i k\alpha}\zeta^{k\alpha}}{\sin \pi k\alpha}$$

and

$$\frac{f_k B^{k\alpha}}{k\alpha - n}$$

would be infinite. If we assume that α is irrational, we still run into trouble for some irrational numbers. By Chapter 11 of Hardy and Wright [1954], the quantities $k\alpha$ will intermittently come as near as desired to being positive integers as k increases. Thus $\sin \pi k\alpha$ and $k\alpha - n$ (for suitable n) will intermittently come nearer and nearer to zero, and will not be bounded away from zero.

A phenomenon of this sort was noted in Lehman [1954]. It turns out that we can escape from this predicament by suitably grouping together some of the terms on the right of (4.25). To this end, we define $T(\tau,w)$ as follows:

Case 1. τ is an integer. Then we take

$$T(\tau,w) = w^{\tau}\{i - \frac{1}{\pi}\log w\} . \qquad [4.26]$$

Case 2. τ is not an integer. Let n be the integer nearest τ. If τ is halfway between two integers, take $n = \tau + \frac{1}{2}$. That is, we take

$$n = [\tau + \tfrac{1}{2}] . \qquad [4.27]$$

Then we take

$$T(\tau,w) = w^{\tau}\{\frac{w^{n-\tau}}{\pi(\tau - n)} - \frac{e^{-\pi i\tau}}{\sin \pi\tau}\} . \qquad [4.28]$$

Lemma 4.1. For fixed τ and $0 \leq \arg w < 2\pi$ and $0 < |w|$, $T(\tau,w)$ is a continuous function of w.

Proof. Obvious.

Lemma 4.2. For fixed τ and $0 \leq \arg w < 2\pi$ and $0 < |w|$,

$$\frac{d^m}{dw^m} T(\tau,w)$$

is a continuous function of w.

Proof. Temporarily define

$$S(\tau,w) = -\frac{1}{\pi} w^{\tau-1}$$

if τ is an integer, and

$$S(\tau,w) = -\frac{1}{\pi} w^{n-1}$$

with n given by (4.27) if τ is not an integer. Then

$$\frac{\partial}{\partial w} T(\tau,w) = \tau T(\tau - 1,w) + S(\tau,w) .$$

Clearly $S(\tau,w)$ and all its derivatives with respect to w are continuous. So our lemma follows from Lemma 4.1 by induction on m.

Lemma 4.3. For arbitrary τ and $0 \leq \arg w < 2\pi$ and $0 < |w| \leq 1$,

$$w^{1-\tau}T(\tau,w) \qquad [4.29]$$

is uniformly bounded.

Proof. If τ is an integer, then the quantity we are considering is

$$w\{i - \frac{1}{\pi} \log w\} .$$

For the specified range of w, this is uniformly bounded. So take τ not an integer. Then the quantity we are considering is the sum of

$$\frac{w}{\pi} \frac{w^{n-\tau} - 1}{\tau - n} \qquad [4.30]$$

$$\frac{w}{\pi} \frac{1 - e^{-\pi i(\tau-n)}}{\tau - n} \qquad [4.31]$$

$$w\{\frac{e^{-\pi i(\tau-n)}}{\pi(\tau - n)} - \frac{e^{-\pi i(\tau-n)}}{\sin \pi(\tau - n)}\} . \qquad [4.32]$$

As $|\tau - n| \leq \frac{1}{2}$, the latter two are clearly bounded. Take

$$w = \frac{e^{i\lambda}}{\sigma}$$

with $0 \leq \lambda < 2\pi$ and $1 \leq \sigma$. Then

$$\log w = -\log \sigma + i\lambda ,$$

$$w^{n-\tau} = 1 + (n - \tau)\log w + \frac{(n - \tau)^2(\log w)^2}{2!} + \frac{(n - \tau)^3(\log w)^3}{3!} + \ldots .$$

As $|n - \tau| \leq \frac{1}{2}$, we get

$$\left|\frac{w^{n-\tau} - 1}{\tau - n}\right| \leq |\log w| \sum_{r=0}^{\infty} \frac{1}{(r + 1)!} \left|\frac{\log w}{2}\right|^r .$$

This gives

$$\left|\frac{w^{n-\tau} - 1}{\tau - n}\right| \leq |\log w| e^{|\log w|/2} . \qquad [4.33]$$

Case 1. $1 \leq \sigma \leq e^{2\pi}$. Then

$$|\log w| \leq 2\sqrt{2}\, \pi .$$

So by (4.33)

$$\left|\frac{w^{n-\tau} - 1}{\tau - n}\right| \leq 2\sqrt{2}\, \pi e^{\pi\sqrt{2}} .$$

So, as $|w| \leq 1$, we have a bound for (4.30).

Case 2. $e^{2\pi} \leq \sigma$.

$$|\log w| \leq \sqrt{2} \log \sigma .$$

So by (4.33)

$$\left|\frac{w^{n-\tau} - 1}{\tau - n}\right| < \sqrt{2}\ (\log \sigma)\sigma^{1/\sqrt{2}} .$$

Then (4.30) is bounded by

$$\frac{\sqrt{2}\ \log \sigma}{\pi\sigma^{1-(1/\sqrt{2})}} .$$

As $e^{2\pi} \leqslant \sigma$, this is bounded.

If we wish to extend the definition of $T(\tau,w)$ by analytic continuation (foregoing the convention that $0 \leqslant \log w < 2\pi$, of course), then indeed, for fixed N, Lemma 4.3 would hold for $-N\pi \leqslant \arg w \leqslant N\pi$ and $0 < |w| \leqslant 1$. The proof would be quite analogous.

Lemma 4.4. Fix $m \geqslant 0$. Then for $-1 \leqslant \tau$, $0 \leqslant \arg w < 2\pi$, and $0 < |w| \leqslant 1$,

$$(2 + \tau)^{-m} w^{1+m-\tau} \frac{\partial^m}{\partial w^m} T(\tau,w)$$

is uniformly bounded.

Proof. Use induction on m, as in the proof of Lemma 4.2.

Theorem 4.2. For $-1 < \tau$, $0 \leqslant \arg \zeta < 2\pi$, and $0 < |\zeta| < B$, we have

$$F(\tau,\zeta) = B^{\tau}T\left(\tau, \frac{\zeta}{B}\right) + \frac{B^{\tau}}{\pi}\left\{\sum_{n=0}^{[\tau-1/2]} \frac{(\zeta/B)^n}{\tau - n} + \sum_{n=1+[\tau+1/2]}^{\infty} \frac{(\zeta/B)^n}{\tau - n}\right\} . \qquad [4.34]$$

Proof. If τ is not an integer, this follows by Theorem 4.1. One can derive the result for τ an integer T by taking the limit as τ approaches T on both sides of (4.34).

We could use (4.34) to extend $F(\tau,\zeta)$ for values with other ranges for $\arg \zeta$.

Theorem 4.3. Let (4.17) hold, and choose B so that

$$\sum_{k=0}^{\infty} f_k B^{k\alpha}$$

converges absolutely. Then for $|\zeta| < B$ and $0 < \arg \zeta < \pi$

$$\frac{\eta}{\pi}\int_0^B \frac{f_0(t^{\alpha})dt}{(\xi - t)^2 + \eta^2} = I \sum_{k=0}^{\infty} f_k B^{k\alpha} T\left(k\alpha, \frac{\zeta}{B}\right) + I\ \frac{1}{\pi} \sum_{n=0}^{\infty} \left(\frac{\zeta}{B}\right)^n C_n \qquad [4.35]$$

where

$$C_n = \sum_{0 \leqslant k < \frac{2n-1}{2\alpha}} \frac{f_k B^{k\alpha}}{k\alpha - n} + \sum_{\frac{2n+1}{2\alpha} < k} \frac{f_k B^{k\alpha}}{k\alpha - n} . \qquad [4.36]$$

Proof. We substitute (4.34) into (4.20), and interchange the order of summation. To see that this interchange is permissible, note that in the series for C_n we have $|k\alpha - n| > \frac{1}{2}$, so that the series for C_n is absolutely convergent. Indeed

$$|C_n| < 2 \sum_{k=0}^{\infty} |f_k B^{k\alpha}| ,$$

so that

$$\sum_{n=0}^{\infty} (\tfrac{\zeta}{B})^n C_n$$

converges absolutely for $|\zeta| < B$, and converges uniformly for $|\zeta| \leq B - \varepsilon$ for $\varepsilon > 0$. Moreover, it can be differentiated term-by-term as often as desired, and the derivative series will have the same convergence properties. The absoluteness of the convergence justifies the interchange of summation.

By Lemma 4.3 there is an M such that for non-negative integer k and for $0 \leq \arg \zeta \leq 2\pi$ and $0 < |\zeta| \leq B$

$$|(\tfrac{\zeta}{B})^{1-k\alpha} T(k\alpha, \tfrac{\zeta}{B})| \leq M .$$

So

$$|f_k B^{k\alpha} T(k\alpha, \tfrac{\zeta}{B})| \leq |\tfrac{B}{\zeta} f_k \zeta^{k\alpha}| M .$$

Thus we conclude that the first series on the right of (4.35) converges absolutely for $0 < |\zeta| \leq B$, and converges uniformly for $\varepsilon \leq |\zeta| \leq B$ for $\varepsilon > 0$. By appealing to Lemma 4.4 we conclude that it can be differentiated term-by-term with respect to ζ as often as desired, and the derivative series will have the same convergence properties. Also by Lemmas 4.1 and 4.2 the series and the derived series are the sums of continuous terms, and so are themselves continuous.

In the same way, we can conclude that if

$$u^{(s)}(r;\theta) = f_\alpha(r)$$

for $\theta = \alpha\pi$ and $0 < r < C$, and

$$f_\alpha(r) = \sum_{k=0}^{\infty} f_k^* r^k$$

for $0 \leq r < D$, and if

$$\sum_{k=0}^{\infty} f_k^* B^{k\alpha}$$

converges absolutely, then for $|\zeta| < B$ and $0 < \arg \zeta < \pi$

$$\frac{\eta}{\pi} \int_{-B}^{0} \frac{f_\alpha((-t)^\alpha)dt}{(\xi - t)^2 + \eta^2}$$

has an expansion in terms of $e^{\pi i}\zeta$ similar to (4.35). Thus for $|\zeta| < B$ and $0 < \arg \zeta < \pi$ we have an expansion for

$$\frac{\eta}{\pi} \int_{-B}^{B} \frac{g(t)dt}{(\xi - t)^2 + \eta^2}$$

with $g(t)$ given by (2.12), with $\mu = 1/\alpha$, $f_\beta = f_0$, and $f_\gamma = f_\alpha$.

If $B^\alpha \geq C$ (in which case we may as well take $C = B^\alpha$), then we have an expansion for $v(\xi,\eta)$. If not we still have

$$\frac{\eta}{\pi} \int_{-C^{1/\alpha}}^{-B} \frac{g(t)dt}{(\xi - t)^2 + \eta^2}$$

and

$$\frac{\eta}{\pi} \int_{B}^{C^{1/\alpha}} \frac{g(t)dt}{(\xi - t)^2 + \eta^2}$$

to account for. Take $B^* < B$. Then for the sector $\rho \leq B^*$ and $0 < \phi < \pi$ in the ζ-plane, each of the above represents a harmonic function which satisfies homogeneous conditions on the bounding radii. So, by the methods of Section 3, each can be expanded as the real (or imaginary) part of a series in ζ. So adding them to (4.35) will only change the values of C_n.

The revised series hold only for $\rho \leq B^* < B$. However, we can do this for each $B^* < B$. As $w(\xi,\eta)$ can have only one such expansion, the series must be valid for $\rho < B$.

We turn to the case of Neumann conditions. The key formula here is (4.24). The first term will come along with no change. We substitute from (4.34) into the second term and wind up with something resembling the right side of (4.35), except that we will be taking the real part of it. From there on the considerations will be quite analogous.

In both the Dirichlet and Neumann cases, when we go back to the original z coordinates, we put $\zeta = z^{1/\alpha}$. We have to add in $u^{(r)}(x,y)$ to get $u(x,y)$. But the $u^{(r)}(x,y)$ are imaginary or real parts of series in $z^{1/\alpha}$. This combines with what results from the final term on the right of (4.35), or its analogue in the Neumann case. So we get a series in $z^{1/\alpha}$ plus results from the first term on the right of (4.35).

What results from the $T(k\alpha,\zeta/B)$ terms could be split into two parts; see (4.26) and (4.28). There would result two series, one in z, multiplied by $\log z$, and one in $z^{1/\alpha}$. This altogether then resembles the series derived in Wasow [1957] and Lehman [1959]. However, these new series are only asymptotic series, and even divergent for some values of α. It was to avoid such difficulties that we combined two terms to form a $T(k\alpha,\zeta/B)$ term. Such terms should be kept combined.

The resulting series in $z^{1/\alpha}$ converge only for $z < B^{\alpha}$. So, unless B is large, we have got a representation of $u(x,y)$ only in the neighborhood of the vertex. For values elsewhere, we probably have to resort to numerical quadrature.

If we have a Dirichlet condition on one radius and a Neumann condition on the other, things work out very much the same, except that we will be using 2α instead of α.

Acknowledgement. Sponsored in part by the United States Army under Contract No. DA-31-124-ARO-D-462, in part by the Science Research Council under Grant B/RG 4121 at Brunel University, in part by the Wisconsin Alumni Research Foundation, and in part by the Rockefeller University. I wish to acknowledge the help given by the sponsors, and also by Dr. John R. Whiteman, who brought these questions to my attention, and took part in stimulating discussions of them. I also wish to thank Dr. M. G. Crandall for some useful suggestions.

REFERENCES

1. G. D. Bergland, "A radix-eight Fast Fourier Transform subroutine for real-valued series," IEEE Trans. on Audio and Electroacoustics, vol. AU-17 (1969), pp. 138-144.

2. G. C. Evans, "The logarithmic potential," A.M.S. Colloquium Publications, vol. VI, 1927.

3. G. H. Hardy and E. M. Wright, "An introduction to the theory of numbers," third edition, Clarendon Press, Oxford, 1954.

4. O. D. Kellogg, "Foundations of potential theory," Dover Publications, Inc., New York, 1953.

5. R. Sherman Lehman, "Developments in the neighborhood of the beach of surface waves over an inclined bottom," Communications on Pure and Applied Mathematics, vol. 7 (1954), pp. 393-439.

6. R. Sherman Lehman, "Developments at an analytic corner of solutions of elliptic partial differential equations," Journal of Mathematics and Mechanics, vol. 8 (1959), pp. 727-760.

7. W. E. Milne, "Numerical solutions of differential equations," John Wiley and Sons, Inc., 1953.

8. Wolfgang Wasow, "Asymptotic development of the solution of Dirichlet's problem at analytic corners," Duke Mathematical Journal, vol. 24 (1957), pp. 47-56.

9. E. T. Whittaker and G. N. Watson, "A course of modern analysis," American Edition, The Macmillan Company, New York, 1946.

SINGULARITIES OF CRACKS WITH GENERALIZED FINITE ELEMENTS

E. Schnack
Institute of Solid Mechanics
Karlsruhe University
7500 Karlsruhe, F.R.Germany

Dedicated to Prof. Dr. Dr. Heinz Neuber for his 78th birthday

SUMMARY

Pian's hybrid method is frequently used for computing stress intensity factors in fracture mechanics. It has been shown, however, that this method has its limitations. In using this method, it is found that the resulting stiffness matrix is often poorly conditioned and it is not possible to determine surface displacements of notch or crack zones accurately. Most of these aforementioned difficulties can be eliminated by using a modified hybrid stress model in combination with the displacement method. The proposed <u>modified hybrid method</u> will be shown to offer some significant advantages for plane, axi-symmetric and three dimensionel problems of fracture mechanics.

INTRODUCTION

The premise of the following treaties is the linear elastic fracture mechanics (LEFM). Actual problems of the LEFM are first the crack bifurcation problems (fig. 1), second the notch-crack problems (fig. 2). In case of the bifurcation problems, the influence of the intensities among themselves is examined. It is very important to compute the stress intensity factors with high accuracy in order to be able to predict the development of crack progress. In case of the notch-crack problems, the stress intensity factors are functions of the notch parameters, as for instance the notch radius and the notch depth. The approximation of screw-nut connections with bodies of revolution is a very interesting case of application (fig. 3). For example it is important to examine the crack propagation from outside surfaces against the middle axis of notched round bars. Of the same importance for fracture mechanics in practice is the determination of stress intenstiy factors along the crack fronts in generally three-dimensional problems (fig. 4).
From the described actual problems of the LEFM result important claims to an optimal numerical algorithm for determination of the stress intensity factors from the characteristic singularities.

1. To be able to solve crack bifurcation problems, the size of special crack elements must be variable. Therefore it is demanded that the characteristic value D (see fig. 5) can change to smaller values (demanded by the geometry through the close

location of two crack-tips) without the condition number of the resulting stiffness matrices leading to bad values and therefore causing a strongly oscillating convergence behavior [1].

2. As a next, decisive point the strong, also known eigen solution should be included into the trial functions. This eigen solution of the crack-tip vicinity is described by the Irwin-Sneddon equation.
3. It is also important for a good approximation of the stress intensity factor to effect a suitable approximation of the surface fiber stresses.
4. A very essential point for reaching high accuracies for stress concentration problems in general is to fulfill exactly the traction condition (fig. 6), i.e.:

$$\vec{t} := \tau_{km}\, n_k\, \vec{e}_m \Big|_{\partial V_s} = \vec{\bar{t}}. \qquad (1)$$

5. In connection with point 4), it is of great importance to fulfill the equilibrium condition inside the volume around the crack-tip exactly, i.e.:

$$\tau_{km,k} + \rho f_m = 0. \qquad (2)$$

6. Not all of the algorithms are able to produce the surface displacement vector [2]. However, this is an important claim, if for example contact problems shall be computed, and for example a crack bifurcation starts from one of the two elastic bodies (fig. 7).
7. The last point considers mainly the notch-crack problems. So it is very important to consider also the influence of the main parameters of the notch theory during the consideration of suited trial functions for notch-crack problems, as for example the influence of the notch curvature and the notch depth (fig. 8).

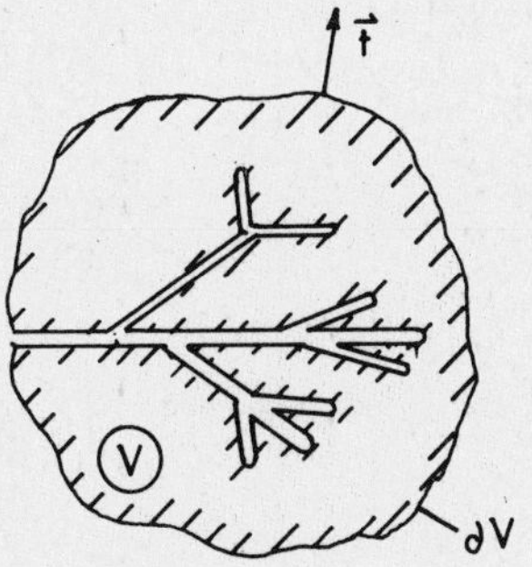

Fig. 1: crack bifurcation

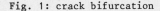

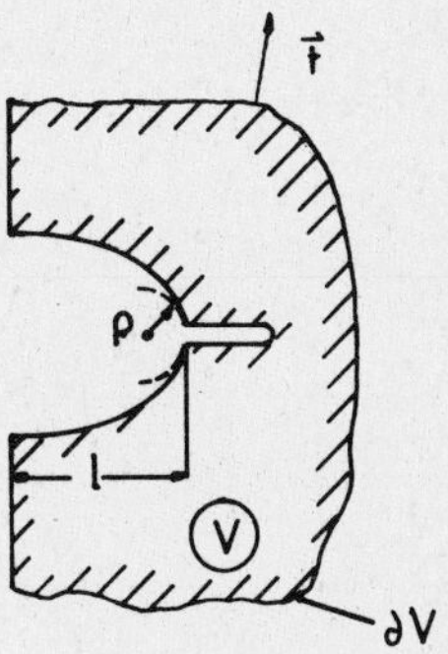

Fig. 2: notch-crack problem

The question arises, which numerical procedures are suited for solving these problems. Because finite bounded bodies generally occuring in practice are based, only a solution of crack or notch-crack problems with numerical methods is possible. So, one time, the boundary element method (BEM) or the class of FEM-methods can be used. When using the FEM-method, the classical Finite-Element-Method can be used without consideration of the singularity in the trial function. Then the possibility is given to work with so-called 'collapsed elements', which are special displacement elements with the $\sqrt{r}$-characteristic by a suited intermediate transformation. Another possibility is to work with hybrid methods, such as the hybrid displacement or the hybrid stress method [2]. To fulfill the above listed seven claims, the most suitable method is a combination of the classical displacement element method with the so-called hybrid assumed stress method. However, the hybrid assumed stress method in its classical form is impossible for solving the problem in consideration of the seven aims.

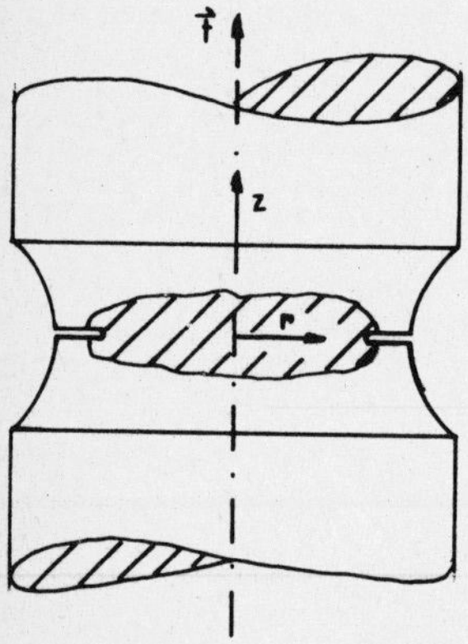

Fig. 3: axi-symmetric notch-crack problem

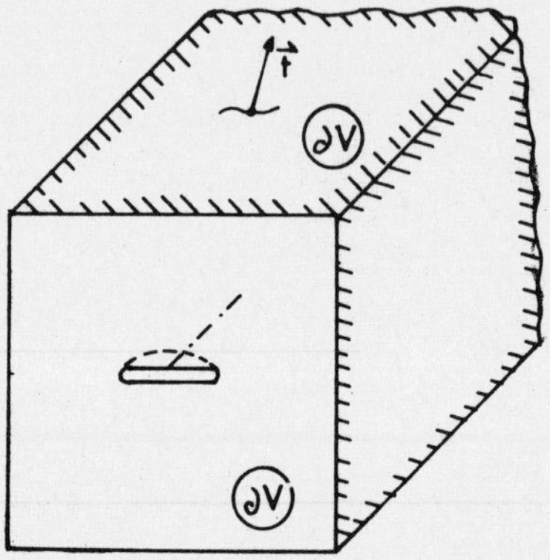

Fig. 4: surface crack

You can often hear in discussions that such hybrid models were not necessary, because the so-called 'collapsed elements' exist. In the following I want to show that the collapsed elements, such as the quarter-point element or others [3], are not at all capable to fulfill the claims from 1-7. For example at first the equilibrium condition is injured:

$$\tau_{km,k} + \rho f_m \neq 0. \tag{3}$$

2. The traction condition is injured:

$$\vec{t} \neq \vec{\bar{t}}. \tag{4}$$

3. The exact eigen solution of the singularity is not incorporated in the trial functions, but only and not more the $\sqrt{r}$-characteristic is. For the quarter-point element, the $\sqrt{r}$-characteristic is fulfilled only for the vertical and the horizontal line, but not in the element. A better element of this group has been developed by M. Stern and E.B. Becker [3], but also in this element, the exact eigen solution is not incorporated in the trial function, but only the $\sqrt{r}$-characteristic is. Conse-

quently the components of the boundary stress vector, e.g. a loadfree crack surface, differ essentially from zero (fig. 9 and fig. 10). If an extrapolation of the normalized stress intensity factor, see fig. 11 and fig. 12, is effected through the crack opening displacement (COD) concept, which means to look for the best angle φ based on a polar co-ordinate system from the crack-tip and to extrapolate with $r \to 0$ against the crack-tip to get the stress intensity factor by the displacement field, one states that the stress intensity factor does not converge against the exact value in front of the crack-tip, but is decreasing considerably.

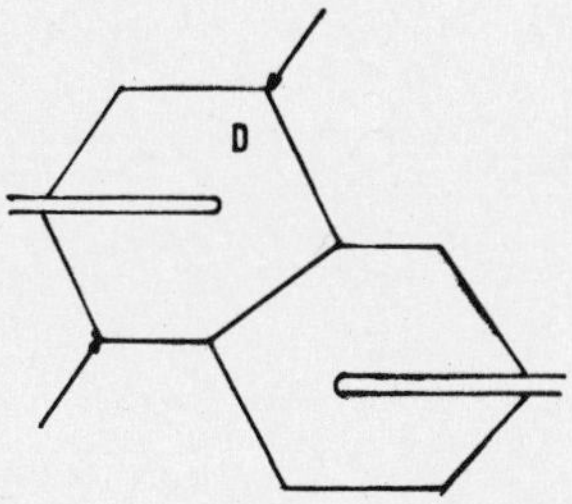

Fig. 5: double-crack problem

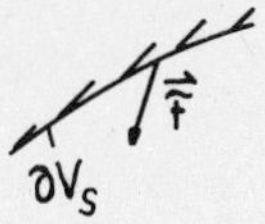

Fig. 6: traction condition

A SURVEY ABOUT HITHERTO KNOWN HYBRID ASSUMED STRESS MODELS

In 1969, Pian [4,5] has created the functional of the modified complementary energy:

$$\Pi^n_{(\{\sigma\},\{u_{\partial V}\})} = \int_{V_n} \hat{W}^{***}\, dV - \int_{\partial V_n} \{t\}^T \{u_{\partial V}\}\, dA + \int_{\partial V_s} \{t_s\}^T \{u_{\partial V}\}\, dA, \qquad (5)$$

(s. equ.(9))

which requires to fulfill a kinematic convergence condition:

$$M \geq N - 1, \qquad \begin{array}{l}(M: = \text{number of stress formulations } \{\sigma\},\\ N: = \text{number of displacement functions } \{u_{\partial V}\},\\ 1: = \text{number of rigid body motions})\end{array} \qquad (6)$$

but this functional can only work effectively in competition with the displacement method, if the numerical computation effort can be reduced. Because with the Pian-formulation one volume integral and two surface integrals must be considered, while with the displacement method only one volume integral and one surface integral in general have to be computed for problems of the same kind. Therefore, an effectivity improvement could be found by E. Schnack [6] for notch mechanics in 1973, and by P. Tong, T.H.H. Pian, S.J. Lasry [2] for problems in fracture mechanics.

A summarized description can be found for example in [7]. That means for problems of notch and fracture mechanics that a further restriction can be required in a manner that the compatibility condition for the interior of the hybrid elements is exactly fulfilled. So an incompatibility tensor formulated as a function of the components of the strain tensor must disappear:

$$i_{rs} = 0. \tag{7}$$

For the Hooke's material law follows consequently:

$$\int_{V_n} \hat{W}^{***}\, dV = \frac{1}{2} \int_{\partial V_n} \{t\}^T \{u_V\}\, dA. \tag{8}$$

Another modification is possible, but not effective, if one is unable to fulfill exactly the traction condition along surfaces. So another term can be added to the functional by the Lagrange-multiplier technique, in order to consider this condition more precisely at least on the average, see papers of J.P. Wolf in 1974 [8] and S.N. Atluri, H.C. Rhee in 1978 [9]. If the FE-formulation will be effected, first the continuum must be devided into a finite number of finite elements with the volume V_n and the surface ∂V_n (fig. 13 and fig. 14). In this process the surface ∂V_n will be devided into two parts:

$$\partial V_n = \partial V_0 \cup \partial V_s. \tag{9}$$

Fig. 7: contact problem

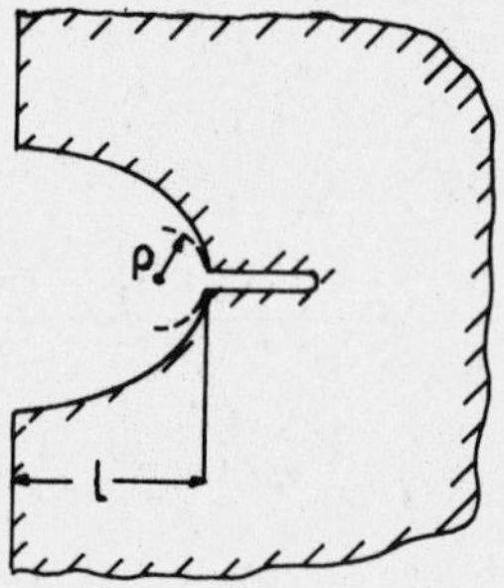

Fig. 8: notch parameters

The predetermined traction vector acts only on the surface ∂V_s, as fig. 14 shows. The constraints for the continuum begin with the equilibrium condition:

1. $\tau_{km,k} = 0\ (-\rho_{fm})$, (10)

provided that no volume forces may act.

2. The equilibrium condition of the boundary must be fulfilled. Written as a tensor is valid:

$$\tilde{t}_m = \tau_{km}\, n_k, \qquad \text{(s. equ. (1))} \tag{11}$$

n_k being the surface unit normal vector. For the original Pian formulation, the following restrictions 3 and 4 are not necessary, however, for the modified hybrid stress method (HSM) they are necessary. So the restrictions 3 must be fulfilled:

$$3.\quad e_{km} = \frac{1}{2}\,(v_{k,m} + v_{m,k}) \tag{12}$$

in the volume element V_n, with full compatibility in V_n, so that the incompatibility tensor within the continuum must disappear as mentioned before (see equ.(7)). Additionally the inverse Hooke's law is necessary as a fourth law:

$$4.\quad e_{km} = \frac{1}{2G}\,(\tau_{km} - \frac{\nu}{1+\nu}\,\delta_{km}\,\tau_{qq}) \tag{13}$$

ν meaning the transverse contraction ratio and G the shear modulus.
In the following the trial functions will be formulated, first for the surface of the element. For ∂V_n the displacement vector is defined:

$$\{u_{\partial V}\} = [N]\{d\}, \tag{14}$$

[N] signifying the form function for the displacement field. {d} are the nodal displacements. For V_n a stress formulation:

$$\{\sigma\} = \{\sigma_0\} + \{\sigma_s\} \tag{15}$$

is set up. For this process it is very important to devide the stress formulation into two parts, i.e. in the homogenious part $\{\sigma_0\}$ and in the inhomogenious part $\{\sigma_s\}$. For the homogenious part the following is set up:

$$\{\sigma_0\} = [P_0]\{\beta_0\}. \tag{16}$$

Through the equilibrium condition of the boundary, see equ. (11), the traction matrix is reached:

$$\{t_0\} = [R_0]\{\beta_0\}. \tag{17}$$

It is important to fulfill the restriction $\{t_0\} = \{o\}$ on that part of the surface where the traction vector is predetermined, that means for ∂V_s.
Because equ. (7) has been fulfilled, and the inverse Hooke's relation is known, the displacement field can be found by integration:

$$\{u_V^0\} = [L_0]\{\beta_0\}. \tag{18}$$

For the inhomogenious stress part a similar formulation can be effected. First a formulation

$$\{\sigma_s\} = [P_s]\{\beta_s\} \tag{19}$$

will be formulated and this leads to:

$$\{t_s\} = [R_s]\{\beta_s\}, \tag{20}$$

which can be computed by the equilibrium condition of the boundary, see equ. (11). It is now very important for a high rate of convergence concerning problems with high stress gradients, that the traction condition is exactly fulfilled for ∂V_s. It is valid:

$$\{t_s\} = \{\tilde{t}_s\} \text{ for } \partial V_s. \tag{21}$$

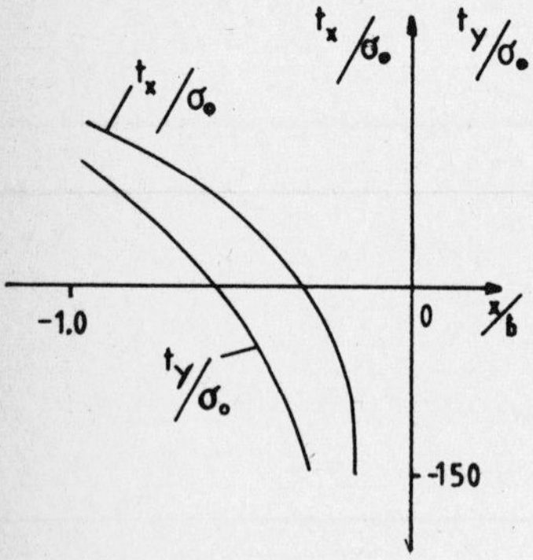

Fig. 9: traction components

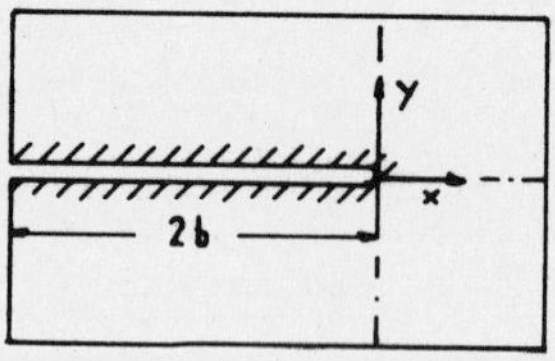

Fig. 10: crack with quarter-point elements

Because $i_{rs} = 0$, see equ. (7), the inhomogenious displacement field in the interior can be computed by integration:

$$\{u_V^s\} = [L_s]\{\beta_s\}. \tag{22}$$

It is known from the principle of virtual work, that the first variation of the functional (equ. (5)) disappears. It is valid:

$$\delta\Pi^n = \underbrace{\int_{\partial V_0} \delta\{t_0\}^T(\{u_V\} - \{u_{\partial V}\})\, dA}_{I} - \underbrace{\int_{\partial V_0} (\{t_0\}^T + \{t_s\}^T)\, \delta\{u_{\partial V}\}\, dA}_{II} = 0. \tag{23}$$

This functional was used during the years 73 to 77 [1,2] to compute crack and notch-crack problems. Part I of $\delta\Pi^n$ represents the compatibility condition in a weak form, while part II is identified by the equilibrium condition as a weak formulation. The trial and test functions can now be inserted and an equation will be reached for part II in form of:

$$\left\langle \{\beta_0\}^T \underbrace{\int_{\partial V_0} [R_0]^T[N]\, dA}_{[T_0]} + \{\beta_s\}^T \underbrace{\int_{\partial V_0} [R_s]^T[N]\, dA}_{[T_{0s}]} \right\rangle \delta\,\{d\} = 0. \tag{24}$$

The interpolation matrices are additionally inserted into the compatibility condition, so that an equation results with three terms for part I:

$$\delta\,\{\beta_0\}^T \left\langle \underbrace{\int_{\partial V_0} [R_0]^T[L_0]\, dA}_{[H_0]} \{\beta_0\} + \underbrace{\int_{\partial V_0} [R_0]^T[L_s]\, dA}_{[H_{0s}]} \{\beta_s\} - \underbrace{\int_{\partial V_0} [R_0]^T[N]\, dA}_{[T_0]^{\cdot}} \{d\} \right\rangle = 0. \tag{25}$$

Hence follows that $[H_0]$ is a positive definite matrix, because $[H_0]$ is the kernel matrix of the strain energy. Consequently $\{\beta_0\}$ can be written as a function of the nodal displacements and of the coefficients of the inhomogenious stress formulation $\{\beta_s\}$:

$$\{\beta_0\} = [H_0]^{-1}\,([T_0]\{d\} - [H_{0s}]\{\beta_s\}). \tag{26}$$

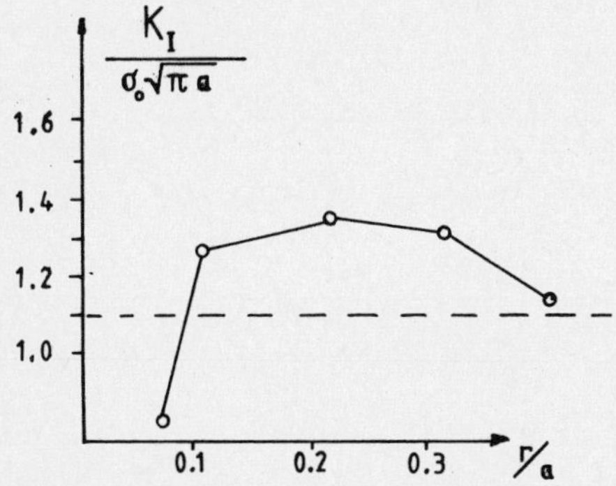

Fig. 11: extrapolation with r → o

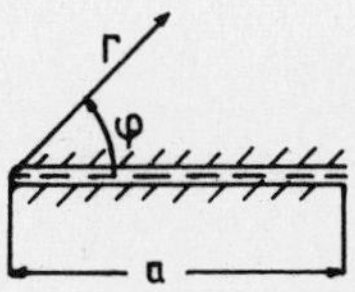

Fig. 12: co-ordinates for COD-concept

After equ. (26) has been inserted into equ. (24), a stiffness formulation results as used by P. Tong et al [2] and M. Wolf [1] for crack and notch-crack problems:

$$\underbrace{[T_0]^T[H_0]^{-1}\,[T_0]}_{[k]}\{d\} = \underbrace{\left\langle [T_0]^T[H_0]^{-1}\,[H_{0s}] - [T_{0s}]^T \right\rangle \{\beta_s\}}_{\{\tilde F\}}. \tag{27}$$

However, a lot of difficulties arise from using this method. So, numerical tests have shown [10], that

1. The procedure shows a strongly oscillating convergence behavior referring to the characteristic values, which are the stress intensity or notch factors. R. Drumm [10] has stated that the background for this is the satisfaction of the kinematic convergence condition of the Pian functional, because the unbalanced equ. (6) must be fulfilled. This unbalanced equation was fulfilled element by element by some investigators until 1982 [1,7,11,12], so that a lot of stress formulations become necessary. This has for consequence that the selectivity between the functions becomes very small. By reason of this, the condition number of the resulting matrix $[H_0]$ gets high values and causes the numerical instabilities and consequently the strong oscillating convergence behavior.
2. The displacement vector $\{u_{\partial V}\}$ is unknown for the part of the surface where the traction vector is predetermined, because equ. (25) has only to be integrated over ∂V_0.
3. The nodal force matrix $\{\tilde{F}\}$, see equ. (27), is numerically unstable, see paper of E. Schnack and R. Drumm in 1981 [13] for solving notch-crack problems with a predetermined traction vector.

Therefore E. Schnack suggested another way in 1981 [13]. It can be derived easily from the principle of virtual work:

$$\overset{v}{W}_a = \{F_\mu\}^T\{u_\mu\} + \{M_\mu\}^T\{\varphi_\mu\} + \int_{\partial V_s} \{t_s\}^T \{u\}\, dA + \int_V \{f\}^T\rho\{u\}\, dV = \int_V \{\sigma\}^T \{\varepsilon\}\, dV. \tag{28}$$

The first four terms represent the external virtual work, while the right-hand term describes the internal virtual work. For the above discussed examples, only a traction vector shall be predetermined, so that the compatibility condition can be written directly in this form:

$$\int_{\partial V_n} \{t\}^T\{u_{\partial V}\}\, dA = \int_{V_n} \{\sigma\}^T \{\varepsilon\}\, dV = \int_{\partial V_n} \{t\}^T \{u_V\}\, dA. \quad (29$$

Because equ. (7) has been fulfilled, the volume integral can be written as a surface integral. It leads to:

$$\int_{\partial V_n} \{t\}^T (\{u_V\} - \{u_{\partial V}\})\, dA = 0. \tag{30}$$

The weight function has become the complete boundary stress vector $\{t\}$ of the homogenious and inhomogenious stress field:

$$\int_{\partial V_n} (\{t_0\}^T + \{t_s\}^T)\, (\{u_V\} - \{u_{\partial V}\})\, dA = 0. \tag{31}$$

This equation represents the so-called generalized compatibility condition, as already used for classical stress calculations to get weak solutions [14]. With definition (s. equ. (9)) and equ. (31) follows, that the displacement vector $\{u_{\partial V}\}$ is now also known for ∂V_s. The generalized equilibrium condition can also be derived from the principle of virtual work, see fig. 14 and equ. (28). With test functions the traction vector can be written:

$$\{t\} = \underbrace{[[R_0][R_s]]}_{[R]} \begin{bmatrix} \{\beta_0\} \\ \{\beta_s\} \end{bmatrix} = [R]\{\beta\}. \tag{32}$$

For the displacement vector inside the element V_n results:

$$\{u_V\} = \{u_V^0\} + \{u_V^s\} = \underbrace{[[L_0][L_s]]}_{[L]} \begin{bmatrix} \{\beta_0\} \\ \{\beta_s\} \end{bmatrix} = [L]\{\beta\}. \tag{33}$$

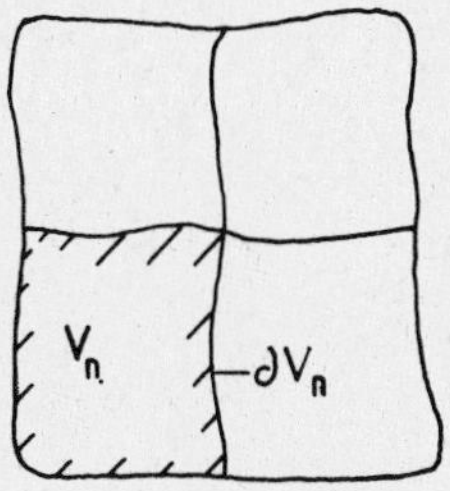

Fig. 13: discretization

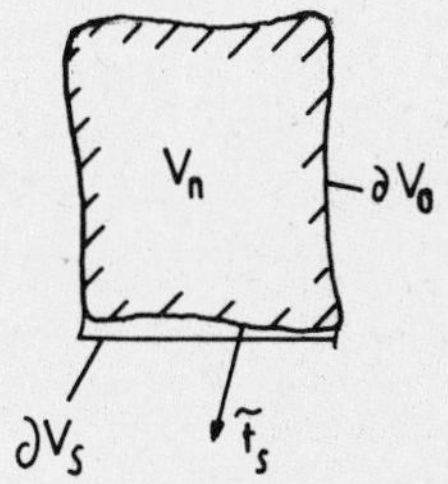

Fig. 14: surface definitions

Independent from this displacement distribution, the displacement distribution for the surface ∂V_n is formulated as:

$$\{u_{\partial V}\} = [N]\{d\}, \tag{34}$$

with the restriction, that an exact compatibility shall exist for the connected displacement elements of the classical FEM. After inserting the test and trial functions in the generalized compatibility condition (s. equ. (31)):

$$\{\beta\}^T \left\langle \underbrace{\int_{\partial V_n} \begin{bmatrix} [R_0]^T \\ [R_s]^T \end{bmatrix} [[L_0][L_s]]\, dA}_{[H]} \{\beta\} - \underbrace{\int_{\partial V_n} \begin{bmatrix} [R_0]^T \\ [R_s]^T \end{bmatrix} [N]\, dA}_{[T]} \{d\} \right\rangle = 0 \tag{35}$$

is reached with the corresponding abreviations. From the generalized equilibrium condition

$$\int_{\partial V_s} \{\tilde{t}\}^T \{u_{\partial V}\}\, dA = \int_{\partial V_n} \{t\}^T \{u_V\}\, dA \tag{36}$$

results:

$$\Big\langle \underbrace{\int_{\partial V_s} \{\tilde{t}\}^T[N]\,dA - \{\beta\}^T[T]}_{\{F\}^T} \Big\rangle \{d\} = 0. \tag{37}$$

It is evident that the surface integral over ∂V_s provides a nodal force matrix from the equivalent form of the displacement method, which is numerically stable. Thus no difficulties arise for the treatment of notch-crack problems with loaded notch boundaries. By insertion of equ. (35) in equ. (37) you get:

$$\{F\} = [T]^T[H]^{-1}[T]\{d\}. \tag{38}$$

In 1982, R. Drumm [15] has shown, that the satisfaction of equ. (6) is totally unnecessary (in combination with displacement elements) for the convergence of the procedure. R. Drumm showed that this condition is neither a necessary nor a sufficient condition. A new formulation can be found in such a form that equ. (6) is replaced. For this purpose, a positive definite test must be made for the residual total stiffness matrix [K] after the elimination of the rigid body motions and the degrees of freedom (the displacements) of only the displacement elements. So, a residual total stiffness matrix is reached only for those degrees of freedom, which are marked by crosses around the special element in fig. 15. The degrees of freedom to be eliminated are marked by small circles. The symbols for the rigid body motions can also be seen in fig. 15. Now the degrees of freedom will be devided along the boundary into a part $\{d_2\}$ (displacement only of ∂V_0, number n) and into a part $\{d_1\}$ (displacements only of ∂V_s, number p), see fig. 16. The stiffness matrix [K] can be devided in two parts, in $[K_H]$ and $[\bar{K}]$, see fig. 17. $[\bar{K}]$ corresponds only to the displacements $\{d_2\}$ and has only coefficients of the displacement elements. Therefore the submatrix [B] of $[\bar{K}]$ is a positive definite stiffness matrix. For $[K_H]$ can be written, see equ.(38):

$$[K_H] = [T]^T[H]^{-1}[T]. \tag{39}$$

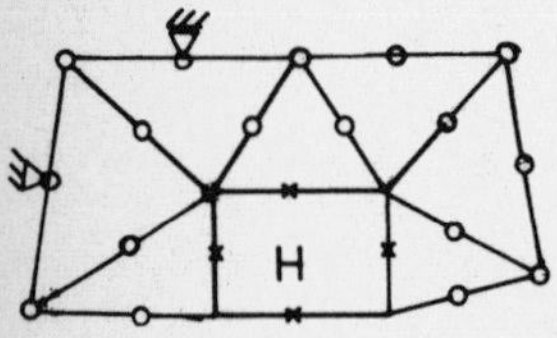

Fig. 15: nodal points for mixed FEM

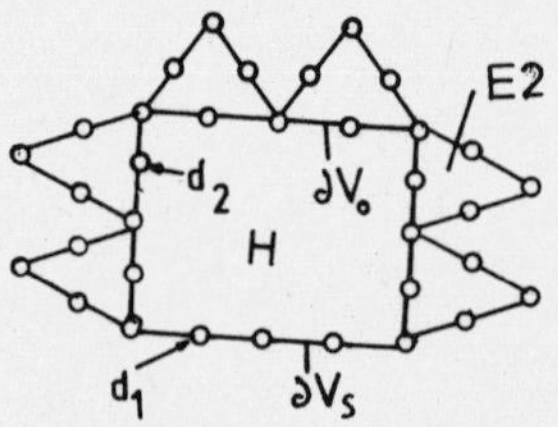

Fig. 16: definition of nodes for hybrids

As it can be seen in fig. 18, [T] is also devided in two parts, in a part $[\bar{T}]$ for the nodal displacements $\{d_1\}$ and in a part from $\{d_2\}$. Therefore a matrix [A] results (see fig. 18):

$$[A] = [\bar{T}]^T [H]^{-1} [\bar{T}]. \tag{40}$$

The superposition of $[K_H]$ and $[\bar{K}]$ produces the matrix $[K]$, as you can see from fig. 17. Therefore the total quadratic form Q can be written in form of:

$$Q = \begin{bmatrix} \{d_1\} \\ \{d_2\} \end{bmatrix}^T [K_H] \begin{bmatrix} \{d_1\} \\ \{d_2\} \end{bmatrix} + \{d_2\}^T [B]\{d_2\}. \tag{41}$$

In the case $m < n + p$ (m is the number of stress formulations and $(n+p)$ the total number of displacement formulations), the rank of $[T]$ is m, and the quadratic form with $[K_H]$ has a simple form with $\{d\}$ instead of $\{d_1\}$ and $\{d_2\}$, as can be seen from equ. (39):

$$\bar{Q} = (\{d\}[T])^T [H]^{-1} \underbrace{([T]\{d\})}_{\{v\}}. \tag{42}$$

In contrary to R. Drumm [10], you can see now, that with a new vector $\{v\}$ follows, that $\bar{Q}$ is generally a positive semi-definite form. For $\{d_2\} \neq \{o\}$, $[B]$ is positive definite, Q is greater than zero and therefore the matrix $[K]$ is positive definite. In case $m \geq n + p$, Q is always positive definite, if the traction functions are different from zero on ∂V_n.
If $\{d_2\}$ is equal zero, $[A]$ (see fig. 18) is a definite form, because the rank of the matrix $[\bar{T}]$ (see fig. 17) is p and it has to be $\tilde{m} \geq p$ with the traction functions (number $\tilde{m}$) different from zero on ∂V_s. Therefore this quadratic form is positive definite, too. Hence follows, that $[K]$ superposed of $[K_H] + [\bar{K}]$ is generally positive definite without fulfilling equ. (6), because Q (see equ. (41)) is always greater zero.
You can see, that equ. (6) is not a sufficient condition, because the elements t_{ij} of $[T]$ would be zero, if the traction functions are zero.
Another problem is to show the uniqueness of $\{\tilde{t}_s\}$, see equ. (21). A unique solution for the predetermined traction vector (see fig. 19) exists, if for one degree of freedom of the stress formulation $\{\sigma_s\}$ exists one degree of freedom of the displacement distribution on ∂V_s. In contrary to the description in [10] it can be shown exactly by separation of equ. (37).

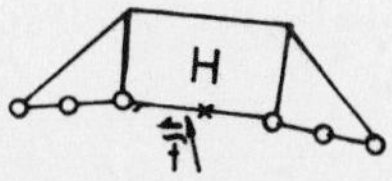

X: 2 nodal displacements
H: 2 degrees of freedom for $\{\sigma_s\}$

Fig. 19: traction condition for hybrids

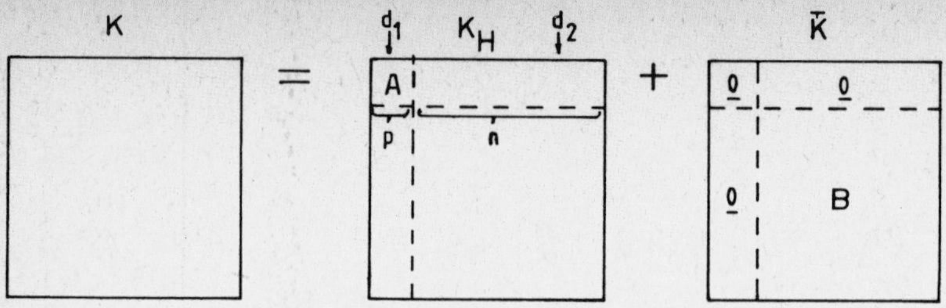

Fig. 17: construction of total stiffness matrix

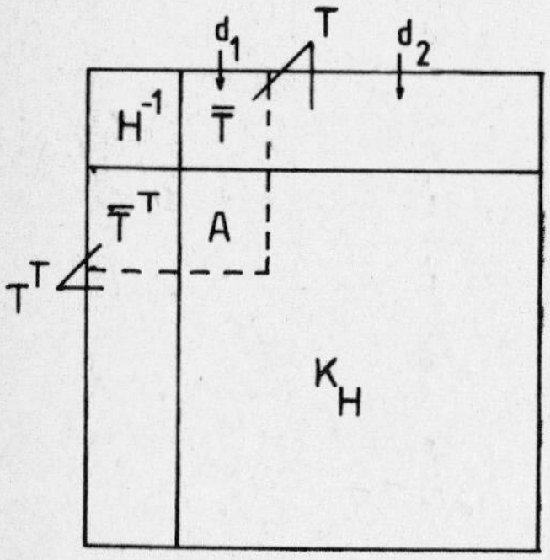

Fig. 18: Falk-scheme

In the following the advantages to other HSM-concepts shall be summoned up:

1. The condition number of the stiffness matrix changes to smaller values, because equ. (6) has not to be fulfilled. The selectivity between the functions in $\{\sigma\}$ is better, as it can be seen from numerical tests. If a special crack-tip element exists, see fig. 20, 31 functions for $\{\sigma\}$ are necessary, see [1,11,12]. But with this concept only four functions are used, if the crack surface is loadfree. Therefore no difficulties arise, if the dimension size at the crack-tip element is going to small values, i.e. the value D.
2. As a consequence to that the strongly oscillating convergence ceases.
3. Now $\{u_{\partial V}\}$ of the part ∂V_s, where the traction vector is acting, is known.
4. The nodal force matrix $\{F\}$ is a numerically stable nodal force matrix, and therefore there are no difficulties to solve notch-crack problems with loaded surfaces.

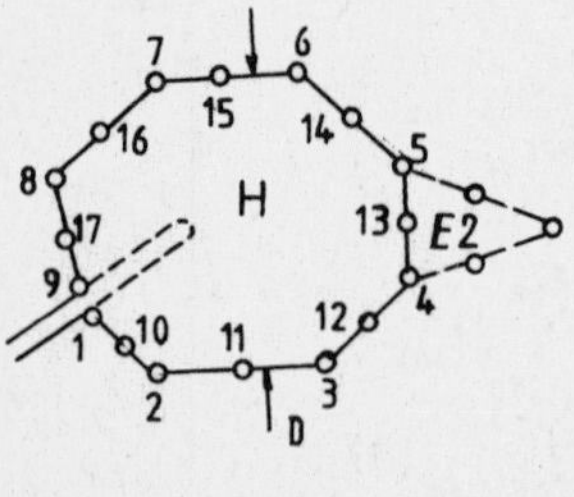

Fig. 20: crack-hybrid

TRIAL FUNCTIONS FOR PLANE PROBLEMS

For crack-tip elements, a homogenious group from $\{\sigma\}$, so-called $\{\sigma_0\}$, and an inhomogenious group, so-called $\{\sigma_s\}$, must be formulated. The eigen solutions of mode I (fig. 21) and Mode II (fig. 22) is incorporated in the homogenious group. The functions themselves are the Irwin-Sneddon-equations

$$\{f_1\} = \frac{K_I}{\sqrt{2\pi r}} \{f_{(\varphi)}\} \tag{43}$$

and

$$\{f_2\} = \frac{K_{II}}{\sqrt{2\pi r}} \{\tilde{f}_{(\varphi)}\} \tag{44}$$

in matrix notation. Therefore the unknowns, the degrees of freedom of the stress formulations, are the stress intensity factors themselves, because $\beta_1 = K_I$ and $\beta_2 = K_{II}$. Other functions with the characteristic of $r^{3/2}$ are formulated, as can be seen from equ. (45) and (46).

$$\{f_3\} = r^{3/2} \{\Phi_{(\varphi)}\} \beta_3, \tag{45}$$

$$\{f_4\} = r^{3/2} \{\psi_{(\varphi)}\} \beta_4. \tag{46}$$

As you know from the chapter before, it is very important to have a good approximation of the surface fiber stresses (related to the x-y-co-ordinate system in fig.23). That can be succeeded in writing the stress functions $\{f_5\}$ and $\{f_6\}$ with a polynome P_1:

$$\{f_5\} = \begin{bmatrix} \sigma_x \\ \sigma_y \\ \tau_{xy} \end{bmatrix} = \begin{bmatrix} 1 \\ 0 \\ 0 \end{bmatrix} \beta_5, \tag{47}$$

$$\{f_6\} = \begin{bmatrix} x \\ 0 \\ -y \end{bmatrix} \beta_6. \tag{48}$$

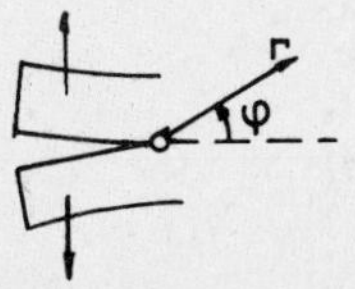

Fig. 21: mode I

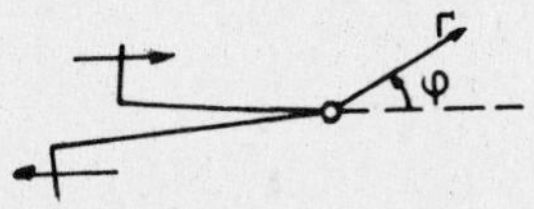

Fig. 22: mode II

For satisfying the traction condition for practical problems with the inhomogenious group the polynome P_1 is sufficient, too. With a formulation in the co-ordinate system from fig. 23 the functions $\{g_1\}$ and $\{g_2\}$ result in:

$$\{g_1\} = \begin{bmatrix} \sigma_x \\ \sigma_y \\ \tau_{xy} \end{bmatrix} = \begin{bmatrix} 0 & 0 \\ 1 & x \\ 0 & 0 \end{bmatrix} \begin{bmatrix} \beta_7 \\ \beta_8 \end{bmatrix} \tag{49}$$

$$\{g_2\} = \begin{bmatrix} 0 & 0 \\ 0 & -y \\ 1 & x \end{bmatrix} \begin{bmatrix} \beta_9 \\ \beta_{10} \end{bmatrix} \tag{50}$$

For notch-crack problems it is necessary to have a good approximation of the stress formulation inside the notch elements by reason of a big influence of notch parameters to the stress intensity factors. Important is the satisfaction of the so-called fade-away-law [16]. The characteristics of these functions can be seen from equ. (51):

$$f_k = a_k \, r^{-p_k} \, f_{(\varphi)} . \tag{51}$$

A shifting of the functions was made to the middle node of the notch-hybrid, see fig. 24. The functions F_k in Airy-formulation are:

$$\begin{bmatrix} F_i \\ F_j \end{bmatrix} = \begin{bmatrix} (c_{1i}\, r^{\lambda_i} + c_{2i}\, r^{-\lambda_i+2} + c_{3i}\, r^{-\lambda_i}) \cos(\lambda_i \varphi) \\ (c_{1j}\, r^{\lambda_j+2} + c_{2j}\, r^{-\lambda_j+2} + c_{3j}\, r^{-\lambda_j}) \sin(\lambda_j \varphi) \end{bmatrix} . \tag{52}$$

You get a symmetric and an antisymmetric term for a good approximation of all possible stress distributions inside a notch element.

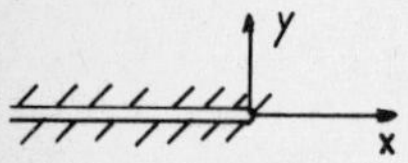

Fig. 23: co-ordinate system

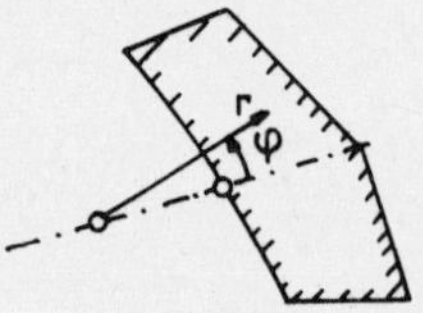

Fig. 24: notch-hybrid

TRIAL FUNCTIONS FOR AXI-SYMMETRIC PROBLEMS

Instead of plane problems, there are now four components of stress values in the co-ordinate system r,z (fig. 25). The equilibrium condition is:

$$[B]\{\sigma\} = 0 \in V. \tag{53}$$

The operator [B] has the form:

$$B = \begin{bmatrix} \frac{\partial}{\partial r}+\frac{1}{r} & -\frac{1}{r} & 0 & \frac{\partial}{\partial z} \\ 0 & 0 & \frac{\partial}{\partial z} & \frac{\partial}{\partial r}+\frac{1}{r} \end{bmatrix}. \tag{54}$$

The best method to formulate trial functions for bodies of revolution is with a three-function-concept formulated by J. Boussinesq, H. Neuber, and P.F. Papkovitch [6,18]. For these problems you get with the ansatz F:

$$F = \Phi_0 + z\Phi_1 \qquad \text{with the restriction} \quad \Delta\Phi_k = 0 \in V. \tag{55}$$

The stress field to be computed can be devided in two groups. The first group includes only functions with high gradients, i.e. the so-called eigen solution of the crack tip vicinity. The second group consists of relatively undisturbant functions. For group I the Laplace operator can be written as:

$$\Delta = \frac{\partial^2}{\partial r^2} + \frac{\partial^2}{\partial z^2} + \frac{1}{r} \cdot \frac{\partial}{\partial r} \tag{56}$$

and tends to $\tilde{\Delta}$ for big values in r and bounded $\frac{\partial}{\partial r}$:

$$\tilde{\Delta} = \frac{\partial^2}{\partial r^2} + \frac{\partial^2}{\partial z^2}. \tag{57}$$

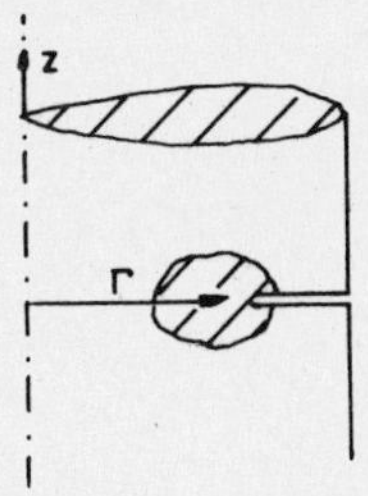

Fig. 25: axi-symmetric crack

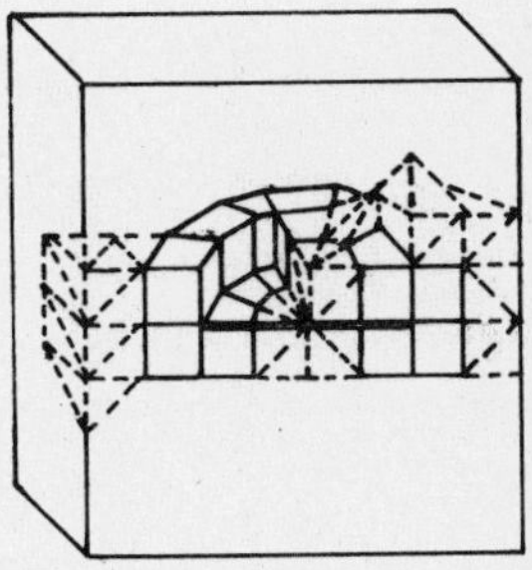

Fig. 26: discretization of 3D-problems

It is evident, that the problem is now the same as in state of plane strain, but only for group I, and you can work with the same eigen solution (equs. (43),(44)). However, as shown before, additionally trial functions are needed for the surface fiber stresses and because of the state of axi-symmetry also for the ring-stress. Stress distributions must be constructed, which leads to the satisfaction of the traction boundary condition. These functions must also be developed in state of plane strain, so that a positive definite stiffness matrix results. This causes automatically an error, because this kind of functions cannot be ranged in the first group. The numerical results produce therefore an error in the fiber and ring-stresses. Numerical tests, however, have shown, that nevertheless the stress intensity factor can be computed with high accuracy for axi-symmetric problems.

COMPUTATION OF STRESS INTENSITY FACTORS FOR THREE-DIMENSIONAL PROBLEMS

Very important in fracture mechanics are the surface cracks (fig. 4). The crack front becomes linearly approximated peacewise and special hybrids are used for the elements around the crack front (see fig. 26 and 27). In the residual domain, the tetrahedron elements of V.Ph. Nguyen [19] will be applied, which work effectively [20]. So it is not a big effort to construct a network for this problem, because the tetrahedron elements can be laid by an automatic mesh generator [19]. It can be seen from fig. 27, that cylinder co-ordinates are inserted now along the crack front for the description of the state of stress. The development of software for 3D-problems according to the theory explained in the passages before, is in progress. Such special hybrid elements exist already from T.H.H. Pian and K. Moriya (1980) [11] and from M. Kuna (1982) [12], which, however, show great disadvantages compared with the theory explained before:

1. Volume integrals must be computed, i.e. high CPU-time.
2. The kinematic convergence condition must be fulfilled, see equ. (6). Therefore 57, respective 63 trial functions become necessary (see [12]) for $\{\sigma\}$-interpolation which causes high CPU-time. A further question arises, if the condition number of the resulting stiffness matrix is the best one possible. Therefore a new concept will be developed which allows 1st only surface integrals for the computation and 2nd leads to a reduction of interpolation-functions. So a better selectivity between the functions occurs and by this a better condition number for the resulting stiffness matrix is reached.
3. Special elements at the edge points on the crack front shall be developed, especially at those points, where the crack front pushes the free surface.

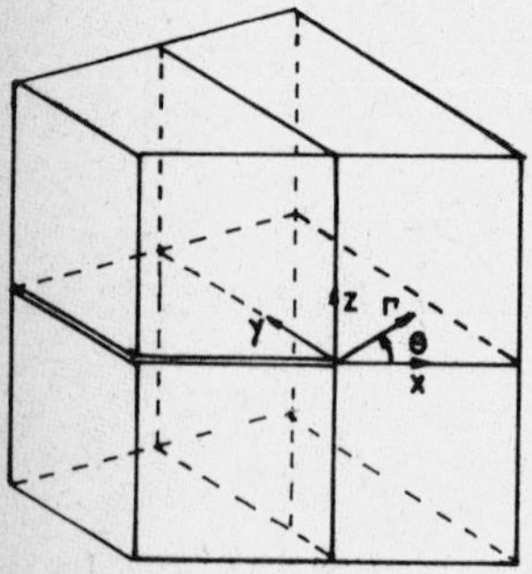

Fig. 27: 3D-hybrid

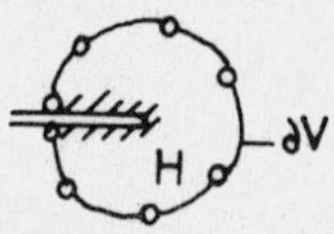

Fig. 28: surface integral

FINAL REMARKS TO THE WHOLE THEORETICAL CONCEPT

For the computation of stress intensity factors in plane, axi-symmetric, and three-dimensional problems, the eigen solution is exactly incorporated in the trial functions. So the displacement field will be computed by the resulting total stiffness matrix under exact consideration of the singulatity. In the second step, however, it is necessary, to determine a surface integral around the crack front (see fig. 28) for computation of the stress intensity factors K_i $(i = 1,2,3)$ by the generalized compatibility condition (see equ. (35)). Therefore an analogy exists to the determination with the J-integral [21]. So additional acceleration of accuracy is possible and random values for K_i of the numerical process can be avoided.

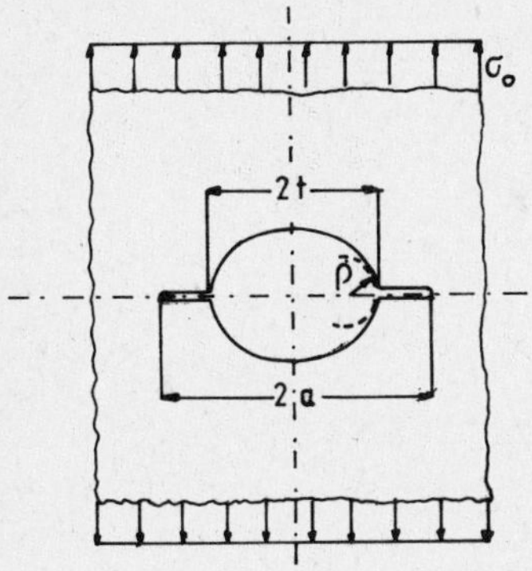

Fig. 29: notch-crack problem

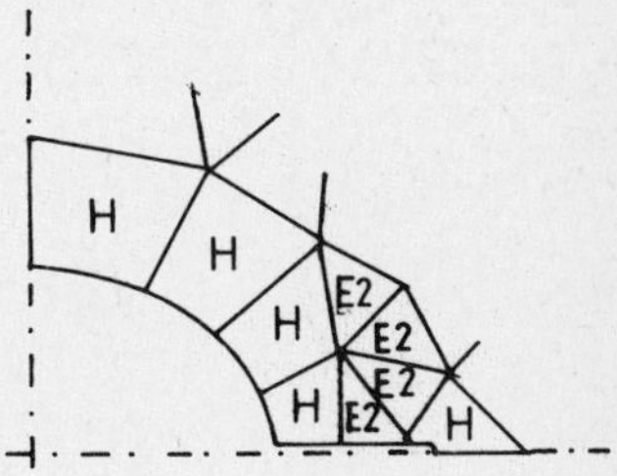

Fig. 30: discretization of problem from fig. 29

NUMERICAL TESTS

Out of the group of plane crack problems, as a first example a plate with an elliptic hole and symmetric exterior cracks will be computed (see fig. 29). Fig. 30 shows that a special hybrid will be inserted at the crack-tip and special hybrids along the notch surface, each exactly satisfying the traction conditions. In the residual domain only the classical triangle displacement elements are used with quadratic displacement distribution. Fig. 31 shows the convergence curves for the normalized stress intensity factors depending from the degrees of freedom (the unknowns of the linear equation system). It is evident that the best one of the classical procedures for computation of stress intensity factors is the crack driving force concept (G/CE-concept) in combination with the so-called collapsed elements (quarter-point-elements). When using these collapsed elements with the COD-concept (COD/CE), the convergence curves are not so good. The same happens, when working with the G-concept and classical elements (G/E2) without singularity incorporation in the displacement distribution of the element. But best of them all works the combination of the modified hybrid assumed stress model with the classical displacement elements (HSM/E2).

The next case shows an axi-symmetric problem for a body of revolution, where a surface crack is running from the outside to the middle of a round bar (see fig. 32). The crack surfaces are under pressure p_0. The best one of the classical methods is to compute the crack closure work by using ring-elements with a quadratic displacement distribution. In about 53 seconds CPU-time the stress intensity factor can be produced with a 2% accuracy. But the best one to work with, is also the combination of hybrid stress elements and classical displacement elements. The stress intensity factor can be produced with a 0.3% deviation to the exact value in only 30 seconds. Further tests can be seen in [10,15].

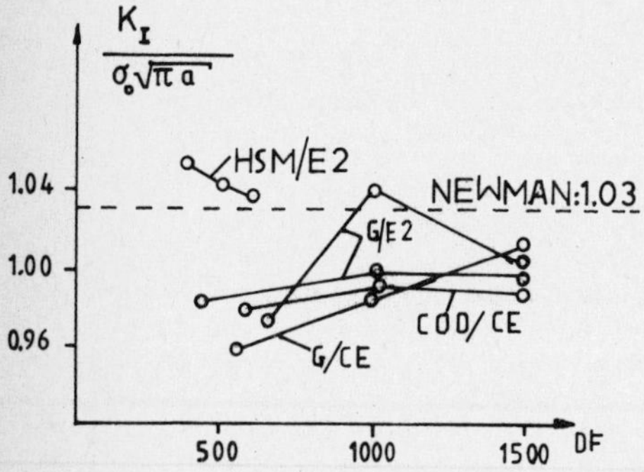

Fig. 31: convergence curves

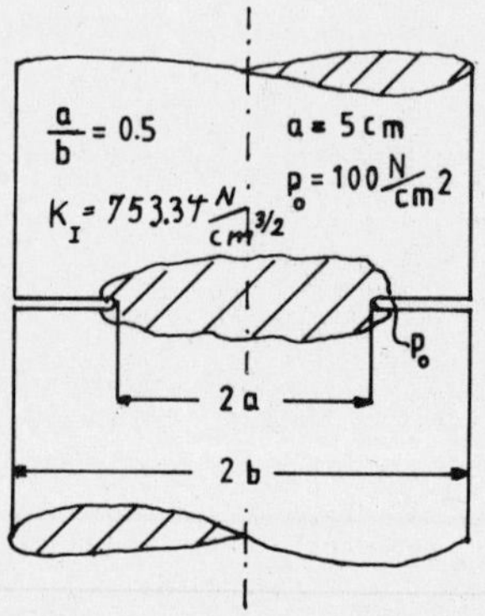

Fig. 32: axi-symmetric crack problem

CONCLUSION

The test examples have shown that the stress intensity factors can be produced for complicated problems. The best way is by combination of the modified hybrid stress model with the classical displacement elements. Because this combination leads to very good values, it is preferred for other open questions in three-dimensional fracture mechanics. So it is possible to solve problems with curved crack fronts, and curved crack surfaces, too. The developments of such algorithms are in progress.

ACKNOWLEDGEMENTS

The author wishes to thank Dr.-Ing. R. Drumm from the Institute of Solid Mechanics at the Karlsruhe University for his detailed excellent tests ([10,15]) out of the field of fracture mechanics.

REFERENCES

[1] M. Wolf: Lösung von ebenen Kerb- und Rißproblemen mit der Methode der finiten Elemente. Diss. Techn. Univ. München, 1977.
[2] P. Tong, T.H.H. Pian, S.J. Lasry: A hybrid element approach to crack problems in plane elasticity. Int. J. Num. Meth. Engng., Bd. 7, 1973, pp. 297-308.
[3] M. Stern, E.B. Becker: A confirming crack-tip element with quadratic variation in the singular fields. Int. J. Num. Meth. Engng., Vol. 12, 1978, pp.279-288.
[4] T.H.H. Pian, P. Tong: Basis of finite element methods for solid continua. Int. J. Num. Meth. Engng., Bd. 1, 1969, pp. 3-28.
[5] P. Tong, T.H.H. Pian: Variational principle and the convergence of a finite element method based on assumed stress distribution. Int. J. Solids Structures, Bd. 5, 1969, pp. 463-472.
[6] E. Schnack: Beitrag zur Berechnung rotationssymmetrischer Spannungskonzentrationsprobleme mit der Methode der finiten Elemente. Diss. Techn. Univ. München, 1973.
[7] E. Schnack, M. Wolf: Application of displacement and hybrid stress methods to plane notch and crack problems. Int. J. Num. Meth. Engng., Vol. 12, No. 6, 1978, pp. 963-975.
[8] J.P. Wolf: Generalized stress models for finite element analysis. Ph.D. Thesis, ETH Zürich, 1974.
[9] S.N. Atluri, H.C. Rhee: Traction boundary conditions in hybrid stress finite element model. AIAA Bd. 16, Nr. 5, 1978, pp. 529-531.
[10] R. Drumm: Zur effektiven FEM-Analyse ebener Spannungskonzentrationsprobleme. Fortschritt-Bericht der VDI-Z, Reihe 18, Nr. 13, Düsseldorf 1983.
[11] T.H.H. Pian, K. Moriya: Three dimensional fracture analysis by assumed stress hybrid elements. in Luxmoore, A.R., Owen, D.R.J.: Proc. 1st and 2nd International Conference on "Numerical Methods in Fracture Mechanics". Pineridge Press, Swansea 1978 und 1980, S. 363-373.
[12] M. Kuna: Konstruktion und Anwendung hybrider Rißspitzenelemente für dreidimensionale bruchmechanische Aufgaben. Techn. Mechanik Bd. 2 (1982), S. 37-43.
[13] E. Schnack, R. Drumm: Zur exakten Erfassung des Randspannungsvektors bei der Hybridmethode. ZAMM Bd. 62, 1982, pp. 167-170.
[14] H. Neuber: Elastostatik und Festigkeitslehre. Springer-Verlag, Berlin, Heidelberg, New York, 1971.
[15] R. Drumm: Zur effektiven FEM-Analyse ebener Spannungskonzentrationsprobleme, Diss. Univ. Karlsruhe, 1982.
[16] E. Schnack: An optimization procedure for stress concentrations by the finite-element-technique. Int. J. Num. Meth. Engng., Vol. 14, No. 1, 1979, pp.115-124.
[17] E. Schnack, M. Wolf: Die Konstruktion von Spannungsansätzen der Hybridspannungsmethode. Forsch. Ing.-Wes., Bd. 44, Nr. 3, 1978, pp. 74-79.
[18] E. Schnack: Zur Berechnung rotationsyymmetrischer Kerbprobleme mit der Methode der finiten Elemente. Forsch. Ing.-Wes. Bd. 42, Nr. 3, 1976, pp. 73-81.
[19] V.Ph. Nguyen: Automatic mesh generation with tetrahedron elements. Int. J. Num. Meth. Engng., Vol. 18, 1982, pp. 273-289.
[20] E. Schnack: Effektivitätsuntersuchung für numerische Verfahren bei Festigkeitsberechnungen. VDI-Z 119, 1977, Nr. 1/2, pp. 43-50.
[21] J.R. Rice: A path independent integral and the approximate analysis of strain concentration by notches and cracks. J. Appl. Mech. Trans. ASME Series, E 35, 2, 1968, pp. 379-386.

THE PRACTICAL TREATMENT OF STRESS CONCENTRATIONS AND SINGULARITIES WITHIN FINITE ELEMENT DISPLACEMENT ALGORITHMS

E. Stein
Institut für Baumechanik und
Numerische Mechanik
Universität Hannover
Callinstr. 32, D-3000 Hannover 1

1. Introduction

The Finite-Element-Method (FEM) in the original version with shape functions for the displacement components is a *Ritz*-method with finite subdomains. In the case of linear elasticity theory, a direct variational process of the *Dirichlet* functional is performed, and one gets a system of linear equations for the *Ritz* parameters /1-4/. In this displacement approach the equilibrium conditions within the finite element domains and at the element boundaries are only approximately fulfilled by the stresses derived from the displacements, using the kinematic relations and the constitutive equations. Chosing linear shape functions for the displacements in plane stress analysis, e.g., the stresses in the element domains are constant, and they have jumps at the element boundaries. From these jumps error indicators for adaptive mesh refinements (h-adaptivity; h: mesh width) can be calculated /5,6/. Note that the stresses are mostly of higher interest for engineering problems than the displacements. It is obvious that the described FEM in the form of the classical displacement approaches is:

i) not efficient to describe strong stress concentrations and
ii) not capable to describe stress singularities.

There exist different strategies to improve the accuracy of stresses within the FE-concept.
Concerning stress concentrations, see i), one can use:

a) shape functions of higher order (p-approximation),

b) hybrid stress method /7/,
c) a postprocessor for the stress calculation following the displacement calculation. This algorithm can be classified as a <u>stepwise hybrid displacement method</u>, called <u>equilibrium method</u> /8,9/.

Concerning the approximation of stress singularities, see ii), it is possible

d) to add singular functions for the singularities in the regions concerned, i.e. working with two regions for the FE-discretization /10/ or with three regions /11/. This procedure can also be interpreted as a combination of *Ritz* and *Trefftz* method, see /12/.
e) to construct special singular elements, mostly within the hybrid stress method.

2. Mechanical and engineering aspects of singularities

2.1 "Artificial" singularities

Most singularities in elastic systems are caused by idealizations

a) of mechanical field properties
b) of the loading
c) of the geometrical and statical boundary conditions.

Examples are:
. The point load acting on a body (Fig. 1).

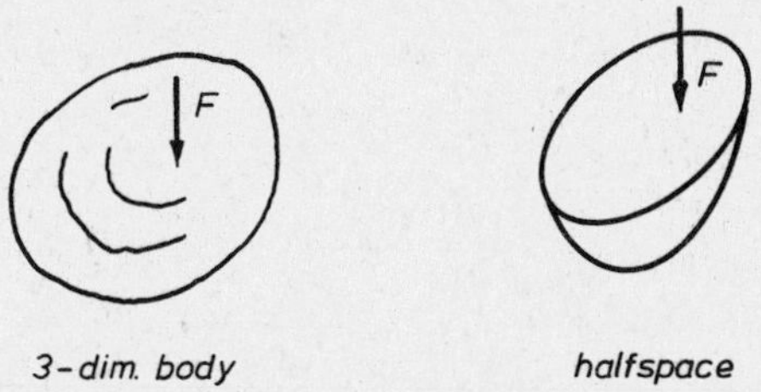

Fig. 1: Point load on a 3d-body and on a half space

· The point or line support for beams, plates, shells.

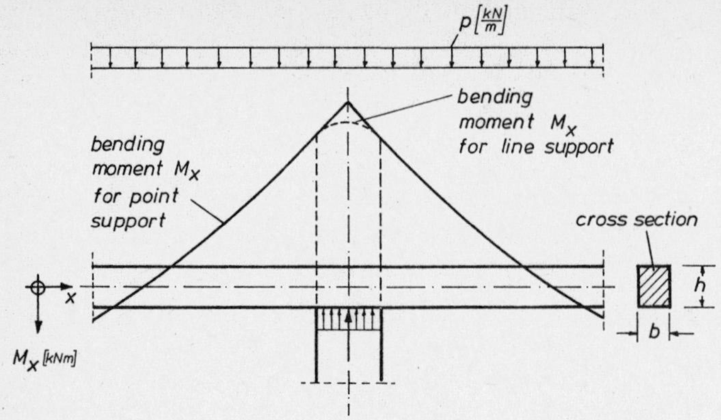

Fig. 2: Bending moment of a beam with a point support and with a line support

· The corner forces of hinged (*Navier*-supported) *Kirchhoff*-plates as singularities of the effective shear forces.

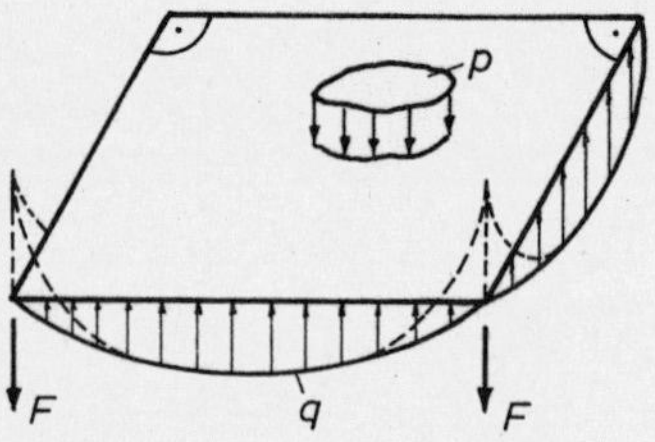

Fig. 3: Effective shear forces of the Navier-supported *Kirchhoff* plate with uniform loading.

2.2 "Existing" singularities

Within the general assumptions for elastic and thermal deformations of *Boltzmann*-continua singularities exist at e.g.

· nodges, slits and crack tips of 2d- and 3d-solids
· edges of heated composites of several materials with different

material properties (e.g. heat elongation α).

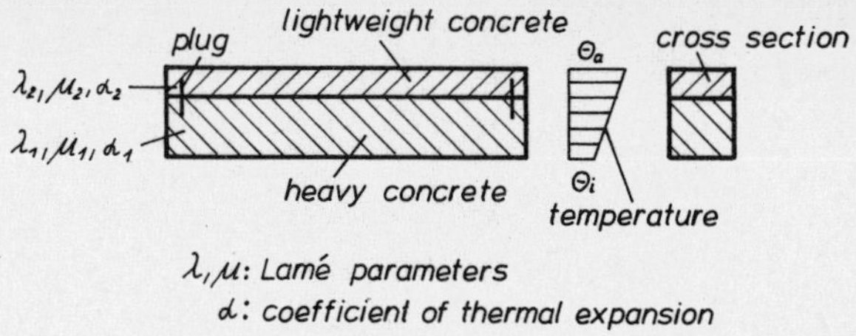

Fig. 4: heating of composite materials

These singularities vanish by regarding plastic material properties, e.g. but the analytical description of the stress concentrations is only possible for special cases.

2.3 Finite Element approach

The FEM allows the approximation of more realistic properties with respect to loading and boundary conditions so that singularities may not arise. On the other hand one has to regard the fact that extreme mesh refinements around ideal elastic or even rigid point supports result in zero reaction forces in 2d- and 3d-elastic systems. (This phenomen was first noticed by *I. Babuska*.) Therefore, the realistic modelling of topological and metric system properties is very important for the efficient and reliable of FEM.

2.4 Practical aspects

The significance of singularities in idealized mechanical systems for engineering problems is to be seen in the judgement of local failure mechanismus, e.g. in fracture mechanics and fatigue analysis.

3. Improved stress calculation in the FE-displacement method - a stepwise hybrid displacement method -

3.1 Concept and scope of the postprocessor "stress"

The method is represented for linear plane stress analysis. It starts with the usual FE-displacement approach. After determining the global displacement vector of the system, shape functions for the stresses are chosen along all element boundaries or in special sections of the system. The nodal values of the stresses are calculated by solving another system of linear equations, expressing the equilibrium conditions between the integrated boundary stresses and the fictitious element nodal forces, resulting from the given loads and the calculated displacement field according to the stiffness properties. This "equilibrium method" was formulated using the principle of virtual work /8/, /9/ and can be interpreted as a stepwise hybrid displacement method.

The local and global matrices are symmetric, the global matrix is positive definite and banded. The computation effort for this stress calculation is less or at most equal to such for the preceding displacement calculation; of course, the effort is higher than such for the usual stress calculation using kinematic relations and constitutive equations.

The scope of the postprocessor "stress" is an improved accuracy of the stresses with respect to the equilibrium conditions at the element boundaries. Numerical comparisons for beams, plane stress problems, plates in bending and shells showed the efficiency of the method especially for polynomial shape functions of low order (linear and cubic (for bending) resp.), see /8/, /9/, /14/.

3.2 The FE-displacement method in matrix notation

We treat linear elastic plane stress analysis with the approximation (for very thin plates) that also plane displacement fields can be considered.

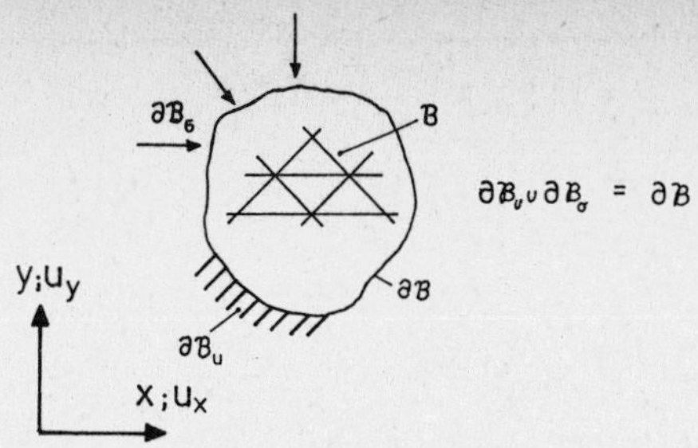

Fig. 5: Elastic plate described in cartesian coordinates

Introducing the displacement vector $\underset{\sim}{u}$ in each point of the domain $\mathcal{B}$

$$\underset{\sim}{u} = u_x(x,y)\underset{\sim}{e}_x + u_y(x,y)\underset{\sim}{e}_y, \tag{3.1a}$$

with the matrix notation for the components

$$\underline{u} = \begin{bmatrix} u_x \\ u_y \end{bmatrix} , \tag{3.1b}$$

the linear kinematic relations - defining the components of the symmetric plane strain tensor $\underset{\sim}{\varepsilon}$ - are

$$\underbrace{\begin{bmatrix} \varepsilon_x \\ \varepsilon_y \\ \gamma_{xy} \end{bmatrix}}_{\underline{\varepsilon}^{(g)}} := \underbrace{\begin{bmatrix} \partial/\partial x & 0 \\ 0 & \partial/\partial y \\ \partial/\partial y & \partial/\partial x \end{bmatrix}}_{\underline{D}} \underbrace{\begin{bmatrix} u_x \\ u_y \end{bmatrix}}_{\underline{u}} \; ; \; \underline{\varepsilon}^{(g)} := \underline{D}\,\underline{u} \;. \tag{3.2}$$

The static equilibrium conditions for the stresses are

$$\underline{D}^T\underline{\sigma} + \rho\underline{b} = \underline{0} \quad \text{with } \underline{\sigma} = \begin{bmatrix} \sigma_x(x,y) \\ \sigma_y(x,y) \\ \tau_{xy}(x,y) \end{bmatrix} ; \; \underline{b} = \begin{bmatrix} b_x(x,y) \\ b_y(x,y) \end{bmatrix} . \tag{3.3}$$

The constitutive equation - here Hooke's law - are

$$\underline{\sigma} = \underline{C}\;\underline{\varepsilon}^{(p)} \;. \tag{3.4}$$

Due to the potential properties of the strain energy density

$$\mathcal{W}_s = \frac{1}{2}\,\underline{\varepsilon}^{(p)T}\,\underline{\sigma} = \frac{1}{2}\,\underline{\varepsilon}^{(p)T}\,\underline{C}\,\underline{\varepsilon}^{(p)} \;, \tag{3.5}$$

the elastic matrix $\underline{C}$ is symmetric and positive definite, and the total differential

$$d\mathcal{W}_s = d\,\underline{\varepsilon}^{(p)^T}\,\frac{\partial \mathcal{W}_s}{\partial \underline{\varepsilon}^{(p)}} = d\,\underline{\varepsilon}^{(p)^T}\,\underline{\sigma} \tag{3.5a}$$

exists. The complementary strain energy density is

$$\overset{*}{\mathcal{W}}_s = \frac{1}{2}\,\underline{\sigma}^T\,\underline{C}^{-1}\underline{\sigma} \;, \tag{3.6}$$

and the total differential becomes

$$d\overset{*}{\mathcal{W}}_s = d\underline{\sigma}^T\,\frac{\partial \overset{*}{\mathcal{W}}_s}{\partial\,\underline{\sigma}} = d\underline{\sigma}^T\,\underline{\varepsilon}^{(p)} \;. \tag{3.6b}$$

The geometric field conditions are

$$\underline{\varepsilon}^{(p)} = \underline{\varepsilon}^{(g)} = \underline{\varepsilon} \;;\; \frac{\partial \overset{*}{\mathcal{W}}_s}{\partial \underline{\sigma}} = \underline{D}\;\underline{u} \;. \tag{3.7}$$

The *Navier-Lamé* differential equations for plane stress analysis arise by introducing (3.2) into (3.7) and (3.3)

$$\underline{\sigma} = \underline{C}\;\underline{D}\;\underline{u} \tag{3.8}$$

and (3.8) into (3.3)

$$\underbrace{\underline{D}^T\,\underline{C}\;\underline{D}}_{\underline{L} = \underline{L}^T}\,\underline{u} + \rho\underline{b} = \underline{O} \quad \text{in } \mathcal{B} \;. \tag{3.9}$$

The (bilateral) geometric boundary conditions are

$$\underline{u} - \underline{\bar{u}} = \underline{O} \quad \text{on } \partial\mathcal{B}_u \;;\; \underline{\bar{u}} = \begin{bmatrix} \bar{u}_x \\ \bar{u}_y \end{bmatrix} \tag{3.10}$$

where the upper bar marks given quantities, and the static boundary conditions can be written as

$$\underline{t} - \underline{\bar{t}} = \underline{O} \quad \text{on } \partial\mathcal{B}_\sigma \;;\; \underline{\bar{t}} = \begin{bmatrix} \bar{t}_x \\ \bar{t}_y \end{bmatrix} \tag{3.11}$$

with the boundary stresses

$$\underline{t} = \underline{N}^T\underline{\sigma} \quad ; \quad \underline{N} = \begin{bmatrix} n_x & 0 \\ 0 & n_y \\ n_y & n_x \end{bmatrix} \quad ; \quad n_x = \tilde{n}\cdot\tilde{e}_x, \quad |\tilde{n}| = 1 \tag{3.12}$$

where $\tilde{n}$ is the outer unit normal vector at the boundary.

The FE approach uses test functions

$$\underline{u}_h^e = \tilde{\underline{u}}^e = \underline{\Omega}\ (x,y)\ \underline{v}^e \tag{3.13}$$

within the element domain $\mathcal{B}^e$. The shape functions $\underline{\Omega} = \begin{bmatrix} \underline{\Omega}_x & \underline{0} \\ \underline{0} & \underline{\Omega}_y \end{bmatrix}$ describe unit displacement states, and $\underline{v}^e = \begin{bmatrix} \underline{v}_x^e \\ \underline{v}_y^e \end{bmatrix}$ is the nodal displacement vector of element $\mathcal{B}^e$.

The shape functions $\underline{\Omega}$ have to be: (1) linear independent, (2) continuously differentiable, (3) rel. complete, (4) rotational invariant (not for rectangles), (5) geometrically "conforming" (i.e. C^0-continuity at the boundaries), (6) containing the rigid body movements and (7) complete in the mechanical sense (i.e. the stresses $\tilde{\underline{\sigma}}^e$ in (3.14b) must provide at least constant values).
Introducing (3.13) into (3.2) with (3.7) and into (3.8) yields

$$\tilde{\underline{\varepsilon}}^e = (\underline{D}\ \underline{\Omega})\underline{v}^e = \underline{H}\ \underline{v}^e \quad \text{and} \quad \tilde{\underline{\sigma}}^e = \underline{C}^e\ \underline{H}\ \underline{v}^e. \tag{3.14a,b}$$

For the mechanical understanding of the element stiffness matrix $\underline{k}^e$, the application of principle of virtual work is very illustrative. It has the form

$$\delta\mathcal{A}^e = \delta\mathcal{W}^e \tag{3.15}$$

and is a sufficient condition for equilibrium if the virtual displacement are geometrical admissible. From (3.15), (3.13) and (3.14) one gets

$$\delta\underline{v}^{eT}\overset{*}{\underline{p}}{}^e = \int_{\mathcal{B}^e} \delta\ \underline{\varepsilon}^T\cdot h\cdot\underline{\sigma}\ dA \quad ; \quad h^e\text{: thickness of the element}$$

$$= \delta\underline{v}^{eT}\ \underbrace{h^e \int_{\mathcal{B}^e} \underline{H}^T\ \underline{C}\ \underline{H}\ dA}_{\underline{k}^e = \underline{k}^{eT}\text{, pos. semidef., } d = 3}\ \underline{v}^e$$

$$\overset{*}{\underline{p}}{}^e = \underline{k}^e\ \underline{v}^e\ ;\ \underline{k}^e = h^e \int_{\mathcal{B}^e} \underline{H}^T\ \underline{C}\ \underline{H}\ dA \tag{3.16}$$

where $\overset{*}{\underline{p}}^e$ is the fictitious nodal force vector according to $\underline{v}^e$.

The global stiffness matrix $\underline{K}$ is derived by implementing geometric continuity conditions at the element boundaries and at the overall boundary according to the differential operator of the bounbary value problem. In our case C^o-continuity is realized by *Boolean* matrices $\underline{a}^e$ of the elements. The global displacement vector of the system is $\underline{V}$, so it holds

$$\underline{v}^e = \underline{a}^e \, \underline{V} \, . \tag{3.17}$$

The principle of virtual work for the whole system then takes the form

$$\delta\mathcal{A} = \delta\mathcal{W} \; ; \quad \delta\mathcal{A} = \sum_{(e)} \delta\mathcal{A}^e \quad \delta\mathcal{W} = \sum_{(e)} \delta\mathcal{W}^e \, . \tag{3.18}$$

Introducing fictitious nodal load vectors $\overline{\underline{p}}^e$ of the elements due to the given loading, the "outer" virtual work is

$$\delta\mathcal{A} = \sum_{(e)} \int_{\mathcal{B}^e} \delta\, \underline{u}^T h \rho \underline{b} \, dA + \sum_{(e)} \int_{\partial\mathcal{B}^e} \delta \underline{u}^T \, h \, \overline{t} \, ds \tag{3.19}$$

and with (3.13) and (3.17)

$$\delta\mathcal{A} = \sum_{(e)} \delta\, \underline{v}^{e^T} \int_{\mathcal{B}^e} \underline{\Omega}^T h\rho \underline{b} \, dA = \sum_{e} \delta\, \underline{v}^{e^T} \tag{3.2o}$$

$$\delta\mathcal{A} = \delta\underline{V}^T \sum_{(e)} \underline{a}^{e^T} \, \overline{\underline{p}}^e = \delta\underline{V}^T \, \overline{\underline{P}} \, . \tag{3.21}$$

The global "inner" virtual work follows from (3.19), (3.16) and (3.17) as

$$\delta\mathcal{W} = \sum_{(e)} \int_{\mathcal{B}^e} \delta\, \underline{\varepsilon}^T \, \underline{C} \, \underline{\varepsilon} \, dV = \sum_{e} \delta\, \underline{v}^{e^T} \, \underline{k}^e \, \underline{v}^e$$

$$= \delta\underline{V}^T \underbrace{\sum_{e} \underline{a}^{e^T} \underline{k}^e \, \underline{a}^e}_{\underline{K}} \, \underline{V} = \delta\underline{V}^T \, \underline{K} \, \underline{V} \tag{3.22}$$

with $\underline{K} = \underline{K}^T$ and pos. def.

The principle (3.19) then results in

$$\delta\underline{V}^T (\underline{K} \, \underline{V} - \overline{\underline{P}}) = 0 \tag{3.23}$$

and therefore in the linear system of equations

$$\underline{K} \, \underline{V} - \overline{\underline{P}} \; ; \quad \underline{V} = \underline{K}^{-1} \, \overline{\underline{P}} \, . \tag{3.24}$$

The usual calculation of stresses yields, see (3.14), (3.17)

$$\underline{\tilde{\sigma}}^e = \underline{C}^e \, \underline{\varepsilon}^e = \underline{C}^e \, \underline{H} \, \underline{v}^e = \underline{C}^e \, \underline{H} \, \underline{a}^e \, \underline{V} \quad . \tag{3.25}$$

The best values are gained in the *Gaussian* integration points. It is also usual in practical engineering work to calculate the mean values at nodal points.

The system of equations, (3.24), can also be derived as a *Ritz* process starting from the *Dirichlet* principle of minimum of total potential energy π . Here potential properties of the inner and outer forces must be demanded. The functional π is given as

$$\pi = \int_{\mathcal{B}} h \, \mathcal{W}_s \, dA - \int_{\mathcal{B}} \underline{u}^T \rho \underline{b} h \, dA - \int_{\partial \mathcal{B}_\sigma} \underline{u}^T h \underline{\bar{t}} \, ds \tag{3.26}$$

The *Ritz* approach according to (3.13) yields, with (3.5), (3.13), (3.14a) ff,

$$\tilde{\pi} = \sum_{(e)} \frac{1}{2} \underline{v}^{e^T} \underline{k}^e \, \underline{v}^e - \sum_{e} \underline{v}^{e^T} \, \underline{\bar{p}}^e$$

$$= \frac{1}{2} \underline{V}^T \, \underline{K} \, \underline{V} - \underline{V}^T \, \underline{\bar{P}} \quad . \tag{3.27}$$

The necessary condition for stationarity is

$$\delta \tilde{\pi} = \delta \, \underline{V}^T \frac{\partial \tilde{\pi}}{\partial \underline{V}} = \underline{V}^T \, (\underline{K} \, \underline{V} - \underline{\bar{P}}) = 0 \ , \tag{3.28}$$

which is equal to (3.23).

Due to the positive definiteness of $\underline{K}$, the second variation of $\tilde{\pi}$ is positive,

$$\delta^2 \tilde{\pi} = \delta \, \underline{V}^T \, \underline{K} \; \delta \underline{V} \quad \begin{matrix} > 0 \text{ for } & \underline{V} \neq \underline{0} \\ = 0 \text{ for } & \underline{V} = \underline{0} \ , \end{matrix} \tag{3.29}$$

and therefore the minimal sequence

$$\tilde{\pi}^{(N)} > \tilde{\pi}^{(N+1)} > \ldots > \tilde{\pi}^{(N \to \infty)} = \pi \tag{3.30}$$

exists where N denotes the total number of kinematical freedoms (*Ritz* parameters).

3.3 The FE stress method

The test functions for the stresses in the element domain are

$$\underline{\sigma}_h^e = \underline{\tilde{\sigma}}^e = \underline{\overset{*}{\Omega}}(x,y)\ \underline{s}^e \quad \text{in} \quad \mathcal{B}^e \tag{3.31}$$

The shape functions $\underline{\overset{*}{\Omega}}$ must fulfil the conditions (1) to (4) of the $\underline{\Omega}$ and additionally the equilibrium conditions in $\mathcal{B}^e$ and on $\partial\mathcal{B}^e$, i.e.

$$\underline{D}^T\ \underline{\tilde{\sigma}}^e + \delta\underline{b} = 0 \quad \text{in} \quad \mathcal{B}^e \tag{3.32a}$$

$$\underline{N}^T\ \Delta\underline{\tilde{\sigma}}^e = \underline{\bar{t}}^e \quad \text{on } \partial\mathcal{B}^e \quad \text{with} \quad \Delta\underline{\tilde{\sigma}}^e = \overset{+}{\underline{\tilde{\sigma}}^e} - \overset{-}{\underline{\tilde{\sigma}}^e} \tag{3.32b}$$

$$\underline{N}^T\ \underline{\tilde{\sigma}}^e = \underline{\bar{t}} \quad \text{on } \partial\mathcal{B}_\sigma\ . \tag{3.32c}$$

Then the principle of complementary virtual work

$$\delta\overset{*}{\mathcal{A}} = \delta\overset{*}{\mathcal{W}}\ ;\ \delta\overset{*}{\mathcal{A}} = \sum_{(e)}\int_{\partial\mathcal{B}_u} h\cdot\delta\underline{t}^T\underline{\bar{u}}\,ds\ ,\ \delta\overset{*}{\mathcal{W}} = \sum_{(e)}\int_{\mathcal{B}^e} h\cdot\delta\underline{\sigma}^T\underbrace{\underline{\varepsilon}^{(p)}}_{\underline{C}^{-1}\underline{\sigma}}\,dA \tag{3.33}$$

can be applied and yields a system of linear equations

$$\underline{F}\ \underline{S} = \underline{\bar{V}} \tag{3.34}$$

for the nodal stress parameters $\underline{S}$ where *Boolean* matrices $\underline{b}^e$ according to

$$\underline{s}^e = \underline{b}^e\ \underline{S} \tag{3.35}$$

realize the conditions (3.32 b,c), and

$$\underline{F} = \sum_{(e)} \underline{b}^{e^T}\ \underline{f}^e\ \underline{b}^e\ ;\ \underline{f}^e = h^e\int_{\mathcal{B}^e}\underline{\Omega}^T\ \underline{C}^{-1}\ \underline{\Omega}\,dA \tag{3.36 a,b}$$

are the global and local flexibility matrix, resp. Note that the element flexibility matrix $\underline{f}^e$ is regular whereas the element stiffness matrix $\underline{k}^e$ is nonregular.

Unfortunately, the shape functions $\underline{\overset{*}{\Omega}}$ are very complicated so that the pure stress method has no practical significance.

Again an equivalent representation with a *Ritz* process for the total complementary energy π can be given.

3.4 The FE hybrid stress method

The shape functions $\overset{*}{\underline{\Omega}}$ for the stresses now only fulfill the static field conditions (3.32a) whereas the static boundary conditions (3.32 b,c) are approximated by the help of *Lagrangian* multipliers $\underline{\lambda}$. The extended principle of complementary work then has the form on the element level

$$\delta \overset{*}{\mathcal{A}}{}^{e} = \delta \overset{*}{\mathcal{W}}{}^{e} + \int_{\partial \mathcal{B}^e} \delta\, \underline{\lambda}^T h(\underline{t} - \underline{\bar{t}})\, ds \tag{3.37a}$$

$$\text{with} \quad \underline{t} = \underline{N}^T \underline{\sigma} \quad \text{and} \quad \underline{\lambda} = \underline{u} \quad \text{on } \partial \mathcal{B}^e . \tag{3.37b}$$

Introducing the element shape functions

$$\underline{\tilde{\sigma}}^e = \overset{*}{\underline{\Omega}}\, \underline{s}^e \quad \text{in } \mathcal{B}^e \tag{3.38a}$$

$$\underline{\tilde{u}}^e = \underline{\Omega}\, \underline{v}^e \quad \text{on } \partial \mathcal{B}^e , \tag{3.38b}$$

equ. (3.37) yields the relations

$$\begin{bmatrix} \underline{0} & \overset{*}{\underline{\ell}}{}^{eT} \\ \overset{*}{\underline{\ell}}{}^{e} & \underline{f}^e \end{bmatrix} \begin{bmatrix} \underline{v}^e \\ \underline{s}^e \end{bmatrix} = \begin{bmatrix} \underline{\bar{p}}^e \\ \underline{0} \end{bmatrix} \tag{3.39}$$

with the regular flexibility matrix $\underline{f}^e$ given in (3.36b) and the *Lagrangian* matrix

$$\overset{*}{\underline{\ell}}{}^{e} = \int_{\partial \mathcal{B}^e} \underline{N}^T \overset{*}{\underline{\Omega}}{}^T \underline{\Omega}\, ds . \tag{3.4o}$$

Because of the regularity of $\underline{f}^e$, the nodal stress vector $\underline{s}^e$ can be elimated on the element level in (3.39), and one gets

$$\underbrace{\overset{*}{\underline{\ell}}{}^{eT}\, \underline{f}^{e^{-1}}\, \overset{*}{\underline{\ell}}{}^{e}}_{\overset{*}{\underline{k}}{}^{e}}\, \underline{v}^e = \underline{\bar{p}}^e \tag{3.41}$$

with the pseudo stiffness matrix $\overset{*}{\underline{k}}{}^{e}$ and the nodal displacement vector $\underline{v}^e$. This method was first given in /7/. The assembling of elements is done according to (3.17) and (3.21-23), i.e. in the scheme of the displacement method.

It is also possible to introduce global vectors $\underline{V}$ and $\underline{S}$ for nodal displacements and nodal stresses from element vectors $\underline{v}^e$ and $\underline{s}^e$ (using *Boolean* matrices for the formulation of continuity conditions), but this procedure has no practical importance. The arising global matrix $\begin{bmatrix} \underline{O} & \overset{*}{\underline{L}}^T \\ \overset{*}{\underline{L}} & \underline{F} \end{bmatrix}$ is not positive definite. Furthermore the order of $\underline{V}$ must be smaller than that of $\underline{S}$ to save regularity. This can not be seen in the orders of the vectors $\underline{v}^e$ and $\underline{s}^e$, i.e. on the element level but is influenced by the location of nodal points (corner - edge- or inner points) of the chosen element and by the mesh including boundary conditions.

Another remark concerns the so-called mixed method, based on the *Hellinger-Reissner* variational principle /15/, with stress and displacement test functions within the element domain. Here the same structure of equations as in (3.39) comes out but the matrix $\overset{*}{\underline{\ell}}^e$ contains integrals over the whole element domain. Again the elimination of stresses on the element level is often used in order to avoid the disadvantages and problems discussed above.

3.5 The FE hybrid displacement method

A complementary method to such in chapter 3.4 can be constructed with displacement shape functions $\underline{\Omega}$ which do not - or only partly - fulfil the geometric continuity conditions at the element boundaries. The extended principle of virtual work with respect to (3.15) than becomes on the element level

$$\delta \mathcal{A}^e = \delta \mathcal{W}^e + \int_{\mathcal{B}^e} \delta \overset{*}{\underline{\lambda}}^T h(\underline{u} - \bar{\underline{u}})\, ds \qquad (3.42a)$$

$$\text{with} \quad \overset{*}{\underline{\lambda}} = \underline{t} = \underline{N}^T \underline{\sigma} \quad \text{on } \partial \mathcal{B}^e . \qquad (3.42b)$$

The test functions then have the form

$$\tilde{\underline{u}}^e = \underline{\Omega}\, \underline{v}^e \quad \text{in } \mathcal{B}^e \qquad (3.43a)$$

$$\tilde{\underline{t}}^e = \overset{*}{\underline{\Omega}}\, \underline{s}^e \quad \text{on } \partial \mathcal{B}^e . \qquad (3.43b)$$

From (3.42) one gets - complementary to (3.39) - the element equations

$$\begin{bmatrix} \underline{O} & \underline{\ell}^{eT} \\ \underline{\ell}^e & \underline{k}^e \end{bmatrix} \begin{bmatrix} \underline{s}^e \\ \underline{v}^e \end{bmatrix} = \begin{bmatrix} \underline{O} \\ \bar{\underline{p}} \end{bmatrix} \qquad (3.44)$$

with $\underline{k}^e$ equal to (3.16) and $\underline{\ell}^e = \overset{*}{\underline{\ell}}^{eT}$, see (3.4o). Because of the nonregularity of the element stiffness matrix $\underline{k}^e$, its elimination in (3.44) - corresponding to that of $\underline{f}^e$ in (3.39) - is not possible. Therefore, a stepwise procedure is reasonable in order to get better stresses than those calculated by differentiation of the displacement field. Furthermore displacement test functions with reduced continuity requirements are possible within this procedure.

3.6 The stepwise FE hybrid displacement method - a postprocessor of the displacement method for improved stress calculation -

After calculating the displacement vector $\underline{V}$ with the displacement method, see (3.23), test functions

$$\underline{\tilde{t}}^e(s) = \underline{\overset{*}{\Omega}}\ \underline{s}^e \quad \text{on} \quad \partial\mathcal{B}^e \cup \partial\mathcal{B} \tag{3.45}$$

are introduced - in the same way as in (3.43 b) - along all interior element boundaries $\partial\mathcal{B}^e$ and $\partial\mathcal{B}$ or in selected sections only.

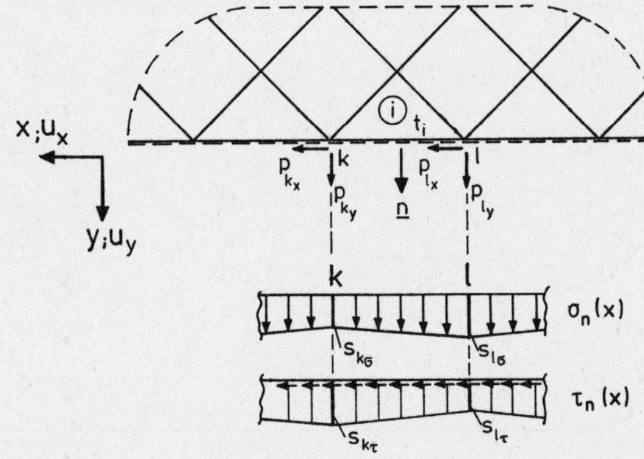

Fig. 6: Linear shape functions for the stresses along element boundaries

In the simplest case the shape functions for normal and shear stresses σ_n and τ_n are assumed to be linear as drawn in fig. 6. For the boundary k-ℓ, e.g., (3.45) can be written as

$$\underbrace{\begin{bmatrix} \tilde{\tau}_n \\ \tilde{\sigma}_n \end{bmatrix}_{k-\ell}}_{\tilde{\underline{t}}_{k-\ell}(x)} = \underbrace{\begin{bmatrix} \overset{*}{\underline{\Omega}}_\tau(x) & \\ & \overset{*}{\underline{\Omega}}_\sigma(x) \end{bmatrix}}_{\overset{*}{\underline{\Omega}}_{k-\ell}(x)} \underbrace{\begin{bmatrix} \underline{s}_\tau \\ \underline{s}_\sigma \end{bmatrix}_{k-\ell}}_{\underline{s}_{k-\ell}} \; ; \; \underline{s}_\tau = \begin{bmatrix} s_k \\ s_\ell \end{bmatrix}_\tau \; ; \; \underline{s}_\sigma = \begin{bmatrix} s_k \\ s_\ell \end{bmatrix}_\sigma . \tag{3.46}$$

Now virtual displacements $\delta\underline{u}_{k-\ell}$ are needed to establish equilibrium conditions by means of principle of virtual work. In order to get symmetric matrices the same test functions as for the stresses are chosen

$$\underbrace{\begin{bmatrix} \delta u_x \\ \delta u_y \end{bmatrix}_{k-\ell}}_{\delta\underline{u}_{k-\ell}(x)} = \underbrace{\begin{bmatrix} \underline{\Omega}_x(x) & \\ & \underline{\Omega}_y(x) \end{bmatrix}}_{\underline{\Omega}(x)} \underbrace{\begin{bmatrix} \delta\overset{*}{\underline{v}}_x \\ \delta\overset{*}{\underline{v}}_y \end{bmatrix}_{k-\ell}}_{\delta\overset{*}{\underline{v}}_{k-\ell}} \quad \text{with} \quad \underline{\Omega} \overset{!}{=} \overset{*}{\underline{\Omega}} . \tag{3.47}$$

Then the virtual work of the assumed stresses (multiplied with the areas) along the virtual displacements is

$$\begin{aligned} \delta\mathcal{W}_{k-\ell} &= \int_{k-\ell} \delta\underline{u}^T \cdot h \cdot \tilde{\underline{t}}\, dx \\ &= \delta\overset{*}{\underline{v}}^T_{k-\ell} \underbrace{\int_{k-\ell} h\, \underline{\Omega}^T\, \underline{\Omega}\, dx}_{\underline{\ell} = \underline{\ell}^T}\, \underline{s}_{k-\ell} . \end{aligned} \tag{3.48}$$

With *Boolean* matrices $\underline{b}_{k-\ell}$ and $\underline{a}_{k-\ell}$, the continuity conditions for the stresses and displacements are established for all elements. So, the global nodal stress vector $\underline{S}$ and the corresponding displacement vector $\delta\overset{*}{\underline{V}}$ can be represented as

$$\underline{s}_{k-\ell} = \underline{b}_{k-\ell}\, \underline{S} \quad ; \quad \delta\overset{*}{\underline{v}}_{k-\ell} = \underline{a}_{k-\ell}\, \delta\overset{*}{\underline{V}} . \tag{3.49}$$

Introducing (3.49) into (3.48) yields

$$\delta\mathcal{W}_{k-\ell} = \delta\overset{*}{\underline{V}}^T (\underline{a}^T\, \underline{\ell}\, \underline{b})_{k-\ell}\, \underline{S} . \tag{3.5o}$$

The total virtual work of the assumed stresses along all boundaries $\partial\mathcal{B}^e$ and $\partial\mathcal{B}$ is

$$\delta \mathcal{W} = \sum_{\partial \mathcal{B}^e \cup \partial \mathcal{B}} \delta \mathcal{W}_{k-\ell} = \delta \overset{*}{\underline{V}}^T \underbrace{\sum_{k-\ell} (\underline{a}^T \; \underline{\ell} \; \underline{b})_{k-\ell}}_{\underline{L}} \underline{S} = \delta \overset{*}{\underline{V}}^T \underline{L} \; \underline{S} \; . \tag{3.51}$$

The righthand side of the wanted system of equations arises from the virtual work done by the fictitious inner and outer nodal forces $\overset{*}{\underline{p}}^e$ and $\overline{\underline{p}}^e$ given in (3.16) and (3.2o) and already calculated in the preceding displacement method

$$\overset{*}{\underline{p}}^e = \underline{k}^e \; \underline{v}^e = \underline{k}^e \; \underline{a}^e \; \underline{V} \; . \tag{3.52}$$

The nodal vectors along the boundary k-ℓ can be represented using *Boolean* matrices $\underline{c}_{k-\ell}$

$$\overset{*}{\underline{p}}_{k-\ell} = \underline{c}_{k-\ell} \; \overset{*}{\underline{p}}^e \quad ; \; \overline{\underline{p}}_{k-\ell} = \underline{c}_{k-\ell} \; \overline{\underline{p}}^e \; . \tag{3.53}$$

In the computation process all the *Boolean* matrices $\underline{a}$, $\underline{b}$, $\underline{c}$ are realized by index vectors in order to save storage and time.

The virtual work of the resulting nodal forces $\overline{\underline{p}}_{k-\ell} - \overset{*}{\underline{p}}_{k-\ell}$ is given by

$$\delta \mathcal{A}_{k-\ell} = \delta \overset{*}{\underline{v}}^T_{k-\ell} \; (\overline{\underline{p}}_{k-\ell} - \overset{*}{\underline{p}}_{k-\ell}) = \delta \overset{*}{\underline{V}}^T \; \underline{a}^T_{k-\ell} \; (\overline{\underline{p}}_{k-\ell} - \overset{*}{p}_{k-\ell}). \tag{3.54}$$

The total virtual work of the nodal forces is given by

$$\delta \mathcal{A} = \sum_{\partial \mathcal{B}^e \cup \partial \mathcal{B}} \delta \mathcal{A}_{k-\ell} = \delta \overset{*}{\underline{V}}^T \underbrace{\sum_{k-\ell} \underline{a}^T \; (\overline{\underline{p}} - \overset{*}{\underline{p}})_{k-\ell}}_{\underline{P}_R} \; . \tag{3.55}$$

For the boundary line k-ℓ it holds

$$\delta \mathcal{A}_{k-\ell} = \delta \mathcal{W}_{k-\ell} \, . \tag{3.56}$$

Introducing (3.53) and (3.48) into (3.56) we get

$$\delta \underline{v}^T_{k-\ell} \; (\overline{\underline{p}}_{k-\ell} - \overbrace{\underline{k}_{k-\ell} \; \underline{v}_{k-\ell}}^{\overset{*}{\underline{p}}_{k-\ell}}) = \delta \underline{v}_{k-\ell} \; \underline{\ell}_{k-\ell} \; \underline{s}_{k-\ell}$$

or

$$\underline{\ell}_{k-\ell} \underline{s}_{k-\ell} + \underline{k}_{k-\ell} \; \underline{v}_{k-\ell} = \overline{\underline{p}}_{k-\ell} \; . \tag{3.57}$$

This equation (3.57) is included in the 2nd equation of (3.44), i.e. the local representation of the hybrid displacement method along the boundary k-ℓ.

For the whole system we arrive from (3.56) at

$$\delta\mathcal{A} \;=\; \delta\mathcal{W}\,. \tag{3.58}$$

Introducing (3.51) and (3.54) we get

$$\delta\underline{\overset{*}{V}}^T\,\underline{P}_R = \delta\underline{\overset{*}{V}}^T\,\underline{L}\,\underline{S} \quad \text{or} \quad \underline{L}\,\underline{S} = \underline{P}_R = \underline{\overline{P}} - \underline{\overset{*}{P}}\;. \tag{3.59}$$

In general $\underline{L}$ is rectangular but column-regular. Therefore, a *Gauß* transformation can be performed to get an approximation for $\underline{S}$ in the form

$$\underbrace{\underline{L}^T\,\underline{L}}_{\underline{A}\,=\,\underline{A}^T,\ \text{regular, pos. def.}}\underline{S} = \underline{L}^T\,\underline{P}_R \;;\quad \underline{S} = \underline{A}^{-1}\,\underline{L}^T\,\underline{P}_R\,. \tag{3.6o}$$

It can be shown, see / 9 /, that the *Gauß* transformation $\underline{L}^T\underline{L}$ can be performed efficiently using the local matrices $\underline{\ell}_{k-\ell}$ so that the computation effort for the postprocessor is nearly the same as such for the displacement vector $\underline{V}$. This holds for the case that all nodal stresses are wanted. In the case of less sections this effort may be considerably reduced. For the described postprocessor the computer program "stress" was developed and implemented in SAP IV, see /14/.

3.7 Examples

A finite strip is loaded by two opposite single forces according to fig. 7.

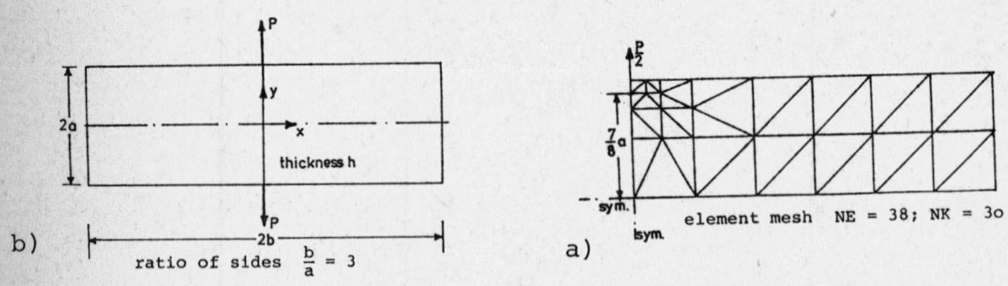

Fig. 7: a) Finite strip with two opposite point loads
b) Discretization with triangular linear displacement elements

The postprocessor "stress", using linear shape functions for the boundary stresses, yields the stress distribution 2 for σ_y in a section y = 7/8 a, i.e. rather near to the singularity, see fig. 8. In the point x = 0, this result - $\sigma_y = 30{,}0 \cdot \sigma_o$ - is equal in the first three digits to such given by *Oliveira* with a *Fourier* series solution. One can see that the conventional stress calculation, curve 1 , gives poor results, and also the so-called conjugate stress method does not improve the result decisively. This conjugate stress method only effects a smoothing of the stresses within the domain of the system (fulfilling the field equilibrium conditions in the middle) but does not include the boundary. Therefore, errors arising from violating the overall equilibrium conditions at the boundary are not reduced.

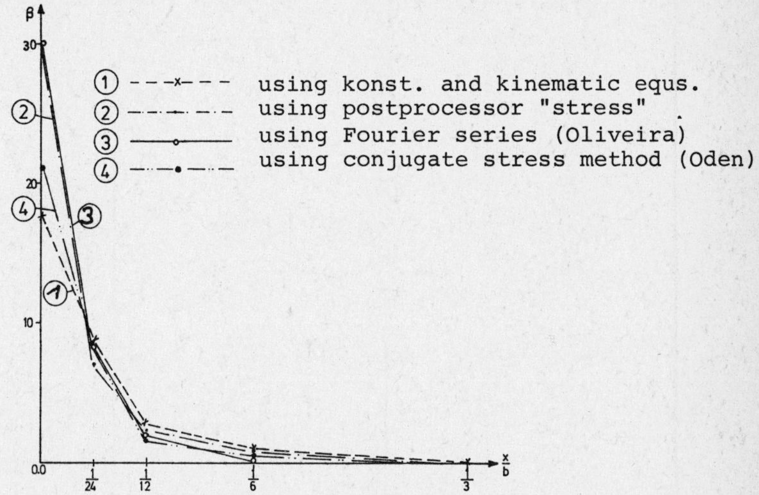

stress ratio $\beta(x) = \frac{\sigma_y(x)}{\sigma_o}$ in section $\frac{y}{a} = \frac{7}{8}$

Fig. 8: Stress distribution $\beta = \frac{\sigma_y}{\sigma_o}$ in section $\frac{y}{a} = \frac{7}{8}$ resulting from four different calculation methods.

4. The FE-displacement method including singular functions in a domain $\mathcal{B}_1$

4.1 Concept of the algorithm

A simple way to extend the FE displacement method to problems with stress singularities is the implementation of singular solutions. In the following the calculation of influence functions for the bending moments of a *Kirchhoff* plate is described, see fig. 9-11. So the boundary value problem for the bi-harmonic equation $\Delta\Delta\, w = 0$ is to be solved. Here, a concept with two domains $\mathcal{B}_1$ (FE's and singular functions) and $\mathcal{B}_2$ (only FE's) is used.

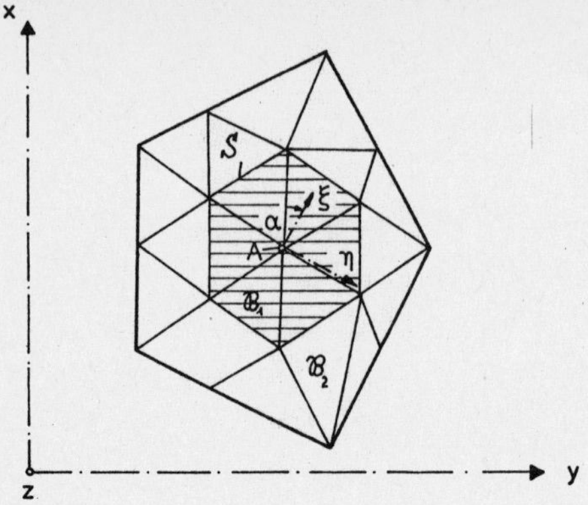

Fig. 9: Domains $\mathcal{B}_1$ (including singularities) and $\mathcal{B}_2$ of a plate

The singularity w_o for the lateral displacement w due to a point load in A is given by a(r^2 ℓn r/r_o)-type function.

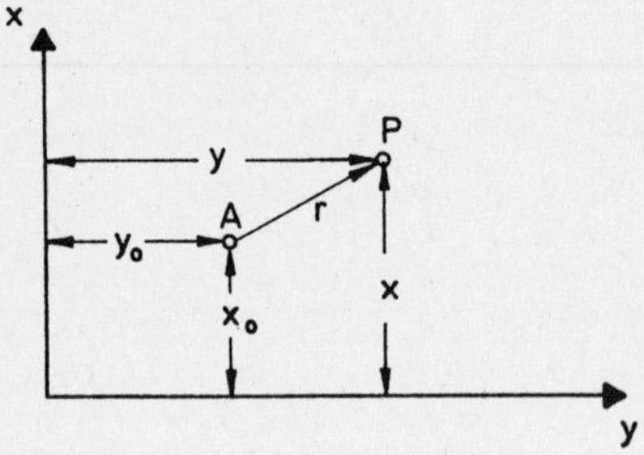

Fig. 1o: Coordinate system of the *Kirchhoff* plate

$$w_o = \frac{1}{8B\pi} r^2 \ln \frac{r}{r_o} \quad ; \quad B = \frac{E\,h^3}{12(1-\nu^2)} \tag{4.1}$$

$$w_o = \frac{1}{16B\pi} \left[(x-x_o)^2 + (y-y_o)^2 \right] \ln \frac{(x-x_o)^2 + (y-y_o)^2}{r_o^2} \; . \tag{4.2}$$

The bending moment m_x is given by

$$m_{x_o} = - B \left(\frac{\partial^2 w_o}{\partial x_o^2} + \nu \frac{\partial^2 w_o}{\partial y_o^2}\right) \tag{4.3}$$

$$= - \frac{1}{8\pi} \left[(1+\nu) \left(\ln \frac{(x-x_o)^2 + (y-y_o)^2}{r_o^2} + 1\right) + 2 \frac{(x-x_o)^2 + \nu(y-y_o)^2}{(x-x_o)^2 + (y-y_o)^2} \right] \; . \tag{4.4}$$

With the transformation

$$\xi = x - x_o \quad ; \quad \eta = y - y_o \tag{4.5}$$

we get the singular solution

$$m_{\xi_o} = w_s = - \frac{1}{8\pi} \left[(1+\nu) \left(\ell n \frac{\xi^2 + \eta^2}{r_o^2} + 1 \right) + 2 \frac{\xi^2 + \nu\eta^2}{\xi^2 + \eta^2} \right]. \tag{4.6}$$

This is also the singular part of the influence function for the moment m_x at point $A(x_o, y_o)$, i.e. the values of the moment m_x at $A(x_o, y_o)$ due to a point load $P = 1$ at $P(x, y)$, plotted at point P. This solution is implemented in all finite elements in the "singular" domain $\mathcal{B}_1$.

The ordinary FE-displacement shape functions - here cubic polynomials without C^1-continuity - are chosen in the domains $\mathcal{B}_1$ and $\mathcal{B}_2$.

The steps of the FE-algorithms are:

STEP 1 Chosing domains $\mathcal{B}_1$ and $\mathcal{B}_2$, Numbering of nodal points, first from outside $\mathcal{B}_2$, then $\partial \mathcal{B}_{1-2} = S$ and $\mathcal{B}_1$, at last point A.

STEP 2 Singular solution $\underline{w}_s$ in elements of $\mathcal{B}_1$, corresponding nodal displacement vector $\underline{\bar{V}}_S$ in the nodal points of $\partial \mathcal{B}_{1-2} = \mathcal{S}$.

STEP 3 $\underline{\overset{*}{P}}_S = \underline{K}_S \; \underline{\bar{V}}_S$; $\underline{\overset{*}{P}}_S$ fictitious nodal forces, $\underline{K}_S$ stiffness matrix of the "macro element" $\mathcal{B}_1$.

STEP 4 FE displacement calculation for $\mathcal{B}_2$ with given $\underline{\bar{V}}_S$ at the border S gives nodal loads $\underline{\bar{P}}_S$ at S and the part w_{s_1} of the influence surface in $\mathcal{B}_2$.

STEP 5 FE displacement calculation in $\mathcal{B}_1 \cup \mathcal{B}_2$ for the residual nodal loads $\underline{P}_S = - (\underline{\overset{*}{P}}_S - \underline{\bar{P}}_S)$ at S gives part w_{s_2} of the influence surface.

STEP 6 Superposition

$$w = w_s + w_{R_2} \quad \text{in } \mathcal{B}_1 \quad ; \quad w = w_{R_1} + w_{R_2} \quad \text{in } \mathcal{B}_2 .$$

On the following page the results for two different boundary conditions and three different domain $\mathcal{B}_1$ are presented and compared with a series solution, see also /1o/.

Another concept uses three domains $\mathcal{B}_1, \mathcal{B}_2, \mathcal{B}_3$ where in $\mathcal{B}_2$ the singular and the regular solution are "sewed" together, see /11/. But this procedure needs more effort in the topological parts of the algorithm and the computer program.

4.2 Examples

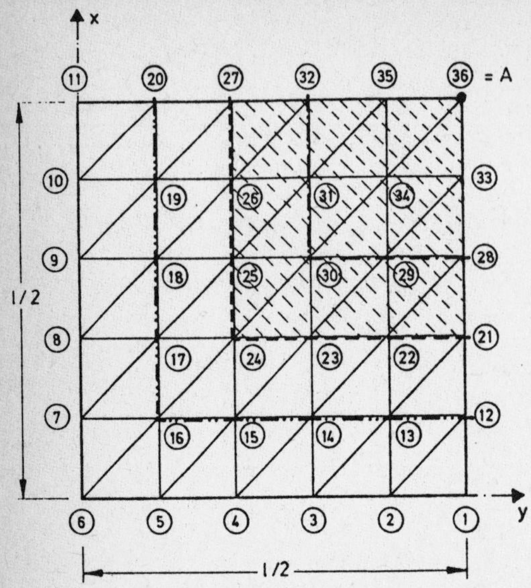

Three different domains $\mathcal{B}_1$ (singular regions):

case a: limit points 28, 29, 3o, 31, 32.

case b: limit points 21, 22,...,26, 27.

case c: limit points 12, 13,...,19, 2o.

case b is marked by hatching.

Fig. 11: Quadratic *Kirchhoff* plate approximated by 5o cubic triangular finite elements without C^1-continuity for the calculation of the influence surface m_x at the centre of the plate.

Table 1 8π-times influence ordinates of m_x for two different boundary conditions and three different singular regions.

	Navier-supported				clamped	
point	series sol.	case a	case b	case c	series sol.	case c
12	0,200	0,234	0,234	0,232	< 0	-0,028
13	0,226	0,249	0,251	0,251	< 0	-0,010
14	0,246	0,267	0,269	0,270	< 0,25	0,015
15	< 0,2	0,231	0,235	0,237	< 0,25	0,019
16	< 0,2	0,132	0,137	0,139	< 0,25	0,007
17	0,28	0,299	0,310	0,314	< 0,25	0,042
18	0,53	0,507	0,522	0,528	< 0,25	0,118
19	0,77	0,721	0,737	0,747	< 0,25	0,228
20	0,85	0,810	0,835	0,848	0,22	0,278
21	0,62	0,571	0,569	0,566	< 0	0,028
22	0,65	0,615	0,618	0,620	< 0,25	0,087
23	0,67	0,646	0,655	0,658	< 0,25	0,151
24	0,57	0,535	0,551	0,553	< 0,25	0,120
25	1,01	0,963	0,991	0,995	0,28	0,367
26	1,50	1,436	1,475	1,488	0,61	0,721
27	1,78	1,652	1,709	1,731	0,81	0,909
28	1,22	1,154	1,157	1,164	0,37	0,387
29	1,27	1,268	1,287	1,296	0,48	0,529
30	1,34	1,269	1,288	1,296	0,50	0,568
31	2,23	2,146	2,187	2,206	1,32	1,325
32	2,81	2,662	2,745	2,773	1,79	1,838
33	2,45	2,339	2,387	2,405	1,45	1,470
34	2,60	2,593	2,645	2,664	1,63	1,739
35	4,30	4,198	4,281	4,304	3,32	3,325
36	∞	(5,261)	(5,332)	(5,353)	∞	

References

/1/ Turner, M. J., Clough, R. W., Martin, H. C. and Topp, L.J.: Stiffness and deflection analysis of complex structures. J. Aeron. Sci. 23 (1956), pp. 8o5 - 823, 854.

/2/ Argyris, J. H.: Energy theorems and structural analysis. Butterworth Scientific Publications, London, 196o.

/3/ Stein, E. and Wunderlich W.: Finite-Element-Methoden als direkte Variationsverfahren der Elastostatik, in: Finite Elemente in der Statik, Edts.: K. E. Buck, D. W. Scharpf, E. Stein, W. Wunderlich, Verlag von Wilh. Ernst u. Sohn, Berlin, 1973.

/4/ Strang, G. and Fix, G. J.: An analysis of the finite element method. Prentice Hall, Inc., Englewood Cliffs, 1973.

/5/ Babuska, I. and Rheinboldt, W. C.: Error estimates for adaptive finite element computations, SIAM J. Num. Anal. 15 (1978), pp. 736 - 754.

/6/ Ciarlet, P. G.: The finite element method for elliptic problems. North-Holland Publ. Comp., Amsterdam, 1978.

/7/ Pian, T. H. H.: Derivation of element stiffness matrices by assumed stress distributions. AIAA J. 2 (1964), pp. 1333 - 1336.

/8/ Bufler, H. and Stein, E.: Zur Plattenberechnung mittels finiter Elemente. Ing. Archiv 39 (197o), pp. 248 - 26o.

/9/ Stein, E. and Ahmad, R.: An equilibrium method for stress calculation using finite element displacement models. Comp. Meth. in Appl. Mech. and Engng. 1o (1977), pp. 175 - 198.

/1o/ Riehle, W. and Stein, E.: Die Berechnung von Zustands- und Einflußflächen für die Schnittgrößen beliebiger Platten mit Hilfe der Methode der finiten Elemente. VDI-Fortschrittbericht, Reihe 4, Nr. 14, 1969.

/11/ Whiteman, J. R. and Barnhill, R. E.: Finite element methods for elliptic mixed boundary value problems containing singularities. Proc. Equadiff 3, Brno, 1972.

/12/ Stein, E.: Die Kombination des Trefftzschen Verfahrens mit der Methode der finiten Elemente, in: Finite Elemente in der Statik, Verlag von Wilh. Ernst u. Sohn, Berlin, 1973.

/13/ Stein, E.: Kriterien für Finite-Element-Programme aus der Sicht prüffähiger statischer Berechnungen, in: Finite Elemente in der Baupraxis, Hrsg. J. Pahl, E. Stein, W. Wunderlich, Verlag von Wilh. Ernst u. Sohn, Berlin, 1978.

/14/ Klöhn, C.: Alternative Spannungsberechnung in Finite-Element-Verschiebungsmodellen, Forschungs- und Seminarberichte aus dem Bereich der Mechanik der Universität Hannover, Bericht Nr. F 82/2, Hannover, 1982.

/15/ Wunderlich, W.: Grundlagen und Anwendung eines verallgemeinerten Variationsverfahrens, in: Finite Elemente in der Statik, Verlag von Wilh. Ernst u. Sohn, Berlin, 1973.

Series Expansions of Biharmonic Functions Around a Slit

J. Steinberg
Department of Mathematics
Technion - Israel Institute of Technology
Haifa 32000 Israel

Introduction. We consider a model problem of fracture mechanics in stress function formulation as follows.

$$\begin{aligned} &\Delta^2 u = 0 \quad , \quad \text{in} \quad D\ ; \\ &u = f \quad ; \quad \frac{\partial u}{\partial n} = g\ , \quad \text{on} \quad \partial D\ ; \qquad (1) \\ &u \quad \text{and} \quad \nabla u \quad \text{bounded in} \quad D, \end{aligned}$$

where D is a bounded domain in the plane, having a straight slit as a part of its boundary ∂D, and the functions f and g are given and vanish on the slit. We shall deal with two geometric situations illustrated in the following figures, where L_o denotes the slit and L the remaining part of ∂D.

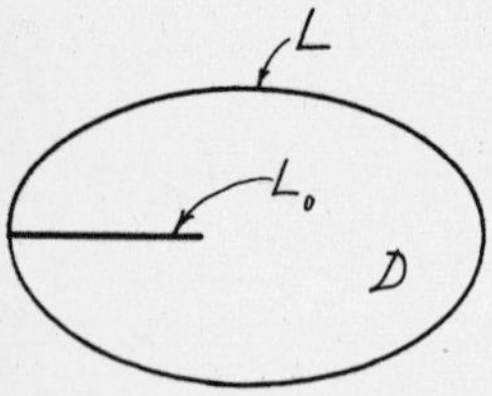

figure 1

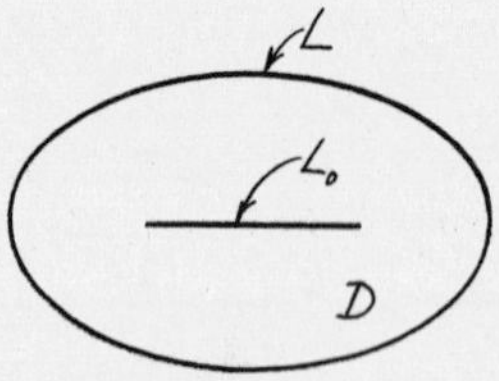

figure 2

The symmetry principle. It is difficult to trace back the origin of the following theorem, see Poritzki [1946].

Let $w(x,y)$ be a biharmonic function in a domain E such that a part of ∂E is a straight segment L_o of the x-axis, and let w satisfy the conditions $\frac{w}{y} \to 0$ as $y \to +0$, then there exists a unique biharmonic extension of w onto the domain $\hat{E}$, symmetric to E with respect to L_o; it is given by

$$w(x,-y) = -w(x,y) + 2y\,\frac{\partial w(x,y)}{\partial y} - y^2 \Delta w(x,y) \equiv \Omega w(x,y)\ ; \quad y \geq 0\ . \qquad (2)$$

It is plain that for the so extended function w, the condition

$w/y \to 0$, as $y \to +0$ takes the form

$$w(x,0) = \frac{\partial w}{\partial y}(y,0) = 0 \quad , \tag{3}$$

on the slit L_o.

In the case of figure 2, for instance, we may divide D into the parts D^+, where $y \geq 0$ and D^-, where $y \leq 0$. We denote by L_1 and L_2 the prolongations of the slit L_o, to the left and to the right in D, see figure 3.

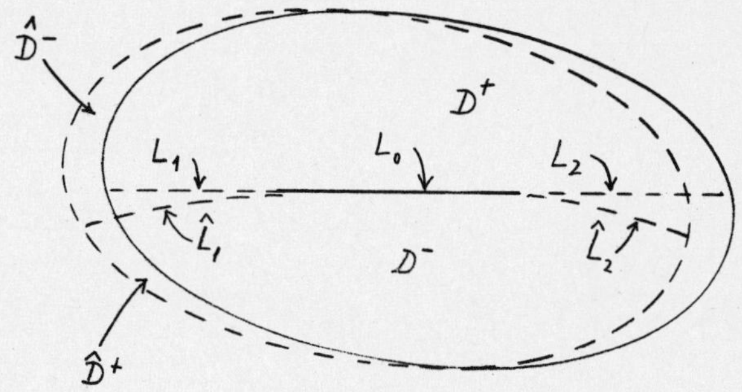

figure 3

Thus there is a continuation of w onto $\hat{D}^+$ across L_o, which is in general different from that of w given in D^- across L_1 and L_2. Likewise, the continuation of w from D^- onto $\hat{D}^-$ across L_o, will differ, in general, from the values given in D^+, and we remark that this continuation is provided by means of the same operator Ω which is invariant if y is changed into $-y$. Now as w belongs to C^∞ in D and its values as well as the values of all its derivatives fit continuously on L_1 and L_2, the same is true of the continuations onto $\hat{D}^+$ and $\hat{D}^-$ along $\hat{L}_1$ and $\hat{L}_2$. Hence if $\hat{D} = \hat{D}^+ \cup \hat{D}^-$, w has a unique extension from D onto $\hat{D}$ across L_o, and so w is defined as a biharmonic function on a two-sheet Riemann surface with the slit L_o. If the slit is $-l \leq x \leq l$, then this surface is the one for the function $(l-z^2)^{1/2}$.

We remark that the continuation of w from $\hat{D}^+$ onto D^+ across L_o gives the original values of w in D^+; thus the continuation operator Ω is involutory for biharmonic functions satisfying (3). In fact Ω is an involution in any set of biharmonic functions; this follows from the identity $\Omega^2=I-y^4\Delta^2$; where I is the identity operator.

From works by Kondratev [1967], Grisward [1976], Blum and Rannacher [1980], it is known that the asymptotic behaviour of w near an end of the slit is given by

$$w = O(r^{3/2}); \quad \frac{\partial w}{\partial r} = O(r^{1/2}) \; ; \; \frac{\partial^2 w}{\partial r^2} = O(r^{-1/2}) \; , \cdots \qquad (4)$$

where r is a polar coordinate.

For simplicity we shall, in the following, restrict ourselves to symmetric biharmonic functions with respect to L_o.

Thus we define the space B as the linear space of all symmetric biharmonic functions in D, vanishing together with the first normal derivative on L_o and possessing the asymptotic property (4). Now, since Ω is an involution in B, it has the only two eigenvalues ± 1. Hence we consider the two following subspaces $B^*, \tilde{B}$.

$$w \in B^* \text{ , iff } w \in B \text{ and } \Omega w = w;$$

$$w \in \tilde{B} \text{ , iff } w \in B \text{ and } \Omega w = -w.$$

Theorem 1. $B=B^* \oplus \tilde{B}$.

Proof. Let $w \in B$ and

$$w^* = \frac{1}{2}(w+\Omega w) \; ; \qquad \tilde{w} = \frac{1}{2}(w-\Omega w)$$

Then

$$\Omega w^* = \frac{1}{2}(\Omega w+\Omega^2 w) = w^* \in B^* \quad ;$$

$$\Omega \tilde{w} = \frac{1}{2}(\Omega w-\Omega^2 w) = -\tilde{w} \in \tilde{B} \; ;$$

$$w = w^* + \tilde{w} \; .$$

That w^* and $\tilde{w}$ belong to B is easy to check.

Series expansion around an end of the slit

In a paper on fracture mechanics, M.L. Williams [1957], set up the following series for the Airy stress function u in the case of symmetric stress conditions, corresponding to mode 1 (pure traction):

$$u = \sum_{j=o}^{\infty} a_j v_j^* + b_j v_j \quad , \qquad (5)$$

where, in polar coordinates r, ϕ,

$$v_j^* = r^{j+2} \, [\cos j\phi - \cos(j+2)\phi] \; ;$$

$$\tilde{v}_j = r^{j+3/2} \left[\frac{\cos(j-1/2)\phi}{j-1/2} - \frac{\cos(j+3/2)\phi}{j+3/2}\right] \tag{6}$$

The convergence of the series (5) can be investigated by putting $u = u^* + \tilde{u}$ according to theorem 1. Then it can be shown that u^* can be expanded into a series of the functions v_j^*, $j=0,1,\ldots$, which belong to B^*, and $\tilde{u}$ can be represented by a series of $\tilde{v}_j$, $j=0,1,\ldots$ which are in $\tilde{B}$. This is done by rearranging these series into ordinary Fourier series and computing the coefficients. The results are

$$a_j = \frac{1}{\pi r_o^{j+2}} \int_o^{\pi} \{f_j^*(\phi) u(r_o,\phi) + g_j^*(\phi) v(r_o,\phi)\} \, d\phi \; ;$$

$$b_j = \frac{1}{\pi r_o^{j+3/2}} \int_o^{\pi} \{\tilde{f}_j(\phi) u(r_o,\phi) + \tilde{g}_j(\phi) v(r_o,\phi)\} d\phi \; , \tag{7}$$

where

$$v = \frac{u}{r} - \frac{r}{2} \Delta u \; ; \tag{8}$$

$$f_j^*(\phi) = -\frac{1}{2}[j \cdot \cos j\phi + (j+2) \cdot \cos(j+2)\phi] \; ;$$

$$g_j^*(\phi) = r_o \sin\phi \, . \, \sin(j+1)\phi \; ;$$

$$\tilde{f}_j(\phi) = (j-1/2)(j+3/2)\cos\phi \, . \, \cos(j+1/2)\phi \; ; \tag{9}$$

$$\tilde{g}_j(\phi) = -r_o[\cos\phi . \cos(j+1/2)\phi + (j+1/2)\sin\phi . \sin(j+1/2)\phi] .$$

In all these formulae r_o is the radius of any circle inside D containing only one end of the slit.

<u>Theorem 2</u>. The series (5) converges absolutely and uniformly in that part of the two-sheet Riemann surface cut off by any circle $r=r_o$. We remark that b_1 is the stress intensity factor, mainly, and that the formulae (7) can be used for the computation of the coefficients if u, $\partial u/\partial r$ and Δu are known on such a circle $r=r_o$. Now in the case of figure 1 it is possible to express a_j and b_j in terms of the data on L, in (1) and Δu, $\partial\Delta u/\partial n$ on L. In order to obtain such

expressions, we first write Green's second formula for the domain obtained by cutting off from D the disc $r \leq r_o$, and take two functions u and w, where u is the solution of (1) and w is any biharmonic function satisfying (3) but not necessarily (4); thus

$$r_o \, S(u,w) = T(u,w) \ , \tag{10}$$

whith the two antisymmetric bilinear forms

$$S(u,w) = \int_o^{\pi} (u \frac{\partial \Delta w}{\partial r} - \frac{\partial u}{\partial r} \Delta w + \Delta u \frac{\partial w}{\partial r} - \frac{\partial \Delta u}{\partial r} w) d\phi \ , \tag{11}$$

$$T(u,w) = \int_L (u \frac{\partial \Delta w}{\partial n} - \frac{\partial u}{\partial n} \Delta w + \Delta u \frac{\partial w}{\partial n} - \frac{\partial \Delta u}{\partial n} w) ds \ . \tag{12}$$

As a function w we may take any of (6) with negative j. Now, the two catenated sequences

$$v_o^*, \ v_1^*, \ldots \quad ; \quad \tilde{v}_o, \tilde{v}_1, \ldots \tag{13}$$

$$v_{-2}^*, \ v_{-3}^*, \ldots \quad ; \quad \tilde{v}_{-1}, \tilde{v}_{-2}, \ \ldots \tag{14}$$

form a biorthogonal system with respect to S(u,w), as can be checked directly. If we put for u its series (5) containing the functions of (13) and for w any function of (14) we obtain the following formulae

$$a_j = - \frac{T(u, v_{-j-2}^*)}{8\pi(j+1)}$$

$$j=0,1,2,\ldots \tag{15}$$

$$b_j = - \frac{(2j-1)(2j+3)}{16\pi(2j+1)} \, T(u, \tilde{v}_{-j-1})$$

For the computation of these coefficients we need the values of Δu, $\partial \Delta u / \partial n$ on L, while u and $\partial u / \partial n$ are given there. We shall consider this problem later.

Series expansion around the whole slit.

For the case of figure 2, we shall use elliptic coordinates on the two-sheet Riemann surface, with the slit $-1 \leq x \leq 1$. They are defined by the relations

$x=\mathrm{ch}u \cos v \quad ; \quad y=\mathrm{sh}u \sin v$

$-\infty<u<\infty \quad ; \quad -\pi<v\leq\pi \quad .$

A basis in B^* is now given by the following functions

$$w_j^*=[\mathrm{ch}(j+2)u-\mathrm{ch}ju][\cos jv-\cos(j+2)v] \quad ; \quad j=0,1,... \tag{16}$$

and a basis for $\tilde{B}$ is

$$\tilde{w}_{-2}=r_o(u) \equiv \frac{1}{2}(\mathrm{sh}2u-2u) \quad ; \quad \tilde{w}_{-1}=(u\mathrm{ch}u-\mathrm{sh}u)\cos v \quad ;$$

$$\tilde{w}_o=2r_o(u)\cos 2v+r_1(u) \quad ; \quad r_1(u) = \int_o^u G(s)ds \quad ; \tag{17}$$

$$G(s)= -8\mathrm{sh}^2u \int_o^u \frac{r_o(s)ds}{\mathrm{sh}^2 s} \quad .$$

$$\tilde{w}_j = [\frac{\mathrm{sh}(j+2)u}{j+2} - \frac{\mathrm{sh}ju}{j}][\frac{\cos jv}{j} - \frac{\cos(j+2)v}{j+2}] \quad ; \quad j=1,2,... \tag{18}$$

Theorem 3. For any biharmonic w in a domain D of figure 2, satisfying (3) and (4), for instance for the solution of (1), we have

$$w = \sum_{j=o}^{\infty} c_j w_j^* + \sum_{j=-2}^{\infty} d_j\tilde{w}_j \tag{19}$$

and this series converges absolutely and uniformly on that part of the two-sheet Riemann surface cut off by any ellipse $u=u_o$ contained in D. (See Steinberg [1982]).

The coefficients in (19) can be represented by formulae analoguous to (7), but integral representations like (15) are not yet available.

Approximation of Δu and $\partial\Delta u/\partial n$ on L.

It seems appropriate to use the boundary integral method in order the obtain approximate values of Δu and its normal derivative on L, as required in (12) and (15). Let $K(P,Q)$, with $P=(r,\phi)$; $Q=(\rho,\theta)$, in polar coordinates, be a fundamental solution of the biharmonic equation, then we have, (see figure 4 and figure 1),

$$u(P) = \int_{L+L_o} \{u(Q) \frac{\partial \Delta_Q K}{\partial n} - \frac{\partial u(Q)}{\partial n} \Delta_Q K + \Delta u(Q) \frac{\partial K}{\partial n} - \frac{\partial \Delta u(Q)}{\partial n} K\} ds \ . \qquad (20)$$

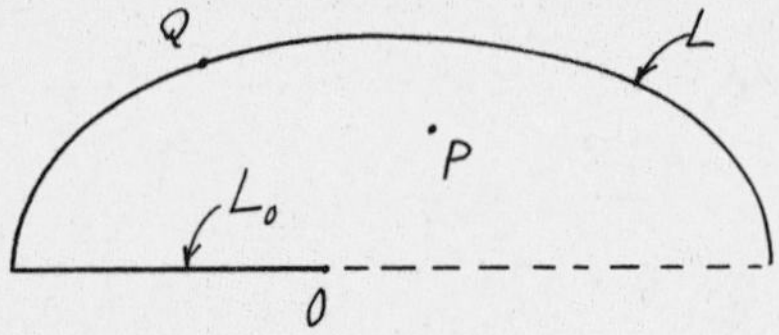

figure 4

In order to avoid integration along the slit L_o, whose end **0** is a singular point of the solution u, we seek K satisfying

$$K(P,Q) = \frac{\partial K}{\partial n}(P,Q) = 0 \quad ; \qquad Q \varepsilon L_o \ .$$

This can be obtained by putting te expressions (15) in the series (5) for u and inverting summation and integration, which is allowed by uniform convergence. The result is (20) without L_o, where

$$K = K^* + \tilde{K} \ ,$$

$$K^* = -\frac{1}{8\pi} \sum_{j=o}^{\infty} \frac{v_j^*(P) v_{-j-2}^*(Q)}{j+1} = -\frac{1}{8\pi} r \sin\phi . \sin\theta \log \frac{r^2+\rho^2-2r\cos(\phi+\theta)}{r^2+\rho^2-2r\cos(\phi-\theta)}$$

$$\tilde{K} = -\frac{1}{16\pi} \sum_{j=o}^{\infty} \frac{(2j-1)(2j+3)}{2j+1} \tilde{v}_j(P) \tilde{v}_{-j-1}(Q) \ .$$

$$= \frac{1}{16\pi}\{-8(r+\rho)\sqrt{r\rho}\ \cos\phi/2 \cdot \cos\theta/2 \ +$$

$$+(r^2+\rho^2-2r\rho\cos\phi.\cos\theta)\log \frac{(r+\rho+2\sqrt{r\rho}\cos\frac{\phi-\theta}{2})(r+\rho+2\sqrt{r\rho}\cos\frac{\phi+\theta}{2})}{(r+\rho-2\sqrt{r\rho}\cos\frac{\phi-\theta}{2})(r+\rho-2\sqrt{r\rho}\ \cos\frac{\rho+\theta}{2})}$$

Letting P tend to a point on L we obtain an integral equation for Δu and $\partial \Delta u/\partial n$ on L. A second integral equation for these two functions is obtained by letting Δ_P operate on (20) and P tend to a point of L.

The system of these two integral equations can be solved numerically by collocation or by a Galerkin procedure. Results for a model problem where D is a rectangle will be publisehd elsewhere.

References

Blum, H.and Rannacher, R. (1980). On the boundary value problem of the biharmonic operator on domains with angular corners. Math. Meth. in Appl. Sci.2, pp.556-581.

Grisvard, P. (1976). Behaviour of the solutions of an elliptic boundary value problem in a polygonal or polyhedral domain. Numerical solution of partial differential equations. III. B. Hubbard, editor. Academic Press, pp.207-274.

Kondratev, V.A. (1967). Boundary problems for elliptic equations in domains with conical or angular points. Trans. Moscow Math. Soc. 16, pp.227-313.

Melzer, H. and Rannacher, R. (1980). Spannungskonzentrationen in Eckpunkten der Kirchhoffschen Platte. Bauingenieur, 55, pp.181-184.

Poritzki, H. (1946). Application of analytic functions to two-dimensional biharmonic analysis. Trans. A.M.S. , vol.59, pp.248-279.

Steinberg, J. (1980, 1982). The symmetry principle for biharmonic functions and its application to a problem in fracture mechanics. Technion preprint series No.MT-492, and No.MT-563.

Williams, M.L. (1957). On the stress distribution at the base of a stationary crack. J. Appl. Mech.24, pp.109-114.

INVARIANCE PROPERTIES AND SPECIAL STRUCTURES NEAR CONICAL BOUNDARY POINTS

Peter Tolksdorf
Abt. Angew. Anal.
Universität Bonn
Beringstr. 6
53 Bonn 1

In this talk, I want to indicate how one can prove theorems concerned with the special structure near conical boundary points, also for the solutions of strongly nonlinear equations, in particular the p-harmonic equation. Before that, I want to recall the situation for the Laplace equation which is typical for all linear elliptic equations with constant coefficients.

1. Laplace - Equation

Here and in the following, S is an open subset of the unit sphere S^{n-1} having a smooth C^∞- boundary ∂S. By Ω, we denote the unit cone

$$\Omega = \{ r\cdot\sigma \mid 0 < r < 1 , \quad \sigma \in S \}.$$

We consider the solution u of the problem

$$- \Delta u = 0 , \quad \text{in } \Omega , \tag{1.1}$$

$$u = 0 , \quad \text{on } \partial\Omega \setminus S , \tag{1.2}$$

$$u = g , \quad \text{on } S . \tag{1.3}$$

Here, g is any given function. In practice, it is given by the values of a solution which is defined in a larger domain Ω' satisfying $\Omega = \Omega' \cap B_1(0)$. The solution u of (1.1) - (1.3) can be represented in the following way.

$$u(r,\sigma) = \sum_{\nu=1}^{\infty} a_\nu \cdot r^{\lambda_\nu} \cdot \varphi_\nu(\sigma) . \tag{1.4}$$

Here, $\lambda_1<\lambda_2<\lambda_3<\dots$ are the eigenvalues, $\varphi_1,\varphi_2,\varphi_3,\dots$ are the eigenfunctions and $a_1,a_2,a_3,\dots$ are the Fourier-coefficients of g with respect to the eigenvalue problem

$$-\Delta_\sigma\varphi = \lambda\cdot(\lambda+n-2)\cdot\varphi , \quad \text{in } S , \tag{1.5}$$

$$\varphi = 0 , \quad \text{on } \partial S , \tag{1.6}$$

posed for the Laplace-Beltrami operator Δ_σ on S^{n-1}.

Remarks. (i) If the first stress intensity factor a_1 does not vanish, then there is an $\varepsilon>0$ such that

$$\nabla^m u(r,\sigma) = a_1\cdot\nabla^m\{r^{\lambda_1}\cdot\varphi_1(\sigma)\} + O(r^{\lambda+\varepsilon-m}) ,$$

for $m=0,1,2,\dots$.

(ii) The condition that the first stress intensity factor a_1 disappears is "unstable". Hence, "in most cases", the solution u does not change sign, near the conical point.

2. The p-Harmonic Equation - General Remarks

The p-harmonic equation ($1 < p < \infty$)

$$- \operatorname{div} \{ |\nabla u|^{p-2} \cdot \nabla u \} = 0 , \quad \text{in } \Omega , \tag{2.1}$$

is the Euler-equation of the variational integral

$$\int_\Omega |\nabla u|^p \, dx .$$

In particular for p=n, it has applications in the theory of quasiconformal mappings [2],[19]. There is also another reason to study it. Namely, for some Non-Newtonian fluids [1],[16;Chap.2.5] , the Laplace operator in the Navier-Stokes equations is replaced by the p-harmonic operator. Therefore, the regularity properties of elliptic and parabolic systems and equations similar to the p-harmonic equation have found much interest. Here, we mention the $C^{1,\alpha}$-regularity results of N.N. Uraltseva [25], L.C. Evans [8], J.E. Lewis [15], E. DiBenedetto [4] and P. Tolksdorf [22], for elliptic equations, of K.Uhlenbeck [24] and P.Tolksdorf [23], for elliptic systems, and the C^1-regularity result of E.DiBenedetto & A.Friedman [5], for parabolic systems. The function $v(x) = |x|^{p/(p-1)}$ solves

$$- \operatorname{div} \{ |\nabla v|^{p-2} \cdot \nabla v \} = p \cdot n/(p-1) , \quad \text{in } \mathbb{R}^2 ,$$

and it does not belong to $C^2(\mathbb{R}^2)$, for $p > 2$. Hence, $C^{1,\alpha}$-regularity is optimal for the p-harmonic equation. This counterexample can be found in the work [10] by R.Glowinski & A.Marroco. Another counterexample for the homogenous p-harmonic equation will be given in the next section.

3. The p-Harmonic Equation - Behavior Near Conical Boundary Points

Theorem. *Suppose that* $u \in H^{1,p}(\Omega)$ *solves the p-harmonic equation* (2.1), *in* Ω , *and that*

$$u > 0 , \quad \text{in } \Omega , \tag{3.1}$$

$$u = 0 , \quad \text{on } \partial\Omega \setminus S . \tag{3.2}$$

Then, there are a constant a , *a function* $\varphi \in C^\infty(\overline{S})$ *and a*

$$\lambda > \max \{ 0 , (p-n)/p \} \tag{3.3}$$

such that

$$\nabla^m u(r,\sigma) = a \cdot \nabla^m \{ r^\lambda \cdot \varphi(\sigma) \} + o(r^{\lambda-m}) , \tag{3.4}$$

for m=0,1,2,... . *Moreover,* λ *and* φ *are determined uniquely by the nonlinear eigenvalue problem*

$$- \operatorname{div}_\sigma \{ (\lambda^2 \varphi^2 + |\nabla_\sigma \varphi|^2)^{(p-2)/2} \cdot \nabla_\sigma \varphi \} =$$

$$\lambda\cdot\{\lambda(p-1)+n-p\}\cdot(\lambda^2\varphi^2+|\nabla_\sigma\varphi|^2)^{(p-2)/2}\cdot\varphi \quad , \quad \textit{in } S \quad , \tag{3.5}$$

$$\varphi \quad = \quad 0 \quad , \qquad \textit{on } \partial S \quad , \tag{3.6}$$

a normation of φ *and the positivity condition*

$$\varphi \quad > \quad 0 \quad , \qquad \textit{in } S \quad . \tag{3.7}$$

Remarks. (i) A function of the form $r^\lambda\cdot\varphi(\sigma)$ solves the p-harmonic equation (2.1), iff λ and φ solve (3.5).

(ii) In the case p=2, (3.5) and (1.5) are identical.

(iii) In a more special form, the nonlinear eigenvalue problem (3.5) and (3.6) has already been studied by I.N.Krol & V.G. Mazja [13] and by I.N.Krol [12].

(iv) It would be nice to have a special structure theorem also without the positivity condition (3.1), though it can be excused by the Remark (ii) of Section 1.

Plane Domains (n=2). M.Dobrowolski [6] showed that one can determine the exponent λ explicitly, if n=2. For this, let $0<\alpha\leqq 2\pi$ be the opening angle of Ω, i.e. $S = \{(\cos\theta,\sin\theta) \mid -\alpha/2<\theta<\alpha/2\}$. Then

$$\lambda \quad = \quad \begin{cases} s + \sqrt{s^2+1/\gamma} \quad , & \text{if} \quad 0<\alpha\leqq\pi \quad , \\ s - \sqrt{s^2+1/\gamma} \quad , & \text{if} \quad \pi\leqq\alpha<2\pi \quad , \\ (p-1)/p \quad , & \text{if} \quad \alpha=2\pi \quad , \end{cases} \tag{3.8}$$

where

$$\gamma \quad = \quad (\alpha/\pi - 1)^2 \quad , \tag{3.9}$$

$$s \quad = \quad \{(\gamma-1)p-2\gamma\}/\{2\gamma(p-1)\} \quad . \tag{3.10}$$

For the details, we refer the reader to [6] and his article contained in these proceedings.

Now, suppose that $\alpha=\pi/2$ and that

$$p \quad > \quad 2 \quad .$$

From the above formulas, one can derive that the exponent λ satisfies

$$1 \quad < \quad \lambda \quad < \quad 2 \quad .$$

By reflection, one can define the function $u(r,\sigma)=r^\lambda\cdot\varphi(\sigma)$, on the whole plane. Thus, we obtain a function which does not belong to $C^2(\mathbb{R}^2)$ and which solves the homogenous p-harmonic equation (2.1), in $\mathbb{R}^2$, for p>2. Another proof of this fact is due to B.Bojarski & T. Iwaniec [3].

4. Proof of the Theorem

The essential tools of the proof are Maximum and Comparison Principles , $C^{1,\alpha}$-regularity results and the Stretching Operator T_R. For every function $w \in L^\infty(\Omega) \setminus \{0\}$ and every $R \in \,]0,1]$, it is defined by

$$T_R\, w\,(x) \quad = \quad c_R \cdot w(R\cdot x) \quad .$$

Here, the factor c_R is positive and chosen in such a way that

$$|T_R w|_{L^\infty(\Omega)} \quad = \quad 1 \; .$$

In the following, we frequently use the abbreviation

$$\Omega_R \quad = \quad \{\; x \in \Omega \;\|\; |x| < R \;\}.$$

<u>Key-Lemma</u>. *Let* $w \in L^\infty(\Omega) \setminus \{0\}$ *satisfy*

$$T_R\, w \quad = \quad w \quad , \quad \forall \;\; R \in \,]0,1] \; .$$

Then, there is a $\lambda \in [0,\infty[$ *such that*

$$w(x) \quad = \quad |x|^\lambda \cdot w(x/|x|) \quad , \quad \forall \; x \in \Omega \quad .$$

<u>Proof</u>. For $R \in]0,1]$, we set

$$f(R) \quad = \quad \operatorname{esssup} \{\; w(x) \;\;|\;\; x \in \Omega_R \;\} \; .$$

We pick $0 < t < T < 1$ and choose $\lambda, \Lambda \in \,]0,\infty[$ such that

$$f(t) \quad = \quad t^\lambda \quad , \qquad f(T) \quad = \quad T^\Lambda \quad .$$

In addition to that, we pick an $m \in \mathbb{N}$ and we choose an $M \in \mathbb{N}$ such that

$$t^{m+1} \quad \leqq \quad T^M \quad \leqq \quad t^m \quad .$$

As f is increasing with respect to R ,

$$t^{\lambda\cdot(m+1)} \quad = \quad f(t^{m+1}) \quad \leqq \quad f(T^M) \quad = \quad T^{\Lambda\cdot M} \quad \leqq \quad t^{\Lambda\cdot m} \quad ,$$

$$t^{\Lambda\cdot(m+1)} \quad \leqq \quad T^{\Lambda\cdot M} \quad = \quad f(T^M) \quad \leqq \quad f(t^m) \quad = \quad t^{\lambda\cdot m} \quad .$$

Now, we let m tend to infinity, in order to obtain that $\lambda = \Lambda$ which proves the Key-Lemma.

The next results are important properties of the solutions of quasilinear elliptic equations. They can be found in [21]. By D, we denote an arbitrary open domain in $\mathbb{R}^n$. The regularity result and the Maximum Principles can be applied to an arbitrary $H^{1,p}$-solution v of

$$-\operatorname{div} \{\; |\nabla v|^{p-2} \cdot \nabla v \;\} \quad = \quad 0 \quad , \quad \text{in} \quad D \; .$$

<u>Regularity</u>. *Let* $B_R = B_R(x_0)$ *be a ball with radius* R. *Suppose that* $\partial D \cap B_{2R}$ *is empty or that it is a regular* C^∞*-surface. Moreover, assume that* $v = 0$, *on* $\partial D \cap B_{2R}$. *Then, there is a constant* k *and an* $\alpha > 0$ *which depend only on* n, p, $\partial D \cap B_{2R}$ *and an upper bound for* $|v|_{L^\infty(D\cap B_{2R})}$ *such that*

$$|v|_{C^{1,\alpha}(\overline{D\cap B_R})} \quad \leqq \quad k \; .$$

In addition to that, the Schauder-estimates [9] *imply that* $v \in C^\infty(\overline{D\cap B_R})$, *provided that* $\nabla v \neq 0$, *in* $\overline{D\cap B_R}$.

<u>Weak Maximum Principle</u>. *The solution* v *satisfies*

$$\operatorname*{essinf}_{\partial D} \; v \quad \leqq \quad v(x) \quad \leqq \quad \operatorname*{esssup}_{\partial D} \; v \quad , \quad \forall \; x \in D \; .$$

Strong Maximum Principle. Suppose that v *is nonnegative and that it does not vanish identically, in* D. *Then,* $v > 0$, *in* D .

Hopf's Maximum Principle. Let B *be a ball contained in* D . *Suppose that* v *is positive, in* B , *and that* $v(x_0) = 0$, *for some* $x_0 \in \partial B$. *Then,* $\nabla v(x_0) \neq 0$.

In the following Comparison Principles, we consider two $H^{1,p}$-functions that satisfy

$$- \operatorname{div} \{ |\nabla v|^{p-2} \cdot \nabla v \} \leq - \operatorname{div} \{ |\nabla w|^{p-2} \cdot \nabla w \} , \quad \text{in } D .$$

Weak Comparison Principle. Let v *and* w *satisfy* $v \leq w$, *on* ∂D . *Then,* $v \leq w$, *also in* D .

Strong Comparison Principle. Assume that $w \in C^1(D)$, $v \in C^2(D)$ *and that* $\nabla v \neq 0$, *in* D . *Moreover, suppose that* v *and* w *are not identical and that* $v \leq w$. *Then,* $v < w$, *in* D .

Hopf's Comparison Principle. Let B *be a ball contained in* D *and assume that* $w \in C^1(\overline{D})$, $v \in C^2(\overline{D})$ *and that* $\nabla v \neq 0$, *in* $\overline{B}$. *Moreover, suppose that* $v < w$, *in* B , *and that* $v(x_0) = w(x_0)$, *for some* $x_o \in \partial B$. *Then,* $\nabla v(x_0) \neq \nabla w(x_0)$.

With the aid of these tools, we can establish the following

Existence Theorem. There exists a $\lambda > 0$ *and a* $\varphi \in C^\infty(\overline{S})$ *satisfying* (3.3) *and* (3.5) - (3.7).

Proof of the Existence Theorem. We set

$$g(x) = \begin{cases} 0 , & \text{if } |x| \leq 3/4 , \\ 4 \cdot (|x|-3/4) , & \text{if } |x| \geq 3/4 . \end{cases}$$

From the theory of monotone operators [11], [16] , we know that there is a $v \in H^{1,p}(\Omega)$ that solves the homogenous p-harmonic equation (2.1), in Ω , and that satisfies $v = g$, on $\partial\Omega$. We note that the Weak Maximum Principle, the above regularity result and the C^α-estimates of [14] imply that $v \in C^0(\overline{\Omega}) \cap C^1(\overline{\Omega} \setminus \{0\})$. We want to prove that there is an $\varepsilon > 0$ such that

$$v(R \cdot x) \leq \{1-\varepsilon \cdot (1-R)\} \cdot v(x) , \quad \forall\, x \in \Omega , \quad \forall\, R \in [1/2,1] . \tag{4.1}$$

For this, we use the function $b(x) = 1 + \alpha^{-2} \cdot (e^{-\alpha} - e^{-\alpha|x|^2})$. By the Strong Maximum Principle, $v < 1$, in $\overline{\Omega}_{1/2}$. Hence, we can choose $\alpha > 0$ so large that

$$- \operatorname{div} \{ |\nabla b|^{p-2} \cdot \nabla b \} \geq 0 , \quad \text{in } \Omega ,$$

$$b \geq v , \quad \text{on } \partial(\Omega \setminus \overline{\Omega}_{1/2}) .$$

This and the Weak Comparison Principle imply that

$$b \geq v \quad \text{in } \Omega \setminus \overline{\Omega}_{1/2} .$$

Consequently, (4.1) holds for some $\varepsilon > 0$, all $x \in \partial\Omega$ and all $R \in [1/2,1]$. Applying the Weak Comparison Principle to $v(R\cdot x)$ and $\{1-\varepsilon\cdot(1-R)\}\cdot v(x)$, we obtain the desired conclusion.

From the above regularity result, the Strong Maximum Principle and Hopf's Maximum Principle, we derive that there is a $k > 0$ such that

$$k \cdot v(x) \leqq v(x/2) \leqq k^{-1} \cdot v(x) , \tag{4.2}$$

for all $x \in \Omega_{1/4}$. This and the Weak Comparison Principle imply that (4.2) holds for all $x \in \Omega_{1/4}$. Consequently, we can use the C^{α}-estimate of [14] and the above regularity result in order to obtain a $v^* \in H^{1,p}(\Omega) \cap C^0(\overline{\Omega}) \cap C^1(\overline{\Omega} \setminus \{0\})$ and a sequence of $R_\nu > 0$ tending to zero such that

$$T_{R_\nu} v \longrightarrow v^* , \tag{4.3}$$

in the sense of $C^0(\overline{\Omega}) \cap C^1(\overline{\Omega} \setminus \{0\})$. Moreover, $v^* \geqq 0$, in Ω , $v^* = 0$, on $\partial\Omega \setminus S$, $|v^*|_{L^\infty(\Omega)} = 1$ and it solves the homogenous p-harmonic equation (2.1), in Ω . By the Strong Maximum Principle and Hopf's Maximum Principle, $v^* > 0$, in Ω , and $\nabla v^* \neq 0$, on $\partial\Omega \setminus \{\{0\} \cup \overline{S}\}$. Moreover, (4.1) and (4.2) hold for v^* and all $x \in \overline{\Omega}$. Consequently

$$\nabla v^*(x) \neq 0 , \quad \forall x \in \overline{\Omega} \setminus \{\{0\} \cup \overline{S}\} . \tag{4.4}$$

Hence, v^* solves an equation which is uniformly elliptic, in each compact subset of $\overline{\Omega} \setminus \{\{0\} \cup \overline{S}\}$. Thus, the Schauder-estimates [9] imply that

$$v^* \in C^\infty(\overline{\Omega} \setminus \{\{0\} \cup \overline{S}\}) . \tag{4.5}$$

We pick an $R \in]0,1]$. By the Key-Lemma and the Remark (i) of Section 3, it is sufficient to prove the existence of a $C_R > 0$ satisfying

$$v^*(R\cdot x) = C_R \cdot v^*(x) , \quad \forall x \in \Omega . \tag{4.6}$$

In order to show (4.6), we set

$$c_{r,R} = \sup \{ c \mid c\cdot u(x) \leqq u(R\cdot x) , \forall x \in \Omega_r \} ,$$
$$C_R = \sup \{ c_{r,R} \mid r \in]0,1]\} .$$

By the Weak Comparison Principle, $c_{r,R}$ is decreasing with respect to $r \in]0,1]$. This and (4.3) imply that

$$C_R\cdot v^*(x) \leqq v^*(R\cdot x) , \quad \forall x \in \Omega . \tag{4.7}$$

Let us suppose that (4.6) is not true. Then, we can use (4.7), the Strong Comparison Principle and Hopf's Comparison Principle to obtain a $\delta>0$ such that

$$(C_R + 2\cdot\delta)\cdot v^*(x) \leqq v^*(R\cdot x) , \quad \forall x \in \partial\Omega_{1/2} .$$

This and (4.3) show that

$$(C_R + \delta)\cdot v(x) \leqq v(R\cdot x) , \quad \forall x \in \partial\Omega_r , \tag{4.8}$$

for some $r > 0$. By the Weak Comparison Principle, (4.8) holds for all $x \in \Omega_r$. This, however, is a contradiction to the definition of C_R. Hence, (4.6) must be true.

Now, it is rather easy to prove the theorem of Section 3. We set

$$b(r,\sigma) = r^{\lambda}\cdot\varphi(\sigma) ,$$

where λ and φ are given by the Existence Theorem and where φ is normed in such a way that $|b|_{L^\infty(\Omega)} = 1$. From the L^∞-estimate of [14], the above regularity result, the Strong Maximum Principle, Hopf's Maximum Principle and the Weak Comparison Principle, one derives that there is a $k > 0$ such that

$$k \cdot b \leqq u \leqq k^{-1} \cdot b \ , \quad \text{in } \Omega_{1/2} \ . \tag{4.9}$$

Now, we pick a sequence of $R_\nu > 0$ tending to zero and suppose that there must exist a subsequence (R_μ) of (R_ν) such that

$$T_{R_\mu} u \longrightarrow b \ , \tag{4.10}$$

in the sense of $C^0(\bar{\Omega}) \cap C^1(\bar{\Omega} \setminus \{0\})$. Then, from (4.9) and the Schauder-estimates, one derives the conclusion of the theorem of Section 3.

Hence, it rests to prove (4.9). For this, we set

$$c_R = \sup \{ c \geqq 0 \mid c \cdot b \leqq u \ , \ \text{in } \Omega_R \} \ .$$

From the Weak Comparison Principle, one derives that c_R is decreasing with respect to R. Consequently, the limit

$$c^* = \lim_{R \to 0} c_R$$

exists and is positive. The C^α-estimates of [14] , the above regularity result and (4.9) imply that there is a subsequence (R_μ) of (R_ν) and a $u^* \in H^{1,p}(\Omega) \cap C^0(\bar{\Omega}) \cap C^1(\bar{\Omega} \setminus \{0\})$ such that

$$T_{R_\mu} u \longrightarrow u^* \ , \tag{4.11}$$

in the sense of $C^0(\bar{\Omega}) \cap C^1(\bar{\Omega} \setminus \{0\})$. Moreover, $u^* = 0$, on $\partial\Omega \setminus S$, u^* solves the homogenous p-harmonic equation, in Ω, and

$$c^* \cdot b \leqq u^* \ , \quad \text{in } \Omega \ . \tag{4.12}$$

Now, suppose that u^* is not identical to $c^* \cdot b$, in Ω . Then, (4.12), the Strong Comparison Principle and Hopf's Comparison Principle imply that there is a $\delta > 0$ such that

$$(c^* + 2\delta) \cdot b \leqq u^* \ , \quad \text{on } \partial\Omega_{1/2} \ .$$

Consequently, by (4.11), there is an $r > 0$ such that

$$(c^* + \delta) \cdot b \leqq u \ , \quad \text{on } \partial\Omega_r \ .$$

The Weak Comparison Principle shows that this is true, also in Ω_r . This, however, is a contradiction to the definition of c^* . Hence, $c^* \cdot b = u^*$, in Ω , and (4.10) is true.

5. Other Nonlinear Equations

In this section, we want to indicate how our results can be generalized to more general nonlinear equations. We also discuss the case p=1. For this, we suppose that Ω is plane and that α is the opening angle of Ω , i.e.

$$S = \{ (\cos\theta \ , \sin\theta) \mid -\alpha/2 < \theta < \alpha/2 \} \ .$$

We will frequently use the polar coordinates r and θ defined by $x = (r \cdot \cos\theta \ , r \cdot \sin\theta)$.

In [21] , it has been shown that one can treat the solutions of the inhomogenous p-harmonic equation having nonzero boundary values, on $\partial\Omega \setminus S$, similarly as in Section 4. The results, however, are not so strong and not yet satisfying.

As a typical example, let us consider a solution $u \in H^{1,p}(\Omega)$ of

$$- \operatorname{div} \{ (1 + |\nabla u|^2)^{(p-2)/2} \cdot \nabla u \} = 0 , \quad (p > 1) \tag{5.1}$$

in Ω , which satisfies (3.1) and (3.2). We have to distinguish beween two cases.

1st *case* $\alpha < \pi$. *In this case, we can show that* (3.4) *holds, for some* $a > 0$, *for all* $m \in \mathbb{N} \cup \{0\}$ *and for*

$$\lambda = \pi / \alpha > 1 , \tag{5.2}$$

i.e. λ *and* φ *are determined by the eigenvalue problem* (1.5) *and* (1.6) *posed for the Laplace-Beltrami operator. Moreover, the results of V.G.Mazja & B.P.Plamenevskii* [17] *imply that there is an* $\varepsilon > 0$ *such that*

$$\nabla^m u(r,\theta) = a \cdot \nabla^m \{ r^\lambda \cdot \varphi(\theta) \} + O(r^{\lambda+\varepsilon-m}) , \tag{5.3}$$

for all $m \in \mathbb{N} \cup \{0\}$.

In order to prove this, we note that

$$- \operatorname{div} \{ (1+|\nabla w|^2)^{(p-2)/2} \cdot \nabla w \} = (1+|\nabla w|^2)^{(p-2)/2} \cdot \{ - \Delta w + (2-p) \cdot (1+|\nabla w|^2)^{-1} \cdot \sum_{i,j=1}^{2} w_{x_i} \cdot w_{x_j} \cdot w_{x_i x_j} \} , \tag{5.4}$$

$$- \Delta \{ r^s \cdot \cos(t \cdot \theta) \} = r^{s-2} \cdot (t^2 - s^2) \cdot \cos(t \cdot \theta) , \tag{5.5}$$

for all $w \in C^2(\Omega)$ and all $s,t \in \mathbb{R}$. Moreover, the L^∞-estimate of [14], the Strong Maximum Principle and the Weak Comparison Principle hold for solutions of (5.1) (cf. [21]). Hence, similarly as in [20], one can use functions of the form $c \cdot r^s \cdot \cos(t \cdot \theta)$ $(c>0)$ as lower and upper barriers. For every $\varepsilon > 0$, this gives a constant $c_\varepsilon > 0$ such that

$$r^{\lambda+\varepsilon} \leqq c_\varepsilon \cdot u(r,\theta) , \quad \text{for} \quad -\alpha/4 \leqq \theta \leqq \alpha/4 \quad \text{and} \quad 0 < r < 1/2 , \tag{5.6}$$

$$u(r,\theta) \leqq c_\varepsilon \cdot r^{\lambda-\varepsilon} . \tag{5.7}$$

From (5.6), one derives similarly as in [20] that

$$|\nabla u(r,\theta)| \leqq c' , \quad \text{in } \Omega_{3/4} , \tag{5.8}$$

for some $c' > 0$. Consequently, u solves an equation which is uniformly elliptic, in $\Omega_{3/4}$. Hence, we can use the $C^{1,\alpha}$-regularity result of E.Miersemann [18] and the Schauder-estimates [9] , in order to see that $u \in C^2(\Omega_{3/4})$ and that

$$r \cdot |\nabla u| + r^2 \cdot |\nabla^2 u| \leqq c''_\varepsilon \cdot r^{\lambda-\varepsilon} , \quad \text{in } \Omega_{1/2} . \tag{5.9}$$

From (5.4) and (5.9), one derives that there is a $c_0 > 0$ and a

$$\lambda_0 > \lambda - 2 \tag{5.10}$$

such that

$$|\Delta u| \leqq c_0 \cdot r^{\lambda_0} , \quad \text{in } \Omega_{1/2} . \tag{5.11}$$

By (5.6), (5.10), (5.11) and Theorem 1.4 of [21] , (3.4) holds for some $a > 0$,

$m=0,1$ and the λ given by (5.2). That (3.4) holds also for $m \geqq 2$ can be derived from the Schauder-estimates [9].

2nd <u>case</u> $\alpha > \pi$. *In this case, we additionally suppose that*

$$p > (1+\sqrt{5})/2 \approx 1.62 \quad . \tag{5.12}$$

We can show that (3.4) *holds for some* $a > 0$, *all* $m \in \mathbb{N} \cup \{0\}$ *and for the* λ *given by* (3.8) - (3.10) , *i.e.* λ *and* φ *are determined by the nonlinear eigenvalue problem* (3.5) *and* (3.6) .

In order to prove this, we note that

$$- \operatorname{div} \{ (1+|\nabla w|^2)^{(p-2)/2} \cdot \nabla w \} = - (1+|\nabla w|^2)^{(p-4)/2} \cdot \Delta w$$
$$- (1+|\nabla w|^2)^{(p-4)/2} \cdot |\nabla w|^{4-p} \cdot \operatorname{div} \{ |\nabla w|^{p-2} \cdot \nabla w \} , \tag{5.13}$$

for all $w \in C^2(\Omega)$ satisfying $\nabla w \neq 0$, in Ω . Moreover, from [20], we know that $b(r,\theta) = r^{\mu} \cdot \cos(\mu \cdot \theta)$ satisfies

$$- \operatorname{div} \{ |\nabla b|^{q-2} \cdot \nabla b \} \leqq 0 , \quad \text{for } -\pi/(2\mu) \leqq \theta \leqq \pi/(2\mu) \text{ and } r > 0, \tag{5.14}$$

if $\mu \in]1/2,1[$ and if $q \leqq 2$. Moreover, for all $q \geqq 2$ and all $\mu \in]1/2,1[$, there is a $\kappa \in]\mu,1[$ such that the function $b(r,\theta) = r^{\kappa} \cdot \cos(\mu \cdot \theta)$ satisfies (5.14), too. These arguments and those given in the 1st case show that we can use functions of the form $c \cdot r^{\kappa} \cdot \cos(\mu \cdot \theta)$ $(c>0)$ as lower barriers, in order to obtain a $c_0 > 0$ and a $\lambda_0 < 1$ such that

$$r^{\lambda_0} \leqq c_0 \cdot u(r,\theta) \quad , \quad \text{for } -\alpha/4 < \theta < \alpha/4 \text{ and } 0 < r < 1/2 \quad .$$

This and Theorem 1.5 of [21] imply that

$$\lim_{R \to 0} T_R u = r^{\lambda} \cdot \varphi(\theta) \quad , \tag{5.15}$$

in the sense of $C^0(\overline{\Omega})$, where λ and φ are determined by (3.5)-(3.7). From the $C^{1,\alpha}$-estimate of [21] , one derives that (5.15) holds also in the sense of $C^1(\overline{\Omega} \setminus \{0\} \setminus \overline{S})$. Consequently, for every $\varepsilon > 0$, there is a constant $c_\varepsilon > 0$ and an $R_\varepsilon > 0$ such that

$$|u| + r \cdot |\nabla u| \leqq c_\varepsilon \cdot r^{\lambda-\varepsilon} \quad , \quad \text{in } \Omega_{1/2} \quad , \tag{5.16}$$
$$r^{\lambda+\varepsilon} \leqq c_\varepsilon \cdot r \cdot |\nabla u| \quad , \quad \text{in } \Omega_{R_\varepsilon} \quad . \tag{5.17}$$

Moreover, $T_R u$ satisfies a uniformly elliptic equarion , in $\Omega_{7/8} \setminus \overline{\Omega}_{1/8}$, if $R > 0$ is sufficiently small. Hence, by (5.16) and the Schauder-estimates [9], for every $\varepsilon > 0$, there is a $c''_\varepsilon > 0$ and an $R''_\varepsilon > 0$ such that

$$r^2 \cdot |\nabla^2 u| \leqq c''_\varepsilon \cdot r^{\lambda-\varepsilon} \quad , \quad \text{in } \Omega_{R''_\varepsilon} \quad . \tag{5.18}$$

From (5.13), (5.16) and (5.18), one derives that there is a $c_1 > 0$, an $R_1 > 0$ and a

$$\lambda_1 > \lambda \tag{5.19}$$

such that

$$| \operatorname{div} \{ |\nabla u|^{p-2} \cdot \nabla u \} | \leqq c_1 \cdot r^{(\lambda_1-1)(p-1)-1} \quad , \quad \text{in } \Omega_{R_1} \quad . \tag{5.20}$$

Theorem 1.4 of [21], (5.17), (5.19) and (5.20) show that (3.4) holds for some $a>0$, $m=0,1$ and the λ given by (3.8)-(3.10). That (3.4) holds also for $m \geqq 2$ can be derived from the Schauder-estimates.

Now, we want to consider the minimal surface equation

$$- \operatorname{div} \{ (1 + |\nabla u|^2)^{-1/2} \cdot \nabla u \} = 0 . \qquad (5.21)$$

It is identical with the equation (5.1), if $p=1$. Going through the proof of the result obtained for $\alpha < \pi$ and $p > 1$, it is easily seen that it is true, also for the minimal surface equation (5.21). The result for $\alpha > \pi$ and $p > 1$, however, cannot be true, also for $p = 1$. This is due to the following

<u>Non-Existence</u> <u>Result</u>. *Suppose that* $0 < \alpha < 2$ *and that* $\alpha \neq \pi$. *Then, there do not exist a function* $\varphi \in C^0([-\alpha/2,\alpha/2]) \cap C^2(]-\alpha/2,\alpha/2[)$ *and a* $\lambda \in \mathbb{R}$ *that satisfy*

$$- \lambda \cdot \varphi \cdot \varphi'' = \lambda^2 \cdot \varphi^2 + (1-\lambda) \cdot (\varphi')^2 , \quad \textit{in }]-\alpha/2,\alpha/2[, \qquad (5.22)$$

$$\varphi > 0 , \quad \textit{in }]-\alpha/2,\alpha/2[, \qquad (5.23)$$

$$\varphi(\alpha/2) = \varphi(\alpha/2) = 0 . \qquad (5.24)$$

<u>Remarks</u>. (i) The nonlinear eigenvalue problem (5.22) and (5.24) is identical with (3.5) and (3.6), if $p = 1$.

(ii) The above non-existence result must be seen in connection with other non-existence results concerned with the Dirichletproblem for the minimal surface equation. For this, we refer to [9] and the literature cited there. The results concerned with parame_ tric minimal surfaces spanning polygonal Jordan arcs [7] make suspect that there does not exist any solution of (5.21), in Ω, that satisfies (3.1) and (3.2).

<u>Proof</u> <u>of</u> <u>the</u> <u>Non-Existence</u> <u>Result</u>. Let us suppose that there exists a $\varphi \in C^0([-\alpha/2,\alpha/2]) \cap C^2(]-\alpha/2,\alpha/2[)$ and a $\lambda \in \mathbb{R}$ that satisfy (5.22)-(5.24). Then, there is a $t \in]-\alpha/2,\alpha/2[$, where φ attains its maximum. The uniqueness result for ordina_ ry differential equations and a reflection argument show that $t=0$. Consequently, we may assume that

$$\varphi(0) = 1 \quad \text{and} \quad \varphi'(0) = 0 . \qquad (5.25)$$

As $\psi(\theta) = (\cos\theta)^\lambda$ satisfies (5.22) and (5.25), $\varphi(\theta) = \psi(\theta)$, in $]-\alpha/2,\alpha/2[$. This, however, is a contradiction to (5.24).

<u>References</u>

1. G. ASTARITA & G. MARRUCI , Principles of non-Newtonian fluid mechanics, Mac Graw Hill (1974).

2. B. BOJARSKI & T. IWANIEC , Analytic foundations of quasiconformal mappings, Ann. Acad. Fenn., to appear.

3. B. BOJARSKI & T. IWANIEC , p-harmonic equation and quasilinear mappings, Pre_ print Nr. 617 of the SFB 72, University of Bonn (1983).

4. E. DiBENEDETTO , $C^{1+\alpha}$ local regularity of weak solutions of degenerate ellip_

tic equations, Nonl. Anal. 7, 827-850 (1983).

5. E. DiBENEDETTO & A. FRIEDMAN , Regularity of solutions of nonlinear degenerate parabolic systems, Preprint.

6. M. DOBROWOLSKI , Nichtlineare Eckenprobleme und Finite Elemente Methode, ZAMM, to appear.

7. G. DZIUK , Über quasilineare elliptische Systeme mit isothermen Parametern an Ecken der Randkurve, Analysis 1 , 63-81 (1981) .

8. L.C. EVANS , A new proof of local $C^{1,\alpha}$-regularity for solutions of certain degenerate elliptic PDE , J. Diff. Equt's 45 , 356-373 (1982).

9. D. GILBARG & N.S. TRUDINGER , Elliptic partial differential equations of second order, Springer-Verlag (1977).

10. R. GLOWINSKI & A. MARROCO , Sur l'approximation, par élements finis d'ordre un, et la résolution par pénalisation-dualité, d'une classe de Dirichlet non linéaires, RAIRO 9 , 41-76 (1975).

11. P. HARTMAN & G. STAMPACCHIA , On some nonlinear elliptic differential-functional equations, Acta Math. 115 , 271-310 (1966).

12. I.N. KROL , On solutions of the equation $\mathrm{div}\{|\nabla u|^{p-2}\cdot\nabla u\}=0$ with a particularity at a boundary point, Trudi Matematiceskovo Instituta Stekłova 125 , 127-139 (1973) [Russian] .

13. I.N. KROL & V.G. MAZJA , On the absense of continuity and Hölder continuity of solutions of quasilinear elliptic equations near a nonregular boundary, Trans. Moscow Math. Soc.26 , 73-93 (1972).

14. O.A. LADYZHENSKAYA & N.N. URALTSEVA , Linear and quasilinear elliptic equations, Academic Press (1968).

15. J.L. LEWIS , Regularity of the derivatives of solutions to certain degenerate elliptic equations, Ind. Math. J. , to appear.

16. J.L. LIONS, Quelques méthodes de résolution des problêmes aux limites non linéaires, Dunod (1969).

17. V.G. MAZJA & B.P. PLAMENEVSKI , The behavior of the solutions of quasilinear elliptic boundary value problems in the neighborhood of a conical point, Journal of Soviet Mathematics 8 , 423-426 (1977).

18. E. MIERSEMANN , Zur Regularität verallgemeinerter Lösungen von quasilinearen elliptischen Differentialgleichungen zweiter Ordnung in Gebieten mit Ecken, Zeitschr. f. Analysis 1 , 59-71 (1982).

19. Yu.G. RESHETNYAK , Index boundedness condition for mappings with bounded distortion, Siberian Math. J. 9 , 281-285 (1968).

20. P. TOLKSDORF , On quasilinear boundary value problems in domains with corners, Nonl. Anal. 5 , 721-735 (1981).

21. P. TOLKSDORF , On the Dirichletproblem for quasilinear equations in domains with conical boundary points, Comm. PDE 8 , 773-310 (1983).

22. P. TOLKSDORF , Regularity for a more general class of quasilinear elliptic equations, J. Diff. Equt's 51 ,126-15o (1984).

23. P. TOLKSDORF , Everywhere-regularity for some quasilinear systems with a lack of ellipticity, Ann. di Mat. p. appl. , to appear.

24. K. UHLENBECK , Regularity for a class of nonlinear elliptic systems, Acta Math. 138 , 219-240 (1977).

25. N.N. URALTSEVA , Degenerate quasilinear systems, LOMI 7 , 184-222 (1968) [Russian].

A COMPARISON OF THE LOCAL BEHAVIOR OF SPLINE L_2-PROJECTIONS, FOURIER SERIES AND LEGENDRE SERIES

Lars B. Wahlbin
Department of Mathematics and
Center for Applied Mathematics
Cornell University
Ithaca, New York 14853, USA

ABSTRACT.

Mechanisms are given which explain, in a quantitative fashion, "pollution" from nonsmooth regions of a function approximated into smooth regions. In spline L_2-projection pollution is negligible. In Fourier series it is governed by a global modulus of continuity in L_1 and in Legendre series by a weighted L_1 global modulus of continuity which accounts for certain boundary effects.

The three cases considered may serve as models for similar effects in approximation of derivatives in the "h-finite element", the "spectral" and the "p-finite element" methods of solving elliptic partial differential equations. However, the case of the "h-method", which is at present reasonably understood also in two space dimensions, shows that the analogy is far from perfect. One notes, though, that in some more complicated situations with the "h-method" the "pollution" effects are never better than in the simple model considered here. Thus, our rather pessimistic results concerning Fourier series and Legendre series may be of some value in delineating the performances of the "spectral" and "p-finite element" methods.

Examples are given that demonstrate sharpness of the "pollution" mechanisms described. Also, examples of de la Vallée-Poussin type summation methods which reduce "pollution" are considered.

1. INTRODUCTION.

Let I be a bounded interval and $x_0 \in I_\delta \subset I$ where I_δ is an interval with $\delta = \text{dist}\,(x_0, I \setminus I_\delta) > 0$. Let $f(x)$, $x \in I$, be a function which is smooth on I_δ but possibly rough on $I \setminus I_\delta$. For simplicity assume that

$$f \equiv 0 \quad \text{on} \quad I_\delta \,. \tag{1.1}$$

In the Fourier series case all intervals and functions are periodically extended. With f_n denoting spline projection or Fourier or Legendre

---

This work was supported by the National Science Foundation, USA.

series approximation we wish to investigate the approximation error at x_0, i.e., $f_n(x_0)$.

In particular we wish to describe the mechanism through which the roughness of f outside I_δ is reflected at x_0, i.e., we wish to elucidate the mechanism responsible for "pollution".

We now define the three classical L_2-projection methods we consider for approximating a given function.

1.i. Spline projections.

Let $I = [0,1]$ and $-1 \leq k < r$, $r \geq 0$, be fixed integers. For n a positive integer let $x_j = j/n$, $I_j = [x_j, x_{j+1}]$ and

$$S_n = S_n(k,r) = \{\chi(x) \in C^k(I) : \chi|_{I_j} \text{ is a polynomial of degree } \leq r,\ j = 0,1,\dots,n-1\}.$$

Let $\mathcal{S}_n f \in S_n$ denote the $L_2(I)$ projection of f into S_n, i.e.,

$$(\mathcal{S}_n f - f, \chi) \equiv \int_I (\mathcal{S}_n f - f)\chi \, dx = 0, \quad \text{for} \quad \chi \in S_n .$$

1.ii. Fourier series.

Here we take (for notational convenience) $I = [-\pi,\pi]$ and set T_n to be the trigonometric polynomials of degree $\leq n$. Then $\mathcal{F}_n f \in T_n$ is defined by

$$(\mathcal{F}_n f - f, \chi) = 0, \quad \text{for} \quad \chi \in T_n .$$

1.iii. Legendre series.

Again for convenience let $I = [-1, 1]$ and let P_n denote the polynomials on I of degree $\leq n$. Then $\mathcal{L}_n f \in P_n$ is given by

$$(\mathcal{L}_n f - f, \chi) = 0, \quad \text{for} \quad \chi \in P_n .$$

We remark, as is well known, that the case of Chebyshev series, $\mathcal{C}_n f \in P_n$ given by

$$\int_{-1}^{1} (\mathcal{C}_n f - f)\chi \frac{dx}{\sqrt{1-x^2}} = 0, \quad \text{for} \quad \chi \in P_n ,$$

reduces to that of Fourier-Cosine series approximation of the function $g(\theta) = f(\cos\theta)$, $0 \leq \theta \leq \pi$. We leave it to the reader to translate our results for Fourier series to this case.

The plan of this paper is as follows. In Section 2 we give a brief discussion of two unsuccessful global measures, the Hausdorff metric and "negative norms", which perhaps at first glance might appear as plausible candidates for explaining the pollution mechanism. In Section 3 we state our results. The parts of the proofs for which we have not found a reference to the literature occupy Sections 4 and 5. Examples of the sharpness of the descriptive mechanisms we have selected are given in Section 6. In Section 7 we give examples of summation methods which reduce the pollution effect.

We finally remark that the reader who so wishes may regard our results as pertaining to the derivatives in $\overset{\circ}{H}{}^1(I)$-projections, thus serving as models for pollution in derivatives in the "h-method", the "spectral method" [8] and the "p-method" [3] for approximating solutions of elliptic differential equations. As for pollution in $\overset{\circ}{H}{}^1$ projections in two or more space dimensions, only the "h-method" is at present reasonably understood, see [17], [18] and [23]. In this respect, cf. Example 6.5 below.

2. STRAWMEN FOR EXPLAINING POLLUTION: THE HAUSDORFF METRIC AND "NEGATIVE NORMS."

It is well known that

$$|\mathcal{S}_n f(x_0)| \le C(k,r) \min_{\chi \in S_n} \|f-\chi\|_{L_\infty(I)}, \tag{2.1.i}$$

$$|\mathcal{F}_n f(x_0)| \le \left(\frac{4}{\pi^2} \ln n + O(1)\right) \min_{\chi \in T_n} \|f-\chi\|_{L_\infty(I)}, \tag{2.1.ii}$$

$$|\mathcal{L}_n f(x_0)| \le C(\ln n + \sigma_n(x_0)) \min_{\chi \in P_n} \|f-\chi\|_{L_\infty(I)} \tag{2.1.iii}$$

where

$$\sigma_n(x_0) = \min\left((1-x_0^2)^{-1/4}, \sqrt{n}\right). \tag{2.2}$$

For (2.1.i) see [5] or [6], for (2.1.ii) [10], [15] or [24] and for (2.1.iii) and (2.2), [1]. Of course, when f is globally unsmooth these bounds will not give sharp estimates. E.g., when f is a stepfunction the L_∞-norms on the right are in general $O(1)$ and not better whereas (as is well known and will be recalled in Examples 2.2 and 6.2 below) for x_0 away from the discontinuities in f the error tends to zero as n tends to infinity.

It has sometimes (and many times in informal discussions with Numerical Analysts) been suggested that the Hausdorff metric might be appropriate for error analysis in problems with certain nonsmooth functions, see e.g. [11]. For J a closed interval and f,g two functions on it one defines

$$h(f,g;J) = \max\{\max_{A\in F}\ \min_{B\in G}\ \mathrm{dist}(A,B),\ \max_{B\in G}\ \min_{A\in F} \mathrm{dist}(A,B)\}.$$

Here F and G denote the augmented graphs of f and g over J, cf. [19], and for $A = (x_A,y_A)$, $B = (x_B,y_B)$,

$$\mathrm{dist}(A,B) = \max(|x_A-x_B|,\ |y_A-y_B|)\ .$$

For our purposes the Hausdorff metric has the attractive feature that on subintervals J where f is smooth, such as I_δ, it is commensurable with the maximum norm, [19, Theorem 1, p. 148]: With $h = h(f,g;J)$,

$$h \le \|f-g\|_{L_\infty(J)} \le h + \omega_f(h)$$

where ω_f is the modulus of continuity of f on J.

Regrettably the projections we consider are not close to being almost best approximations in the Hausdorff metric. We offer two examples.

Example 2.1.

In the spline case, take discontinuous piecewise constants, i.e., $k = -1$, $r = 0$. Let $f(x) = nx(1-nx)$, $x \in I_0$ and vanish for $x \ge 1/n$. Then it is easily calculated that $\mathcal{S}_n f = 1/6$ on I_0 and zero otherwise. Thus, $h(f, \mathcal{S}_n f;I) = O(1)$ and no better whereas a good approximation to f is $\tilde{\chi} = 1/4$ on I_0, $\tilde{\chi}$ vanishing outside, so that $h(f,\tilde{\chi};I) = O(n^{-1})$.

Example 2.2.

Let H denote the stepfunction

$$H(x) = \begin{cases} 0, & -\frac{\pi}{2} \le x \le \frac{\pi}{2} \\ 1, & -\pi \le x < -\frac{\pi}{2}\ , \text{ or } \frac{\pi}{2} < x \le \pi\ . \end{cases}$$

Then $\mathcal{F}_n H(x) = \frac{1}{2} - \frac{2}{\pi}(\cos x - \frac{\cos 3x}{3} + \frac{\cos 5x}{5} - \dots)\ .$

By the classical Gibbs phenomenon $h(H,\mathfrak{F}_n H;I) = O(1)$ and not better whereas, [19, Theorem 6, p. 153], there is a $\tilde{\chi} \varepsilon T_n$ such that

$$h(H,\tilde{\chi};I) = O(\frac{\ell n\ n}{n}) .$$

A similar example is easily constructed for Legendre series.

In the spline case it is well known that "negative norms" are of importance in describing the local influence of global effects. Let, for $s \geq 0$,

$$[\![u]\!]_{-s} = \sup_{\|\phi\|_{\mathbb{C}^s(I)} = 1} (u,\phi).$$

Then, [13], for any $s \geq 0$ and $\delta \geq C_s\ n^{-1}$,

$$|\mathcal{S}_n\ f(x_0)| \leq C\ \delta^{-s-1}\ [\![f-\mathcal{S}_n\ f]\!]_{-s} . \tag{2.3}$$

However, a similar result cannot hold in the Fourier (or Legendre) case. For, using that $(f-f_n,\phi) = (f,\ \phi-\phi_n)$, (2.1) and the classical facts that, cf. [10], [15],

$$\min_{\chi} \|\phi-\chi\|_{L_\infty(I)} \leq \begin{cases} C\ n^{-(r+1)}\ \|\phi\|_{\mathbb{C}^{r+1}(I)} ,\ \chi\ \varepsilon\ S_n\ , \\ C\ n^{-s}\ \|\phi\|_{\mathbb{C}^s(I)} ,\ \text{any } s,\ \chi\ \varepsilon\ T_n,\ P_n, \end{cases} \tag{2.4}$$

one obtains, respectively,

$$[\![f-f_n]\!]_{-s} \leq \begin{cases} C\ n^{-(r+1)}\ \|f\|_{L_1(I)} ,\ s \geq r+1, \\ C_s\ (\ell n\ n)\ n^{-s}\ \|f\|_{L_1(I)}\ , \\ C_s\ n^{-s+1/2}\ \|f\|_{L_1(I)}\ . \end{cases}$$

Since in Example 2.2 one has $|\mathfrak{F}_n\ H(0)| \geq cn^{-1}$, $c > 0$, the result corresponding to (2.3) cannot hold with $s > 1$ in the Fourier case. (By Example 6.1 one sees that it cannot hold for any $s > 0$).

In the next section we shall describe our results. It turns out that for Fourier series the "correct" global quantity is related to the best global approximation in L_1, $\min_{x \varepsilon T_n} \|f-x\|_{L_1(I)}$. In the Legendre series a slightly more complicated L_1-based norm enters whereas in the spline case the error is negligible.

3. STATEMENTS OF RESULTS.

3.i. The spline case.

Note first that in the case of discontinuous piecewise polynomials, $k = -1$, $\mathcal{S}_n f$ in a meshinterval is influenced only by the values of f on that meshinterval. Thus if $\delta > n^{-1}$, $\mathcal{S}_n f(x_0) = 0$. In general we have the following rapid decrease of outside influences, [6, (4.7)].

Theorem 3.1.

Assume that (1.1) holds. There exist positive constants C and c depending only on r and k such that

$$|\mathcal{S}_n f(x_0)| \leq C\, n\, e^{-c\delta n} \|f\|_{L_1(I)} .$$

From this it follows in particular that, given any q, for $\delta \geq \frac{(q+1)}{c} \cdot \frac{\ell n\, n}{n}$,

$$|\mathcal{S}_n f(x_0)| \leq Cn^{-q} \|f\|_{L_1(I)} . \tag{3.1}$$

We regard this type of error spread from the unsmooth parts as negligible and shall not further consider the spline case in this paper.

3.ii. Fourier series.

The question of local versus global influences to convergence of Fourier series was studied in [16]. Riemann's localization principle states the following.

If f and $g \in L_1$ and $f \equiv g$ in a neighborhood of x_0, then $\mathcal{F}_n f(x_0)$ and $\mathcal{F}_n g(x_0)$ converge or diverge simultaneously.

Note however that absolute convergence is not locally determined, see Example 2.2 where $\mathcal{F}_n H(0)$ is not absolutely convergent. Then again, uniform convergence in a neighborhood is locally determined by the Riemann-Lebesgue principle, [24, p. 52]. But clearly the rate of convergence at a point x_0 depends greatly on outside influences; arbitrarily small rates are possible while $f \equiv 0$ in a fixed neighborhood of x_0 and $\|f\|_{L_1} \leq 1$, see Example 6.1 below.

To state a theorem introduce, for $p > 0$ an integer, the pth modulus of continuity in a Banach space B,

$$\omega^P(f,\tau;B) = \sup_{|t| \leq \tau} \|\Delta_t^P f\|_B \tag{3.2}$$

where $\Delta_t^P f(x) = (T_t - I)^P f(x)$, $T_t f(x) = f(x+t)$.

Theorem 3.2.

Assume that (1.1) holds and let p and q be given nonnegative integers. There exist constants C_1, $C_2 = C_2(q)$, $C_3 = C_3(p)$, $C_4 = C_4(p,q)$ such that for $\delta \geq C_4 \frac{\ell n\ n}{n}$,

$$|\mathfrak{F}_n f(x_0)| \leq C_1\ \delta^{-1}\ \omega^p(f, C_2 \frac{\ell n\ n}{n}; L_1(I)) + C_3 n^{-q} \|f\|_{L_1(I)}, \tag{3.3}$$

for $p \geq 1$,

$$|\mathfrak{F}_n f(x_0)| \leq C_1\ \delta^{-1} \|f\|_{L_1(I)}, \text{ for } p = 0. \tag{3.3)'}$$

The rather elementary proof will be given in Section 4.

3.iii. Legendre series.

As is well known, cf. [12], [22], when investigating approximations by polynomials in x on $[-1,1]$, it is often technically advantageous to consider instead even trigonometric approximations to

$$g(\theta) := f(\cos\theta), \ -\pi \leq \theta \leq \pi \tag{3.4}$$

In particular, this procedure makes it easier to account for certain boundary effects, e.g., the fact that polynomial approximation is often capable of handling singularities at the boundary points $x = \pm 1$ better than interior singularities of the same type, at least when the singularities are not too sharp, cf., Examples 6.3 and 6.4 below.

Our result will be stated in terms of the function $g(\theta)$. Recall that

$$\sigma_n(x_0) = \min\left((1-x_0^2)^{-1/4}, \sqrt{n}\right).$$

Theorem 3.3.

Assume that (1.1) holds and let p and q be given nonnegative integers. There exist constants C_1, $C_2 = C_2(q)$, $C_3 = C_3(p)$, $C_4 = C_4(p,q)$ such that for $\delta \geq C_4 \frac{\ell n\ n}{n}$,

$$|\mathcal{L}_n f(x_0)| \le C_1 \delta^{-1} \sigma_n(x_0) \omega^p(g, C_2 \frac{\ell n\, n}{n}; L_1(|\sin|^{1/2})) \qquad (3.5)$$

$$+ C_3 n^{-q} \|g\|_{L_1}, \text{ for } p \ge 1,$$

$$|\mathcal{L}_n f(x_0)| \le C_1 \delta^{-1} \sigma_n(x_0) \|g\|_{L_1(|\sin|^{1/2})}, \text{ for } p = 0 . \qquad (3.5)'$$

Here the norm in $L_1(|\sin|^{1/2})$ is

$$\|g\|_{L_1(|\sin|^{1/2})} = \int_{-\pi}^{\pi} |g(\theta)| \, |\sin \theta|^{1/2} \, d\theta . \qquad (3.6)$$

The proof of Theorem 3.3 will be given in Section 5. It includes ideas from the proof of Theorem 3.2 while also relying on classical properties of Legendre polynomials.

Notice the blend of three boundary effects in the estimate (3.5). For functions with algebraic singularities at the boundary of type $(1-x^2)^\alpha$, $g(\theta) = |\sin \theta|^{2\alpha}$, which is helpful in the modulus of continuity if $\alpha > 0$ but not if $\alpha < 0$. The weight factor $|\sin \theta|^{1/2}$ is always helpful for boundary singularities. However, if the approximation point x_0 approaches ± 1 the factor $\sigma_n(x_0)$ deteriorates to $\sqrt{n}$, cf. Gronwall [9]. It is illuminating in this respect to consider Examples 6.3 and 6.4 below.

We note, incidentally, that the weighted norm $L_1(|\sin|^{1/2})$ on $g(\theta) = f(\cos \theta)$ is probably appropriate for Legendre series. For, an example by Stieltjes, cf. [7, p. 100-103], viz., the Legendre series for $(1-x)^{-\alpha}$, diverges at <u>all</u> points in $(-1,1)$ if $\frac{3}{4} \le \alpha$. Since $g(\theta) = (1-\cos \theta)^{-\alpha} = 2^\alpha (\sin \theta/2)^{-2\alpha}$ belongs to $L_1(|\sin|^{1/2})$ for $\alpha < 3/4$ but not for $3/4 \le \alpha$, we have some indication that our weighted norm is not merely a byproduct of our proof.

4. <u>PROOF OF THEOREM 3.2.</u>

In the proof C will denote generic absolute constants. We have

$$(\mathcal{F}_n f)(x) = \int_{-\pi}^{\pi} D_n(x-y) \, f(y) dy \qquad (4.1)$$

where the Dirichlet kernel D_n is given by

$$D_n(t) = (2\pi)^{-1} \sin((n+1/2)t)/\sin(t/2).$$

From this, the estimate (3.3)' for $p = 0$ is clear.

For $p \geq 1$ we shall compare $\mathfrak{F}_n f(x_0)$ to a suitable $\mathcal{K}_n f \in T_n$ constructed by means of a sharper kernel than D_n. By [14] or [4, p. 270] there exists a trigonometric polynomial $k_n(t)$ of degree n such that with

$$C_2 = 16 (q+1)e, \tag{4.2}$$

$$|k_n(t)| \leq \frac{C}{C_2} \frac{n}{\ell n\ n}, \text{ all } t, \tag{4.3}$$

$$|k_n(t)| \leq C\, n^{-q-1} \quad \text{for} \quad C_2 \frac{\ell n\ n}{n} \leq |t| \leq \pi \tag{4.4}$$

$$\int_{-\pi}^{\pi} k_n(t)dt = 1, \tag{4.5}$$

$$\int_{-\pi}^{\pi} |k_n(t)|dt \leq C . \tag{4.6}$$

(In [4, Lemma, p. 270] take "δ", in her notation, to equal $C_2\ \ell n\ n/n$.) The kernel k_n is, for our purposes, somewhat sharper than the usual generalized Jackson kernel [10, Ch. 4, §3]. Now define

$$\mathcal{K}_n f(x) = -\int_{-\pi}^{\pi} k_n(t) \sum_{\ell=1}^{p} (-1)^{\ell} \binom{p}{\ell} f(x + \ell t)dt \tag{4.7}$$

which belongs to T_n, see e.g. [10, p. 58].

Since $\mathfrak{F}_n \chi_n = \chi_n$ for $\chi_n \in T_n$ we have

$$\mathfrak{F}_n f(x_0) = \mathcal{K}_n f(x_0) + \mathfrak{F}_n[f-\mathcal{K}_n f](x_0) . \tag{4.8}$$

Let

$$C_4 = 2\, C_2\, p = 32p(q+1)e \tag{4.9}$$

so that $\delta/p \geq 2\, C_2\ \ell n\ n/n$. From the support properties of f, using (4.4) we have

$$|\mathcal{K}_n\, f(x_0)| \le \int_{|pt|\ge\delta} |k_n(t)| \sum_{\ell=1}^{p} \binom{p}{\ell} |f(x_0+\ell t)|dt$$

$$\le C\, n^{-q-1}\; 2^p\, \|f\|_{L_1} \;. \tag{4.10}$$

Further, by (4.1), (4.5) and (4.6),

$$\mathcal{F}_n(f-\mathcal{K}_n\, f)(x_0) = \int_{-\pi}^{\pi} D_n(x_0-y)\Big(\int_{-\pi}^{\pi} k_n(t)\,\Delta_t^p\, f(y)dt\Big)dy$$

$$= \int_{-\pi}^{\pi} k_n(t)\Big(\int_{-\pi}^{\pi} D_n(x_0-y)\,\Delta_t^p\, f(y)dy\Big)dt \tag{4.11}$$

$$= \int_{|t|\le C_2 \frac{\ell n\, n}{n}} + \int_{|t|\ge C_2 \frac{\ell n\, n}{n}} \equiv I_1 + I_2 \;.$$

In I_1 the inner integrand $\Delta_t^p\, f(y)$ vanishes unless $|x_0-y \pm pt| \ge \delta$ (modulo 2π). By (4.9) this can happen only if $|x_0-y| \ge \delta/2$. Then $|D_n(x_0-y)| \le C\,\delta^{-1}$ and hence, by (4.6),

$$|I_1| \le \int_{|t|\le C_2 \frac{\ell n\, n}{n}} |k_n(t)|\; C\delta^{-1}\, \|\Delta_t^p\, f\|_{L_1}\, dt$$

$$\le C\delta^{-1}\, \omega^p(f,\, C_2\, \tfrac{\ell n\, n}{n};\, L_1) \;. \tag{4.12}$$

For I_2 we have by (4.3) and since $|D_n(t)| \le \frac{(n+1/2)}{\pi}$,

$$|I_2| \le Cn^{-q-1}\cdot n\cdot 2^p\, \|f\|_{L_1} = Cn^{-q}\, 2^p\, \|f\|_{L_1} \;. \tag{4.13}$$

Combining (4.8), (4.10), (4.11), (4.12) and (4.13),

$$|\mathcal{F}_n\, f(x_0)| \le C\delta^{-1}\, \omega^p(f,\, C_2\, \tfrac{\ell n\, n}{n};\, L_1)$$

$$+\, C\, 2^p\, (n^{-q} + n^{-q-1})\, \|f\|_{L_1}$$

$$\le C_1\, \delta^{-1}\, \omega^p(f,\, C_2\, \tfrac{\ell n\, n}{n};\, L_1) + C_3(p)n^{-q}\, \|f\|_{L_1} .$$

The dependences on p and q for C_2 and C_4 are displayed in

(4.2) and (4.9).

This proves Theorem 3.2.

5. PROOF OF THEOREM 3.3.

Let p_n denote the L_2-normalized Legendre polynomials, $\int_{-1}^{1} p_n\, p_m = \delta_{nm}$, and let q_n denote those normalized by $q_n(1) = 1$ so that $p_n(x) = \sqrt{\frac{2n+1}{2}}\, q_n(x)$. We refer to [20] and [21] for basic definitions and properties of Legendre polynomials.

The Christoffel-Darboux formula, [20, p. 122] or [21, Theorem 3.2.2, p. 43] gives

$$\mathcal{L}_n f(x_0) = \int_{-1}^{1} \Pi_n(x_0,y)\, f(y)dy$$

where

$$\Pi_n(x_0,y) = \sum_{j=0}^{n} p_j(x_0)p_j(y) = \frac{n+1}{2}\left(\frac{q_{n+1}(x_0)q_n(y)-q_{n+1}(y)q_n(x_0)}{x_0-y}\right). \tag{5.1}$$

Introducing $g(\theta) = f(\cos\theta)$ we have

$$\mathcal{L}_n f(x_0) = \int_0^{\pi} \Pi_n(x_0, \cos\theta)\, g(\theta)\sin(\theta)\, d\theta\ . \tag{5.2}$$

Note that $g(\theta) = 0$ for $|\theta-\theta_0| \le \delta/(\max|\sin\tilde{\theta}|)$, $\tilde{\theta} \in \mathrm{Int}[\theta,\theta_0]$. Thus,

$$g(\theta) = 0 \quad \text{for } |x_0-\cos\theta| \le \delta, \text{ in particular for} \quad |\theta_0-\theta| \le \delta\ . \tag{5.3}$$

By the Stieltjes-Bernstein inequality, [20, Theorem 4.3, p. 131], [21, Theorem 7.3.3, p. 165], combined with the estimate $|q_n| \le 1$, [20, Theorem 4.1, p. 128], [21, (7.21.1), p. 164], we have

$$|q_n(x)| \le C\, n^{-1/2}\ \sigma_n(x) \tag{5.4}$$

where $\sigma_n(x) = \min\left((1-x^2)^{-1/4}, \sqrt{n}\right)$. We note in particular that

$$\sigma_n(\cos\theta) \le |\sin\theta|^{-1/2}\ . \tag{5.5}$$

By (5.1) and (5.4), (5.5),

$$|\Pi_n(x_0, \cos\theta)| \le \begin{cases} \dfrac{C \quad \sigma_n(x_0)}{|x_0-\cos\theta||\sin\theta|^{1/2}} \\ \\ C\ n^2 . \end{cases} \tag{5.6}$$

For $p = 0$ we obtain via (5.2), (5.3) and (5.6),

$$|\mathcal{L}_n\ f(x_0)| \le C \int_0^\pi \frac{\sigma_n\ (x_0)}{\delta \sin^{1/2}\theta} \ |g(\theta)| \sin\theta\ d\theta$$

$$\le \frac{C\ \sigma_n(x_0)}{\delta} \|g\|_{L_1(|\sin|^{1/2})}$$

which establishes (3.5)'.

For $p \ge 1$ we introduce $\mathcal{K}_n\ g(\theta)$ as in (4.7). Since, [10, p. 58], $\mathcal{K}_n\ g(\theta)$ is an even trigonometric polynomial, $K_n\ f(x) := \mathcal{K}_n\ g(\theta)$, $x = \cos\theta$, is a polynomial in x. Therefore, since $\mathcal{L}_n\,\chi_n = \chi_n$ for $\chi_n \in P_n$,

$$\begin{aligned} \mathcal{L}_n\ f(x_0) &= K_n\ f(x_0) + \mathcal{L}_n(f-K_n\ f)(x_0) \\ &= \mathcal{K}_n\ g(\theta_0) + \int_0^\pi \Pi_n\ (x_0, \cos\theta)(g(\theta)-\mathcal{K}_n\ g(\theta)) \sin\theta\ \ d\theta . \end{aligned} \tag{5.7}$$

Using (5.3) we obtain as in (4.10), with C_2 as in (4.2), for $\delta \ge C_2\ \ell n\ n/n$,

$$|\mathcal{K}_n\ g(\theta_0)| \le C\ n^{-q-1}\ 2^p\ \|g\|_{L_1} . \tag{5.8}$$

Further,

$$\begin{aligned} \mathcal{L}_n(f-K_n\ f)(x_0) &= \int_0^\pi \Pi_n(x_0, \cos\theta)(\int_{-\pi}^{\pi} k_n(t)\ \Delta_t^p\ g(\theta)dt)\sin\theta\, d\theta \\ &= \int_{-\pi}^{\pi} k_n(t)(\int_0^\pi \Pi_n\ (x_0,\cos\theta)\ \Delta_t^p\ g(\theta)\ \sin\theta\ \ d\theta)dt \\ &= \int_{|t| \le C_2 \frac{\ell n\ n}{n}} + \int_{|t| \ge C_2 \frac{\ell n\ n}{n}} \equiv I_1 + I_2 . \end{aligned} \tag{5.9}$$

By arguments as in Section 4, with C_4 as in (4.9) and using (5.3), (5.6) and (4.6),

$$|I_1| \le C \int_{|t| \le C_2 \frac{\ell n\, n}{n}} |k_n(t)| \left(\int_0^{\pi} \frac{\sigma_n(x_0)}{\delta |\sin\theta|^{1/2}} |\Delta_t^p\, g(\theta)| \sin\theta \, d\theta \right) dt$$

$$\le \frac{C\,\sigma_n(x_0)}{\delta}\, \omega^p(g, C_2 \frac{\ell n\, n}{n}; L_1(|\sin|^{1/2})) . \tag{5.10}$$

Also, by (5.6) and (4.4) (changing q to $q+1$),

$$|I_2| \le C\, n^{-q-2}\, n^2\, 2^p\, \|g\|_{L_1} = Cn^{-q}\, 2^p\, \|g\|_{L_1} . \tag{5.11}$$

Combining (5.8)-(5.11) with (5.7), Theorem 3.3 obtains.

6. EXAMPLES.

We have collected an ensemble of examples of Fourier and Legendre series (and one Chebyshev series). In all of these examples, Theorem 3.2 or 3.3 will be seen to be sharp, modulo (weak) logarithmic factors. Since the examples are quite varied in nature, this gives some confidence that we have (almost) found the "correct" mechanisms for explaining pollution. In this respect, cf. again the last paragraph of Section 3.

Most of the examples involve functions which are analytic, but not identically vanishing, around the point x_0 of interest. To apply our results, simply subtract a local smooth function which is approximable to arbitrary high order by classical well known results, cf., (2.1) and (2.4).

We feel that some of the examples (which are mainly classical) are very interesting in their own right, not merely as vehicles for demonstrating that Theorems 3.2 and 3.3 are sharp. In particular, Examples 6.4 and 6.5 could be connected with a computationally observed severe loss in accuracy in stress calculations in the "p-version" of the finite element method, [2].

Example 6.1. Arbitrary low rates of convergence possible in Fourier series in fixed regions where $f \equiv 0$ while $\|f\|_{L_1} \le 1$.

Let

$$h(x) = \begin{cases} 1, & x \le -\frac{3\pi}{4} \text{ or } x \ge \frac{3\pi}{4}, \\ 0, & \text{for } |x| \le \frac{\pi}{2} \end{cases}$$

while $0 \le h(x) \le 1$, $h \in C^\infty(I)$. For $0 < \alpha < 1$, set

$$f_\alpha(x) = h(x)\left(\sum_{j=1}^{\infty} e^{i2^j x}\, 2^{-j\alpha}\right).$$

Clearly, then $f_\alpha(x) \equiv 0$ for $|x| \le \pi/2$ and $\|f_\alpha\|_{L_1} \le C/\alpha$.

We next use Theorem 3.2 to predict the rate of convergence of $\mathfrak{F}_n f_\alpha(0)$ to 0 at $x = 0$.

Since $h(x)$ is smooth we have upon using that $|e^{i2^j(x+t)} - e^{i2^j x}| \le 2^j t$ for $2^j t \le 1$, while ≤ 2 for $2^j t \ge 1$,

$$\omega^1(f_2, C_2 \frac{\ell n\ n}{n}; L_1) \le C\, n^{-\alpha}(\ell n\ n)^\alpha .$$

Thus, a rather unsophisticated (apparently) prediction from Theorem 3.2 is that

$$|\mathfrak{F}_n f_\alpha(0)| \le C n^{-\alpha}(\ell n\ n)^\alpha \quad . \qquad \text{(Pred)}$$

To best this against reality, calculate the (2^{j_0})th Fourier coefficient of f_α. If $h(x) \sim \sum_{-\infty}^{\infty} c_j e^{ijx}$ where the c_j are rapidly decreasing, one readily finds that it has the magnitude $c_0\, 2^{-j_0\alpha}$ and since $c_0 > 0$ our prediction (Pred) is sharp apart from the weak logarithmic factor.

A similar example is easily constructed for Legendre series using the Stieltjes series mentioned at the end of Section 3.

Our next example is the well known one of stepfunctions in Fourier and Legendre series. The goal here is merely to show that Theorems 3.2 and 3.3 give sharp predictions, modulo logarithmic factors.

Example 6.2. Step-functions in Fourier and Legendre series.

6.2.a. Fourier series.

Here we take H as in Example 2.2. It is easily calculated that

$$\omega^1\left(H,\ C_2\left(\frac{\ell n\ n}{n}\right);L_1\right) \le 2C_2\ \frac{\ell n\ n}{n}$$

and so our prediction is

$$|\mathfrak{F}_n\ H(0)| \le C\ \frac{\ell n\ n}{n}\ . \qquad \text{(Pred)}$$

The Fourier series given in Example 2.2 shows that our prediction is sharp, again modulo a logarithmic factor.

6.2.b. Legendre series.

Here it is somewhat more convenient to choose

$$H(x) = \begin{cases} 0, & -1 \le x \le 0, \\ 1, & 0 < x \le 1\ . \end{cases}$$

For the corresponding $g(\theta) = H(x),\ x = \cos\theta$, we have

$$g(\theta) = \begin{cases} 0, & \frac{\pi}{2} \le \theta \le \pi \\ 1, & 0 \le \theta < \frac{\pi}{2}\ . \end{cases}$$

Extending $g(\theta)$ evenly it is trivially calculated that

$$\omega^1\left(g,\ C_2\ \frac{\ell n\ n}{n};\ L_1(|\sin|^{1/2})\right) \le C_2\ \frac{\ell n\ n}{n}\ .$$

Thus, Theorem 3.3 predicts e.g.,

$$|\ \text{error}\ \left(\frac{1}{\sqrt{2}}\right)| \le C\ \frac{\ell n\ n}{n} \qquad (\text{Pred})_1$$

$$|\ \text{error}\ (1)| \le C\ \frac{\ell n\ n}{\sqrt{n}}\ . \qquad (\text{Pred})_2$$

To check these predictions against the true state of affairs, we note that, [20, p. 163], the Legendre series for H is

$$H(x) = \frac{1}{2} + \sum_{k=0}^{\infty} a_{2k+1}\ q_{2k+1}\ (x)$$

where

$$a_{2k+1} = \frac{(-1)^k}{2}\left(\frac{(2k-1)!!}{(2k)!!} + \frac{(2k+1)!!}{(2k+2)!!}\right) = \frac{1}{2}(q_{2k}(0) - q_{2k+2}(0))\ .$$

These coefficients are alternating, decreasing in magnitude and satisfy, by Stirling's formula (or Laplace's formula, [21, (8.21.2), p. 194]),

$$|a_{2k+1}| \simeq \frac{1}{2\sqrt{\pi}} \left(\frac{1}{\sqrt{k}} + \frac{1}{\sqrt{k+1}}\right) \geq \frac{c}{\sqrt{k}}, \quad c > 0 \quad .$$

Testing convergence at $x_0 = 1/\sqrt{2}$ Laplace's formula gives

$$q_\ell\left(\frac{1}{\sqrt{2}}\right) = \frac{2^{3/4}}{\sqrt{\pi \ell}} \cos\left(\frac{\ell\pi}{4} - \frac{1}{8}\right) + O(\ell^{-3/2})$$

so that

$$q_\ell\left(\frac{1}{\sqrt{2}}\right) \geq c\ \ell^{-1/2} \quad .$$

Thus,

$$\left|\ \text{error}\ \left(\frac{1}{\sqrt{2}}\right)\right| \geq cn^{-1}$$

for infinitely many n and the prediction $(\text{Pred})_1$ is sharp modulo a logarithmic factor.

At the endpoint $x_0 = 1$, $q_n(1) = 1$, and so $|(\text{error})(1)| \geq c\ n^{-1/2}$ and again our prediction, $(\text{Pred})_2$, is almost sharp.

As we have already remarked at the end of Section 3, the estimates for Legendre series of Theorem 3.3 involve certain boundary effects. We shall now give an example of a "weak" algebraic singularity placed at an interior point, or at a boundary point, and contrast the two cases. Our Theorem 3.3 will predict the correct rate, apart from weak logarithmic factors.

<u>Example 6.3.</u> <u>"Weak" singularities of algebraic type, in the interior and at the boundary, in Legendre series</u>. <u>Less pollution if the singularity is at a boundary point</u>

<u>6.3.a.</u> Interior singularity.

Let $f(x) = \sqrt{|x|}$, $-1 \leq x \leq 1$.

Then $g(\theta) = \sqrt{|\cos\theta|}$ with singularities at $\theta = \pm\pi/2$. Splitting the interval $(-\pi,\pi)$ into pieces of length $4C_2 \frac{\ell n\ n}{n}$ centered at

$\pm\,\pi/2$ and their complements, one finds since $\sqrt{|\cos\theta|} \le \sqrt{|\theta \pm \pi/2|}$,

$$\omega^2(g,\ C_2\,\tfrac{\ell n\, n}{n};\ L_1(|\sin|^{1/2}))$$

$$\le \quad C\int_{|\theta\pm\frac{\pi}{2}| \le 4C_2 \frac{\ell n\, n}{n}} \sqrt{\frac{\ell n\, n}{n}}\, d\theta \quad + \quad C\int_{|\theta\pm\frac{\pi}{2}| \ge 4C_2 \frac{\ell n\, n}{n}} \left(\frac{\ell n\, n}{n}\right)^2 \cdot \frac{1}{|\cos\theta|^{3/2}}\, d\theta$$

$$\le C\, n^{-3/2}\, (\ell n\, n)^{3/2} \ .$$

Thus, our prediction for x_0 a "unit" distance away from 0 is

$$|\ \text{error}\ (x_0)| \le\ C\sigma_n(x_0) n^{-3/2}\, (\ell n\, n)^{3/2} \ . \qquad \text{(Pred)}$$

In order to test the sharpness of this prediction one adjusts the procedure in [20, p. 165-167] from the case $|x|^{-p}$, $p > 0$, to the one at hand. One readily finds that $\mathcal{L}_n f = \sum a_\ell\, q_\ell(x)$ where

$$a_{2\ell} \simeq n^{-1} \ .$$

Therefore, taking e.g., $x_0 = \frac{1}{\sqrt{2}}$ and $x_0 = 1$, cf. Example 6.2.b, our prediction is sharp (modulo logarithimc factors).

6.3.b. Boundary singularity.

This time let $f(x) = \sqrt{1-x}$, $-1 \le x \le 1$. Then $g(\theta) = \sqrt{1-\cos\theta}$ which is smooth except when $\theta = 0$. Writing $g(\theta)$ as $\frac{|\sin\theta|}{\sqrt{1+\cos\theta}}$ near $\theta = 0$ and considering $|\sin\theta| \ge C\ \ell n\, n/n$ and its complement, we find that

$$\omega^3(g,\ C_2\ \frac{\ln n}{n};\ L_1(|\sin|^{1/2}))$$

$$\leq \quad C\int_{|\theta| \leq 16C_2 \frac{\ln n}{n}} |\sin\theta||\sin\theta|^{1/2}\, d\theta$$

$$+ \quad C\int_{|\theta| \geq 16C_2 \frac{\ln n}{n}} C_2\left(\frac{\ln n}{n}\right)^3 d\theta \ \leq C(\ln n)^{2.5}\ n^{-2.5} \quad .$$

Thus, for $x_0 \neq 1$ and separated by a "unit" distance from $+1$ we predict

$$|\mathrm{error}(x_0)| \leq C\ \sigma_n(x_0)\ n^{-2.5}\ (\ln\ n)^{2.5}. \qquad \text{(Pred)}$$

Calculating the Legendre series for f, [20, p. 163], one has

$$\sqrt{1-x} \simeq \frac{4}{3\sqrt{2}}\ q_0(x) - \frac{4}{\sqrt{2}} \sum_{n=1}^{\infty} \frac{q_n(x)}{(2n-1)(2n+3)} \quad .$$

Thus, our prediction is again sharp modulo logarithmic factors.

For certain singularities of algebraic type sharper than those appearing in Example 6.3, the advantage of placing the singularity at the boundary is lost.

Example 6.4. "Sharp" singularities of algebraic type in the interior and at the boundary in Legendre series. Same amount of pollution from both.

6.4.a. Interior singularity.

Let $f(x) = |x|^{-1/2}$.

It is easily calculated that

$$\omega^1(g,\ C_2\ \frac{\ln n}{n};\ L_1(|\sin|^{1/2})) \ \leq\ C\ \sigma_n(x_0) n^{-1/2} (\ln\ n)^{1/2}$$

and thus we predict for x_0 away from the origin,

$$|\mathrm{error}\ (x_0)| \leq C\ \sigma_n(x_0) n^{-1/2}\ (\ln\ n)^{1/2}. \qquad \text{(Pred)}$$

The calculation of [20, p. 165-167] show that the Legendre coefficients in $\sum a_n q_n(x)$ are of unit size and not better; thus our prediction is sharp modulo, a logarithimc factor.

6.4.b. Boundary singularity.

Let $f(x) = \dfrac{1}{\sqrt{1-x^2}}$

Here $g(\theta) = \dfrac{1}{|\sin \theta|}$. We use a cut-off function around $\theta = 0$,

$$k(\theta) = \begin{cases} 1, & |\theta| \leq 2C_2 \,\ell n\, n/n , \\ 0, & |\theta| \geq 4C_2 \dfrac{\ell n\, n}{n} \end{cases}$$

and similarly around $\theta = \pm \pi$. It is easily arranged that

$|k'(\theta)| \leq \dfrac{C\, n}{\ell n\, n}$ and one finds that

$$\|gk\|_{L_1(|\sin|^{1/2})} \leq C(\ell n\, n)^{1/2}\, n^{-1/2} ,$$

$$\omega^1(g(k-1), C_2 \frac{\ell n\, n}{n}; L_1(|\sin|^{1/2})) \leq C\, \ell n^{1/2}\, n^{-1/2} .$$

Thus, from Theorem 3.3, our prediction is for x_0 away from ± 1,

$$|\text{error } (x_0)| \leq C\, (\ell n\, n)^{1/2}\, n^{-1/2} . \qquad \text{(Pred)}$$

By [20, p. 164],

$$\frac{1}{\sqrt{1-x^2}} = \frac{\pi}{2} \sum_{n=0}^{\infty} (4n+1) \left[\frac{(2n-1)!!}{n!\; 2n}\right]^2 q_{2n}(x)$$

and Stirling's formula shows that our prediction is sharp (apart from the usual logarithimic factor).

Thus, in this example of a "sharp" singularity, the Legendre series does not handle it better (from the point of view of pollution) if it is placed at the boundary than if it is placed in the interior.

In the next example we show that the Legendre series for the derivative of a function compared to the Legendre series for the function itself may loose an order n^{-2} in the rate of convergence in smooth regions. Since (as is easily verified) the derivative of the $\overset{\circ}{H}{}^1(I)$ projection of f into P_{n+1} equals $\mathcal{L}_n f'$, this is of interest in connection with the question of the derivative errors in the "p-version" of the finite element method, cf. [2]. Again, our theory will give (almost) sharp predictions.

Example 6.5. The Legendre series of a function f versus the Legendre series for f': $O(n^{-2})$ loss of accuracy in $\mathcal{L}_n f'$ in smooth regions compared to the rate in $\mathcal{L}_n f$. Boundary singularity.

Let $f(x) = \arcsin x$. Then $g(\theta) = \arcsin(\cos\theta) = \frac{\pi}{2} - \theta$, $0 \le \theta \le \pi$. Thus, since g is Lipschitz continuous on $(-\pi,\pi)$, and $\Delta_t^2 g \equiv 0$ on most of the interval, one easily finds that

$$\omega^2(g, C_2 \frac{\ell n\, n}{n}; L_1(|\sin|^{1/2})) \le C(\ell n\, n)^{2.5} \quad n^{-2.5} \quad .$$

Hence our predicted rate of convergence at $x_0 \neq \pm 1$ is

$$|\text{error in } f(x_0)| \le C(\ell n\, n)^{2.5} \quad n^{-2.5} . \quad \text{(Pred)}$$

Since [20, p. 164],

$$\arcsin x = \frac{\pi}{2} \sum_{n=0}^{\infty} \left[\left(\frac{(2n-1)!!}{n!\ 2^n}\right)^2 - \left(\frac{(2n-3)!!}{(n-1)!\ 2^{n-1}}\right)^2\right] q_{2n+1}(x)$$

an application of Stirling's formula shows that our predicted rate is sharp, apart from a logarithmic factor.

Since $f'(x) = \dfrac{1}{\sqrt{1-x^2}}$, we have already seen in the previous Example 6.4.b that, away from $x = \pm 1$, the rate of convergence in $\mathcal{L}_n f'$ is not better than $O(n^{-1/2})$. Thus, we see the $O(n^{-2})$ loss in accuracy in Legendre series between the derivative of the function and the function itself.

We finally consider two examples where the singularity is of exponential boundary layer type at the endpoints. In this respect, cf. [18] for pollution effects in the h-finite element method for elliptic singularly perturbed problems.

Example 6.6. Exponential boundary layers in Fourier series.

Let

$$f(x) = e^{\frac{x-\pi}{\varepsilon}} + e^{-\frac{(x+\pi)}{\varepsilon}}, \quad 0 < \varepsilon \leq 1, \ -\pi \leq x \leq \pi .$$

If $\varepsilon \leq 1/n$ take $p = 0$ in Theorem 3.2. Then $\|f\|_{L_1} = O(\varepsilon)$ so that we have for $x_0 \neq \pm\pi$,

$$|\text{error } (x_0)| \leq C\varepsilon, \ \varepsilon \leq 1/n \quad . \qquad (\text{Pred})_1$$

For $\varepsilon \geq 1/n$ take $p = 2$ in Theorem 3.2. Splitting the interval $[-\pi,\pi]$ into pieces of length $C \frac{\ell n\, n}{n}$ around the ends and their complements, a prediction estimate for $x_0 \neq \pm\pi$ of

$$|\text{ error } (x_0)| \leq C \frac{\ell n^2\, n}{n^2\, \varepsilon}, \ \varepsilon \geq 1/n \qquad (\text{Pred})_2$$

ensues. Combining $(\text{Pred})_1$ and $(\text{Pred})_2$ we have

$$|\text{error } (x_0)| \leq C(\ell n\, n)^2 \cdot \frac{\varepsilon}{1+n^2\, \varepsilon^2}, \ x_0 \neq \pm\pi . \qquad (\text{Pred})$$

Elementary calculus shows that the Fourier coefficients of f are of size $\varepsilon/(1+n^2\,\varepsilon^2)$ so that our prediction is sharp, apart from a logarithmic factor.

For our last example we are motivated by Figure 3.10 of [8], reproduced here with the kind permission of SIAM.

We adapt our theory to this case and shall see that, in spite of the asymptotic nature of our theory, it explains qualitatively the salient features of the figure.

Example 6.7. "Explaining" Figure 3.10 in [8].

We consider Chebyshev and Legendre expansions of $f(x) = e^{(x-1)/\varepsilon}$, $-1 \leq x \leq 1$. The Chebyshev series is given by the Fourier series for $g(\theta) = e^{(\cos\theta-1)/\varepsilon}$. Thus, for $x_0 \neq +1$, Theorem 3.2. predicts that

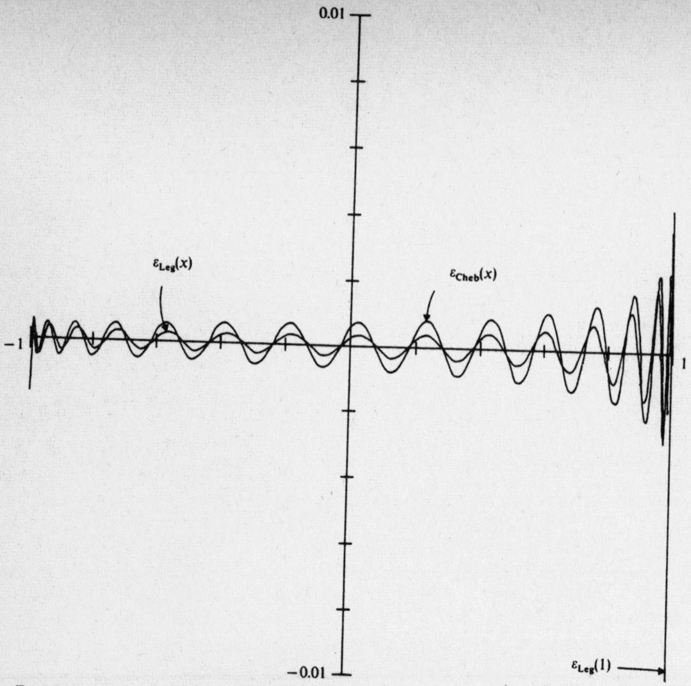

FIG. 3.10. *A plot of the errors $\varepsilon(x) = Q_N(x) - \exp[100(x-1)]$ in the Chebyshev and Legendre polynomial expansions of* $\exp[100(x-1)]$ of degree $N=29$. *Observe that the Legendre expansion has smaller errors away from the boundaries $x = \pm 1$ but that the Chebyshev expansion has a maximum pointwise error that is about 3 times smaller than the Legendre expansion. Here we denote the Chebyshev polynomial approximation errors by $\varepsilon_{\mathrm{Cheb}}(x)$ and the Legendre approximation errors by $\varepsilon_{\mathrm{Leg}}(x)$.*

$$|\text{error}\ (x_0)| \le \begin{cases} C\ \dfrac{\ell n\ n}{n}, & \text{for}\ \dfrac{\ell n^2\ n}{n^2} \le \varepsilon << 1, \\ C\ \sqrt{\varepsilon}, & \text{for}\ \varepsilon \le \dfrac{\ell n^2\ n}{n^2}. \end{cases} \qquad (\text{Pred})_1$$

For the Legendre series we have similarly by Theorem 3.3., for $x_0 \neq +1$,

$$|\text{error}\ (x_0)| \le \begin{cases} C\ \sigma_n(x_0)\ \dfrac{\ell n\ n}{n}\ \varepsilon^{1/4}, & \dfrac{\ell n^2\ n}{n^2} \le \varepsilon << 1, \\ C\ \sigma_n(x_0)\ \varepsilon^{3/4}, & \varepsilon \le \dfrac{\ell n^2\ n}{n^2}. \end{cases} \qquad (\text{Pred})_2$$

In [8, Figure 3.10] we have $\varepsilon = 10^{-2}$, $N = 29$. Thus, $\frac{\ell n^2 n}{n^2} \simeq 1.3 \cdot 10^{-2}$ so that the figure is in the regime of $\ell n^2 n/n^2 \simeq \varepsilon$. Hence our theory predicts that in the interior of $(-1,1)$, the Legendre series error is smaller by a factor of $\varepsilon^{-1/4} \simeq 3$, which fits qualitatively. At $x = -1$, our theory predicts that the error in the Chebyshev series is smaller by a factor of $\sqrt{n}\, \varepsilon^{1/4} \simeq 2$ than that of the Legendre series; again this fits the picture. (The constants in Theorems 3.2 and 3.3 are close to one another. This accounts for the present success.)

7. TWO EXAMPLES OF POLLUTION-DIMINISHING SUMMATION METHODS.

We consider two summation methods, for the Fourier and Legendre case, respectively, and show that they reduce the pollution effect.

7i. Fejér-de la Vallée-Poussin summation in Fourier series.

Given $\mathcal{F}_j f$, $j=1,2,3,..$, the Fejér sum $\sigma'_n f$ is defined as

$$\sigma'_n f = \frac{1}{n} \sum_{j=1}^{n} \mathcal{F}_j f .$$

We recall that Fejér summation has a low, $O(n^{-1})$, saturation order, [10, p. 98]. Therefore, to apply our theory to functions f which are smooth on I_δ, but not identically zero, we would not be able to repeat our argument of the second paragraph of Section 6 of subtracting off a smooth part. Hence we consider instead the de la Vallée-Poussin operator,

$$\mathcal{V}'_n f = 2\sigma'_{2n} f - \sigma'_n f = \frac{1}{n} \sum_{j=n+1}^{2n} \mathcal{F}_j f \quad ; \tag{7.1}$$

in this case the subtraction argument easily works.

We can now show the following.

Theorem 7.1.

Assume that (1.1) holds and let p and q be given nonnegative integers. There exist constants C_1, C_2, C_3 and C_4 such that for $\delta \geq C_4 \,\ell n\, n/n$,

$$|\mathcal{V}_n' f(x_0)| \le C_1 \delta^{-2} n^{-1} \omega^p(f, C_2 \frac{\ell n\, n}{n}; L_1) \tag{7.2}$$

$$+ C_3 n^{-q} \|f\|_{L_1}, \quad \text{for} \quad p \ge 1,$$

$$|\mathcal{V}_n' f(x_0)| \le C_1 \delta^{-2} n^{-1} \|f\|_{L_1}, \quad \text{for} \quad p = 0. \tag{7.2'}$$

Comparing this with Theorem 3.2 we predict a reduction in pollution by $O(n^{-1})$ via the use of the summation method.

Proof.

With $\mathcal{K}_n f$ as in (4.7) we have $\mathcal{K}_n f = \mathcal{F}_j \mathcal{K}_n f$ for $j \ge n$ so that

$$\mathcal{V}_n' f(x_0) = \mathcal{V}_n' (f-\mathcal{K}_n f)(x_0) + \mathcal{K}_n f(x_0) . \tag{7.3}$$

Here, by (4.10),

$$|\mathcal{K}_n f(x_0)| \le C n^{-q-1} 2^p \|f\|_{L_1} .$$

For the first term in (7.3) we consider only a part of it,

$$J := \sigma_n' (f-\mathcal{K}_n f)(x_0);$$

it will be obvious that the estimates for $2\sigma_{2n}' (f-\mathcal{K}_n f)(x_0)$ are completely analogous.

With the Fejér kernel $F_n(t) = (n+1)^{-1} t^{-2} \sin^2 (nt)$ one has, cf., (4.11),

$$J = \int_{-\pi}^{\pi} F_n(x_0-y) \left(\int_{-\pi}^{\pi} k_n(t) \Delta_t^p f(y)dt\right)dy$$

$$= \int_{-\pi}^{\pi} k_n(t) \left(\int_{-\pi}^{\pi} F_n(x_0-y) \Delta_t^p f(y)dy\right)dt .$$

One proceeds as in (4.11) et. seq. and using that

$$|F_n(t)| \le \begin{cases} (n+1)^{-1} t^{-2} , \\ n, \end{cases}$$

one easily obtains

$$|J| \leq |J_1| + |J_2|$$

$$\leq C\,\delta^{-2}(n+1)^{-1}\,\omega^{p}(f,\, C_2\,\frac{\ell n\, n}{n};\, L_1)$$

$$+ Cn^{-q}\quad 2^{p}\,\|f\|_{L_1}\quad .$$

This completes the proof.

7.ii. Second-order Cesàro-de la Vallée-Poussin summation in Legendre series.

As remarked by Fejér [7, p. 76], in Legendre series, the "analogues" of Fejér sums in Fourier series are, naturally, the second Cesàro (or Hölder) sums. We shall consider the Cesàro case, perturbed in the vein of de la Vallée-Poussin to allow our argument of subtracting smooth, but not identically vanishing, functions on I_δ .

We note that, insofar as one is interested in this example as a model for the finite element p-method, the "hierarchic property" of the basis functions, [3, p. 542], makes summation methods feasible.

Let $\mathcal{L}_j$ f, j=0,1,2,.. be the usual Legendre series and

$$\sigma''_n\, f := \frac{2}{(n+1)(n+2)} \sum_{j=0}^{n} (n-j+1)\,\mathcal{L}_j\, f$$

their second Cesàro means. Form then the gliding average

$$\mathcal{V}''_n\, f = \alpha_n\,\sigma''_{4n}\, f + \beta_n\,\sigma''_{2n}\, f + \gamma_n\,\sigma''_n\, f \qquad (7.4)$$

where

$$\alpha_n = \frac{(4n+1)(2n+1)}{3n^2},\quad \beta_n = \frac{-(n+1)(2n+1)}{n^2},\quad \gamma_n = \frac{(n+1)(n+2)}{3n^2}\ .$$

One then has that

$$\mathcal{V}''_n\, f = \sum_{j=n+1}^{4n} \frac{(4n-j+1)}{3n^2}\,\mathcal{L}_j\, f - \sum_{j=n+1}^{2n} \frac{(2n-j+1)}{n^2}\,\mathcal{L}_j\, f$$

so that the "subtraction argument" for smooth but nonvanishing functions on I_δ is, clearly, valid.

Our result for this example is the following. Recall that

$g(\theta) = f(\cos\theta)$.

Theorem 7.2.

Assume that (1.1) holds and let p and q be given nonnegative integers. There exist constants C_1, C_2, C_3 and C_4 such that for $\delta \geq C_4 \ \ell n\ n/n$,

$$|\mathfrak{V}_n'' f(x_0)| \leq C_1 \delta^{-2} n^{-1} \omega^p(g, C_2 \frac{\ell n\ n}{n} L_1(|\sin|)) \qquad (7.5)$$

$$+ C_3 n^{-q} \|g\|_{L_1}, \quad \text{for } p \geq 1,$$

$$|\mathfrak{V}_n'' f(x_0)| \leq C_1 \delta^{-2} n^{-1} \|g\|_{L_1(|\sin|)}, \text{ for } p = 0. \qquad (7.5)'$$

Compared with Theorem 3.3, we have gained (at least) an $O(n^{-1})$ reduction in pollution. Further, the weight in the L_1-norm has changed from $|\sin|^{1/2}$ to $|\sin|$ which is certainly beneficial when singularities are at the boundary. Thus, e.g., for $f(x) = (1-x^2)^{-1/2}$, away from ± 1 we have by Theorem 7.2 (using the technique of Example 6.4.b) a predicted pollution in $\mathfrak{V}_n'' f$ of $O(n^{-2} \ \ell n^2\ n)$ whereas the Legendre series itself exhibited $O(n^{-1/2})$ pollution.

There is no reason to believe that the estimates of Theorem 7.2 are particularly sharp.

Proof.

It is clear that once we have suitable estimates for the Cesàro kernels of σ_n'', σ_{2n}'' and σ_{4n}'' we could use them in our previous manner. Such estimates were given by Fejér [7]; for the convenience of the reader I will briefly describe them here.

Considering the Legendre series as a special case of the Laplace series, one has

$$(\mathcal{L}_n f)(\cos\theta_0) = \sum_{j=0}^{n} \frac{(2j+1)}{4\pi} \int_0^{\pi} \left(\int_0^{2\pi} (q_j(\cos\gamma) d\phi')\right) g(\theta) \sin\theta \, d\theta$$

where γ is the angle between the rays (θ,ϕ') and $(\theta_0,\pi/2)$ so that $\cos\gamma = \cos\theta_0 \cos\theta + \sin\theta_0 \sin\theta \sin\phi'$. Fejér then displays the kernel for σ_n'' as

$$\sigma_n''(\theta_0,\theta) = \frac{1}{2\pi(n+1)(n+2)} \int_0^{2\pi} (U_0(\gamma)\, V_n'(\gamma) + U_1(\gamma)\, V_{n-1}'(\gamma) + \ldots + U_n(\gamma)\, V_0'(\gamma))\,d\phi' \tag{7.6}$$

where

$$U_j(\gamma) = \sum_{\ell=0}^{j} q_\ell(\cos\gamma),$$

$$V_j'(\gamma) = \left(\frac{\sin((j+1)\gamma/2)}{\sin(\gamma/2)}\right)^2 . \tag{7.7}$$

He then utilizes the formula

$$U_j(\gamma) = \frac{2}{\pi} \int_\gamma^\pi \frac{\sin^2((j+1)t/2)}{\sin(t/2)\sqrt{2(\cos\gamma - \cos t)}}\, dt$$

(which, in particular, shows that $\sigma_n''(\theta_0,\theta)$ is positive) to calculate that

$$|U_j(\gamma)| \le C\,\gamma^{-3/2} ,$$

It is, however, easy to sharpen this to

$$|U_j(\gamma)| \le C\,\gamma^{-1} .$$

Combining this with (7.7) into (7.6),

$$\sigma_n''(\theta_0,\theta) \le C n^{-1} \quad \delta^{-2}$$

and the desired result obtains.

REFERENCES

1. S.A. Agahanov and G.I. Natanson; The Lebesgue function in Fourier-Jacobi sums, Vestnik Leningrad University, Mathematics, 1968, No. 1, 11-23.

2. I. Babuška; Lecture at the Finite Element Circus, in Knoxville, Tennessee, October 1983.

3. I. Babuška, B.A. Szabo and I.N. Katz; The p-version of the finite element method, SIAM J. Numer. Anal. 18, 1981, 515-545.

4. N.K. Bary; A Treatise on Trigonometric Series, Volume 2, Macmillan, New York, 1964.

5. J. Descloux; On finite element matrices, SIAM J. Numer. Anal. 9, 1972, 260-265.

6. J. Douglas Jr., T. Dupont and L.B. Wahlbin; Optimal L_∞ error estimates for Galerkin approximations to solutions of two-point boundary value problems, Math. Comp. 29, 1975, 475-483.

7. L. Fejér; Über die Laplacesche Reihe, Math. Annal. 67, 1909, 76-109.

8. D. Gottlieb and S.A. Orszag; Numerical Analysis of Spectral Methods: Theory and Applications, SIAM Regional Conference Series in Applied Mathematics, no. 26, SIAM, Philadelphia, Pennsylvania, 1980.

9. T.H. Gronwall; Über die Laplacesche Reihe, Math. Annal. 74, 1913, 213-270.

10. G.G. Lorentz; Approximation of Functions, Holt, Rinehart and Winston, New York, New York, 1966.

11. J.J. Moreau; Approximation en graphe d'une évolution discontinue, RAIRO, Analyse Numerique 12, 1978, 75-84.

12. S.M. Nikol'skii; On the best approximation of functions satisfying a Lipschitz's condition by polynomials, Izvestia Akad. Nauk. SSSR, Ser. Mat. 10, 1946, 295-322.

13. J. Nitsche and A.H. Schatz; On local approximation properties of L_2 projection on spline subspaces, Applicable Anal. 2, 1972, 161-168.

14. M.E. Noble; Coefficient properties of Fourier series with a gap condition, Math. Annal. 128, 1954, 55-62 (Correction: 256).

15. M.J.D. Powell; Approximation Theory and Methods, Cambridge University Press, Cambridge, 1981.

16. B. Riemann; Über die Darstellbarkeit einer Funktion durch eine trigonometrische Reihe (1854), Collected Works, 2nd Ed., Leipzig, 1892, 227-271.

17. A.H. Schatz and L.B. Wahlbin; Interior maximum norm estimates for finite element methods, Math. Comp. 31, 1977, 414-442.

18. A.H. Schatz and L.B. Wahlbin; On the finite element method for singularly perturbed reaction-diffusion problems in two and one dimensions, Math. Comp. 40, 1983, 47-89.

19. B. Sendov; Some questions of the theory of approximations of functions and sets in the Hausdorff metric, Russian Math. Surveys 24, 1969, 143-183.

20. P.K. Suetin; Classical Orthonormal Expansions, Nauka, Moscow, 1979.

21. G. Szegö; Orthogonal Polynomials, 4th Ed., AMS, Providence, Rhode Island, 1975.

22. A.F. Timan; A strengthening of Jackson's theorem on the best approximation of continuous functions by polynomials on a finite interval of the real axis, Doklady Akad. Nauk. SSSR, 78, 1951, 17-20.

23. L.B. Wahlbin; On the sharpness of certain local estimates for $\overset{\circ}{H}{}^1$ projections into finite element spaces: Influence of a reentrant corner, Math. Comp. January 1984, to appear.

24. A. Zygmund; Trigonometric Series, 2nd Ed., (Reprinted), Cambridge University Press, Cambridge, 1968.

Vol. 980: L. Breen. Fonctions thêta et théorème du cube. XIII, 115 pages. 1983.

Vol. 981: Value Distribution Theory. Proceedings, 1981. Edited by I. Laine and S. Rickman. VIII, 245 pages. 1983.

Vol. 982: Stability Problems for Stochastic Models. Proceedings, 1982. Edited by V. V. Kalashnikov and V. M. Zolotarev. XVII, 295 pages. 1983.

Vol. 983: Nonstandard Analysis-Recent Developments. Edited by A. E. Hurd. V, 213 pages. 1983.

Vol. 984: A. Bove, J. E. Lewis, C. Parenti, Propagation of Singularities for Fuchsian Operators. IV, 161 pages. 1983.

Vol. 985: Asymptotic Analysis II. Edited by F. Verhulst. VI, 497 pages. 1983.

Vol. 986: Séminaire de Probabilités XVII 1981/82. Proceedings. Edited by J. Azéma and M. Yor. V, 512 pages. 1983.

Vol. 987: C. J. Bushnell, A. Fröhlich, Gauss Sums and p-adic Division Algebras. XI, 187 pages. 1983.

Vol. 988: J. Schwermer, Kohomologie arithmetisch definierter Gruppen und Eisensteinreihen. III, 170 pages. 1983.

Vol. 989: A. B. Mingarelli, Volterra-Stieltjes Integral Equations and Generalized Ordinary Differential Expressions. XIV, 318 pages. 1983.

Vol. 990: Probability in Banach Spaces IV. Proceedings, 1982. Edited by A. Beck and K. Jacobs. V, 234 pages. 1983.

Vol. 991: Banach Space Theory and its Applications. Proceedings, 1981. Edited by A. Pietsch, N. Popa and I. Singer. X, 302 pages. 1983.

Vol. 992: Harmonic Analysis, Proceedings, 1982. Edited by G. Mauceri, F. Ricci and G. Weiss. X, 449 pages. 1983.

Vol. 993: R. D. Bourgin, Geometric Aspects of Convex Sets with the Radon-Nikodým Property. XII, 474 pages. 1983.

Vol. 994: J.-L. Journé, Calderón-Zygmund Operators, Pseudo-Differential Operators and the Cauchy Integral of Calderón. VI, 129 pages. 1983.

Vol. 995: Banach Spaces, Harmonic Analysis, and Probability Theory. Proceedings, 1980–1981. Edited by R.C. Blei and S.J. Sidney. V, 173 pages. 1983.

Vol. 996: Invariant Theory. Proceedings, 1982. Edited by F. Gherardelli. V, 159 pages. 1983.

Vol. 997: Algebraic Geometry – Open Problems. Edited by C. Ciliberto, F. Ghione and F. Orecchia. VIII, 411 pages. 1983.

Vol. 998: Recent Developments in the Algebraic, Analytical, and Topological Theory of Semigroups. Proceedings, 1981. Edited by K. H. Hofmann, H. Jürgensen and H. J. Weinert. VI, 486 pages. 1983.

Vol. 999: C. Preston, Iterates of Maps on an Interval. VII, 205 pages. 1983.

Vol. 1000: H. Hopf, Differential Geometry in the Large, VII, 184 pages. 1983.

Vol. 1001: D.A. Hejhal, The Selberg Trace Formula for PSL(2, ℝ). Volume 2. VIII, 806 pages. 1983.

Vol. 1002: A. Edrei, E.B. Saff, R.S. Varga, Zeros of Sections of Power Series. VIII, 115 pages. 1983.

Vol. 1003: J. Schmets, Spaces of Vector-Valued Continuous Functions. VI, 117 pages. 1983.

Vol. 1004: Universal Algebra and Lattice Theory. Proceedings, 1982. Edited by R. S. Freese and O. C. Garcia. VI, 308 pages. 1983.

Vol. 1005: Numerical Methods. Proceedings, 1982. Edited by V. Pereyra and A. Reinoza. V, 296 pages. 1983.

Vol. 1006: Abelian Group Theory. Proceedings, 1982/83. Edited by R. Göbel, L. Lady and A. Mader. XVI, 771 pages. 1983.

Vol. 1007: Geometric Dynamics. Proceedings, 1981. Edited by J. Palis Jr. IX, 827 pages. 1983.

Vol. 1008: Algebraic Geometry. Proceedings, 1981. Edited by J. Dolgachev. V, 138 pages. 1983.

Vol. 1009: T. A. Chapman, Controlled Simple Homotopy Theory and Applications. III, 94 pages. 1983.

Vol. 1010: J.-E. Dies, Chaînes de Markov sur les permutations. IX, 226 pages. 1983.

Vol. 1011: J. M. Sigal. Scattering Theory for Many-Body Quantum Mechanical Systems. IV, 132 pages. 1983.

Vol. 1012: S. Kantorovitz, Spectral Theory of Banach Space Operators. V, 179 pages. 1983.

Vol. 1013: Complex Analysis – Fifth Romanian-Finnish Seminar. Part 1. Proceedings, 1981. Edited by C. Andreian Cazacu, N. Boboc, M. Jurchescu and I. Suciu. XX, 393 pages. 1983.

Vol. 1014: Complex Analysis – Fifth Romanian-Finnish Seminar. Part 2. Proceedings, 1981. Edited by C. Andreian Cazacu, N. Boboc, M. Jurchescu and I. Suciu. XX, 334 pages. 1983.

Vol. 1015: Equations différentielles et systèmes de Pfaff dans le champ complexe – II. Seminar. Edited by R. Gérard et J. P. Ramis. V, 411 pages. 1983.

Vol. 1016: Algebraic Geometry. Proceedings, 1982. Edited by M. Raynaud and T. Shioda. VIII, 528 pages. 1983.

Vol. 1017: Equadiff 82. Proceedings, 1982. Edited by H. W. Knobloch and K. Schmitt. XXIII, 666 pages. 1983.

Vol. 1018: Graph Theory, Łagów 1981. Proceedings, 1981. Edited by M. Borowiecki, J. W. Kennedy and M. M. Sysło. X, 289 pages. 1983.

Vol. 1019: Cabal Seminar 79–81. Proceedings, 1979–81. Edited by A. S. Kechris, D. A. Martin and Y. N. Moschovakis. V, 284 pages. 1983.

Vol. 1020: Non Commutative Harmonic Analysis and Lie Groups. Proceedings, 1982. Edited by J. Carmona and M. Vergne. V, 187 pages. 1983.

Vol. 1021: Probability Theory and Mathematical Statistics. Proceedings, 1982. Edited by K. Itô and J.V. Prokhorov. VIII, 747 pages. 1983.

Vol. 1022: G. Gentili, S. Salamon and J.-P. Vigué. Geometry Seminar "Luigi Bianchi", 1982. Edited by E. Vesentini. VI, 177 pages. 1983.

Vol. 1023: S. McAdam, Asymptotic Prime Divisors. IX, 118 pages. 1983.

Vol. 1024: Lie Group Representations I. Proceedings, 1982–1983. Edited by R. Herb, R. Lipsman and J. Rosenberg. IX, 369 pages. 1983.

Vol. 1025: D. Tanré, Homotopie Rationnelle: Modèles de Chen, Quillen, Sullivan. X, 211 pages. 1983.

Vol. 1026: W. Plesken, Group Rings of Finite Groups Over p-adic Integers. V, 151 pages. 1983.

Vol. 1027: M. Hasumi, Hardy Classes on Infinitely Connected Riemann Surfaces. XII, 280 pages. 1983.

Vol. 1028: Séminaire d'Analyse P. Lelong – P. Dolbeault – H. Skoda. Années 1981/1983. Edité par P. Lelong, P. Dolbeault et H. Skoda. VIII, 328 pages. 1983.

Vol. 1029: Séminaire d'Algèbre Paul Dubreil et Marie-Paule Malliavin. Proceedings, 1982. Edité par M.-P. Malliavin. V, 339 pages. 1983.

Vol. 1030: U. Christian, Selberg's Zeta-, L-, and Eisensteinseries. XII, 196 pages. 1983.

Vol. 1031: Dynamics and Processes. Proceedings, 1981. Edited by Ph. Blanchard and L. Streit. IX, 213 pages. 1983.

Vol. 1032: Ordinary Differential Equations and Operators. Proceedings, 1982. Edited by W. N. Everitt and R. T. Lewis. XV, 521 pages. 1983.

Vol. 1033: Measure Theory and its Applications. Proceedings, 1982. Edited by J. M. Belley, J. Dubois and P. Morales. XV, 317 pages. 1983.